HANDBOOK OF

Cytology, Histology,
and
Histochemistry
of
Fruit Tree Diseases

HANDBOOK OF

Cytology, Histology,
and
Histochemistry
of
Fruit Tree Diseases

Edited by
ALAN R. BIGGS, Ph.D.
Associate Professor
University Experiment Farm
West Virginia University
Kearneysville, West Virginia

CRC Press
Boca Raton   Ann Arbor   London   Tokyo

Library of Congress Cataloging-in-Publication Data

Handbook of cytology, histology, and histochemistry of fruit tree
  diseases/editor, Alan R. Biggs.
      p. cm.
    Includes bibliographical references and index.
    ISBN 0-8493-2939-6
    1. Fruit—Diseases and pests. 2. Fruit trees—Cytology. 3. Fruit
trees—Histochemistry. I. Biggs, A. R. (Alan R.)
  SB608.F8H36 1993
  634.0493—dc20   92-11641
                                                                         CIP

Direct all inquiries to CRC Press, Inc. 2000 Corporate Blvd., N.W., Boca Raton, Florida, 33431.

© 1993 by CRC Press, Inc.

International Standard Book Number 0-8493-2939-6

Library of Congress Card Number 92-11641

Printed in the United States of America 1 2 3 4 5 6 7 8 9 0

Printed on acid-free paper

PREFACE

When I was asked initially by the publisher to consider preparing a volume on cytology, histology, and histochemistry of fruit tree diseases, I recall thinking that this would be a difficult, although achievable task, given the abundance of literature that has been published over the years. Optimistically, I endeavored in my mind to divide a generic fruit tree into its component parts, i.e., roots (primary and secondary), shoots (xylem, phloem, succulent, woody), leaves, flowers, and, of course, fruits. I then overviewed the literature to determine the most important diseases among the various classes of disease-causing organisms. Next, I chose representative diseases found attacking the various plant organs, with the idea in mind that this volume could present, in essence, the principles of cytology, histology, and histochemistry of fruit tree diseases by focusing on specific plant organ-pathogen type interactions. With diseases of fruit trees being incited by fungi, bacteria, viruses, nematodes, mycoplasma-like organisms, fastidious prokaryotes, and adverse environment, combined with a minimum of five plant organs to address, the notion of 30 or more chapters seemed forbidding as well as impractical and unnecessary.

The process of narrowing the field of candidate chapters down to manageable size turned out to hinge not so much upon what I or others felt was important to include, but rather on the individuals in our scientific community who were (1) willing, (2) able, and (3) available to present this information. The field of practicing pathological anatomists, histologists, and cytologists is comparatively small in the profession of plant pathology as a whole and, when this limited field must be narrowed to those working with fruit trees, I discovered that if this book was going to materialize, I would have to either write it myself or gain the cooperation of those few fruit tree pathologists who employ an anatomical approach to their specific disease problems. Being daunted by the first option, I tried the second. During the process of securing commitments from potential authors, I realized that I had gathered together a unique group of people from all over the world—unique in the sense that, for them, the practice of plant pathological anatomy and histology is not only a science, not only an art, but rather a combination of the two fueled by an extraordinary level of fervor. I think this is apparent in most of the chapters in this book.

The literature abounds with papers and monographs on the cytology, histology, and histochemistry of various diseases and disorders of deciduous and nondeciduous fruits. Many of these studies originated in the first decades of this century with the belief that an under-standing of disease process could lead to improved control of the disease in question. With the advent of effective chemicals for controlling such diseases, especially those caused by fungi, investigations on host-pathogen relationships became less of a priority. This book is not a compendium of histopathological investigations on fruit trees. Instead, I have asked the contributing authors to focus on disease problems in which researchers remain active in elucidating the host-pathogen interaction, examining systems for which no effective controls are available, investigating new diseases of recent importance, or looking for novel ap-proaches to old problems. The various approaches taken by the different authors are wide ranging, reflecting their varying areas of interest and expertise, with topics ranging from cytological investigation of pathogens (the pear rust fungus, for example) to studies of plant cell wall fine structure in relation to cold temperature injury. Some of the chapters address single disease problems in great depth (the chapters on apple scab, citrus blight, and fire blight); whereas others are more comprehensive of classes of pathogens or type-diseases (diseases caused by MLOs, *Phytophthora* spp., *Monilinia* spp., and systemic invasion of fruit trees by bacteria). The focus of this book is clearly on fungal and bacterial diseases, with additional chapters on MLOs, cold temperature injury, and citrus blight, a disease of unknown etiology.

The volume is incomplete in that representatives of all the classes of disease-causing organisms are not included. This is due neither to oversight nor design, but rather to those factors listed above. Specific chapters on nematode diseases and virus diseases are not included, even though there are many excellent scientific articles available in the literature.

The reasons for putting this volume together were to consolidate much of the recent histopathological literature into one well-illustrated resource, to provide in-depth or comprehensive chapters which could serve as springboards for future investigations, and to demonstrate the continuing importance of histopathological investigations (and the skills required in order to conduct such studies) to our discipline. It is hoped that all three objectives have been met and that the present volume will serve to increase understanding of host-pathogen interactions and also to stimulate new interests that inspire continued research into important areas of tree fruit pathology.

THE EDITOR

Alan R. Biggs, Ph.D., is Associate Professor of Plant Pathology in the College of Agriculture and Forestry and Extension Specialist in Tree Fruit Pathology at West Virginia University.

Dr. Biggs received his B.S. degree in Forest Science from the Pennsylvania State University in 1976. He obtained his M.S. and Ph.D. degrees in 1978 and 1982, respectively, from the Department of Plant Pathology at the Pennsylvania State University. He was appointed to the position of Research Scientist in Plant Pathology at the Agriculture Canada Research Station in Vineland, Ontario in 1983, with primary responsibilities for diseases of fruit trees. In 1987, he became Head of the Plant Pathology Section at the Vineland Research Station. Dr. Biggs began his present position with West Virginia University in September 1989.

Dr. Biggs is a member of the American Association for the Advancement of Science, the American Phytopathological Society, the Canadian Phytopathological Society, and the International Association of Wood Anatomists. He served as Associate Editor of *Phytopathology* (1988 to 1990) and will serve as Secretary (1992) and Chair (1993) of the Deciduous Tree Fruit Workers of the American Phytopathological Society.

Dr. Biggs has been the recipient of research grants from the Natural Sciences and Engineering Research Council, the Ontario Ministry of Agriculture and Food, the International Society of Arboriculture Research Trust, and private industry. He has published more than 50 research articles in cytology, histology, physiology, biochemistry, epidemiology, tissue culture, and genetics of fruit tree diseases. He recently co-edited a book entitled *Defense Mechanisms of Woody Plants Against Fungi*. His current research interests are in defense mechanisms of woody plants and sustainable fruit production.

ACKNOWLEDGMENTS

This volume is the fruit of the labors of many individuals in addition to those who kindly authored the various chapters. I extend my appreciation to the following individuals who provided either reviews of chapters or helpful discussions during the course of preparing this book: C. Lee Campbell, North Carolina State University; Robert E. Davis, U.S. Department of Agriculture-Agricultural Research Service, Beltsville; Edward Durner, Rutgers University; William Merrill, Pennsylvania State University; William Olien, Clemson University; Paul Peterson, North Carolina State University; David F. Ritchie, North Carolina State University; Paul W. Steiner, University of Maryland; Wayne F. Wilcox, Cornell University; and Keith Yoder, Virginia Polytechnic Institute and State University.

CONTRIBUTORS

Rajeev Arora, Ph.D.
Plant Physiologist
Appalachian Fruit Research Station
Agricultural Research Service
U.S. Department of Agriculture
Kearneysville, West Virginia

Christopher M. Becker, Ph.D.
Research Associate
Department of Plant Pathology
Cornell University
New York State Agricultural Experiment
  Station
Geneva, New York

Ronald H. Brlansky, Ph.D.
Professor and Plant Pathologist
Citrus Research and Education Center
University of Florida
Lake Alfred, Florida

Suzanne Bullock, M.Sc.
Professional Officer
School of Biological Science
University of New South Wales
Kensington, N.S.W.
Australia

Sharon M. Douglas, Ph.D.
Research Plant Pathologist
Department of Plant Pathology and
  Ecology
Connecticut Agricultural Experiment
  Station
New Haven, Connecticut

M. J. Hattingh, D.Sc.
Professor
Plant Pathology Department
University of Stellenbosch
Stellenbosch
South Africa

Mitsuru Kohno, Dr.Agr.
Chief Researcher
Department of Development and
  Planning of Research
Mie Prefectural Agricultural
  Research Center
Ichishi-gun
Japan

Hitoshi Kunoh, Ph.D., Dr.Agr.
Professor, Faculty of Bioresources
Mie University
Tsu-city
Japan

E. Lucienne Mansvelt, Ph.D.
Plant Biotechnology and Pathology
INFRUITEC
Stellenbosch
South Africa

Hideo Nasu, Dr.Agr.
Researcher
Division of Plant Protection
Okayama Prefecture Agricultural
  Experiment Station
Akaiwa-gun, Okayama
Japan

Darren P. Phillips, Ph.D.
Bureau of Rural Resources
Department of Primary Industries and
  Energy
Canberra
Australia

Isabel M.M. Roos, Ph.D.
Plant Biotechnology and Pathology
INFRUITEC
Stellenbosch
South Africa

Henk J. Schouten, Ph.D.
Center for Plant Breeding and
 Reproduction Research
 (CPRO-DLO)
Department of Cultivar Strategy
Wageningen
The Netherlands

Esther Shedletzky, Ph.D.
Researcher
Department of Botany
Hebrew University of Jerusalem
Jerusalem, Israel

Haydn J. Willetts, Ph.D.
Visiting Professor
School of Biological Science
University of New South Wales
Kensington, N.S.W.
Australia

Michael Wisniewski, Ph.D.
Plant Physiologist
Appalachian Fruit Research Station
Agricultural Research Service
U.S. Department of Agriculture
Kearneysville, West Virginia

TABLE OF CONTENTS

Chapter 1

HISTORICAL ASPECTS OF THE PATHOLOGICAL ANATOMY OF FRUIT TREE DISEASES

Alan R. Biggs

TABLE OF CONTENTS

0-8493-2939-6/93/$0.00 + $.50

I. INTRODUCTION

There are many excellent treatises on the histories of cytology, staining, histochemistry, plant anatomy, plant pathology, and fruit growing. All of these are separate and distinct monographs within disciplines or on particular areas of specialty. This chapter presents in a short and very general form the many seemingly similar and disparate interest that ultimately converged and gave rise to the specialty area addressed by this book. What fruit diseases were prevalent when Robert Hooke published "Micrographia" in 1665? What did Grew and Malpighi, the founders of plant anatomy in the 1680s, contribute to studies on the pathological anatomy of fruit trees? Who wrote the first monograph on diseases of fruit trees? Who laid the foundations for investigations into pathological plant anatomy? The history of fruit culture predates recorded history, with some of the earliest writings from Theophrastus dating prior to 285 B.C. The starting point for this synopsis is the beginning of the 17th century and coincides with the invention of the microscope.

II. THE EARLY FRUIT INDUSTRY AND PLANT DISEASE

At the time of the settlement of the earliest colonies in North America, apples were grown extensively from southern Europe northward to the Scandinavian peninsula. The various waves of immigrants, the French to Canada; the English to New England, Maryland, Virginia, the Carolinas, and Georgia; the Dutch to New York; the Swedes to New Jersey; and peoples from all parts of Europe to Pennsylvania, brought young apple trees, seeds, or scions, from which were started the early apple trees of North America. The early history of fungus diseases of fruit in North America is a mixture of allegory and myth. In 1639, John Josselyn visited New England and enjoyed "half a score of very fair pippins" harvested from trees planted by Governor Endicott on Governor's Island in Boston Harbor. In discussing disease problems on apples he revealed that "Fruit trees are subject to two diseases, the measles which is when they are burned and scorched with the sun, and lowsiness, when the woodpeckers jab holes in the bark; the way to cure them when they are lowsie is to bore a hole in the main root with a auger, and pour in a quantity of Brandie or Rhum, and then stop it up with a pin made of the same tree." It seems plant medicine and human medicine were on a similar course in the New World. In the Old World, the talk was of resistant varieties, when R. Austen, in 1657 recorded that "crab trees...are usually free from canker" (presumably Nectria canker).[2]

The fruit industry grew in the 18th century, albeit slowly, in North America. Fruit were largely grown for personal or local consumption or for making beverages or animal feed. However, large apple plantings and thriving nursery businesses were not uncommon at this time. For example, most of the early plantations in Virginia included orchards of apples and other fruits. In New York, the Prince Nursery was active as early as 1732 and their catalogs, from 1815 to 1850, were standard horticultural publications in the U.S. Commercial orcharding in these early days depended on convenient proximity to markets or satisfactory transportation. Indeed, commercial shipments of apples from Pennsylvania to England were occurring as early as 1763. Peter Kalm, perhaps unknowingly portending the future importance of plant diseases in the fruit industry, wrote in 1748 (while traveling from Trenton to Princeton, New Jersey): "Near almost every farm was a spacious orchard full of peach and apple trees, and in some of them the fruit had fallen from the trees in such quantities as to cover nearly the whole surface. Part of it they left to rot...". Coxe in 1817 wrote the first American book on fruit growing and included in it references to peach yellows and fire blight of pear.[3] In the unpublished second edition of this work (1829), mention is made of a peach variety, Morris's Large White Rareripe, as being not subject to rot (presumably brown rot). As some growers began to experiment with clonal rootstock and scion material in the first quarter of the 19th century, The American Pomological Society was founded in 1848, although it was not formally recognized by that name until 1852. Fisher and Upshall's account of the fruit industry on a state by state basis is fascinating reading for those interested in the history of fruit culture in North America.[1]

III. TREE FRUIT DISEASES IN NORTH AMERICA PRIOR TO 1880

Prior to 1850 there was little concern about fruit tree diseases in North America. It was only around 1860, as fruit began to be grown for fresh market purposes, that blemishes were no longer regarded as an inescapable act of "Providence". Spraying orchards to control insects began around 1870 to 1880 with the use of Paris Green, London Purple, and other arsenicals.[4] Spraying for the control of fungus diseases of fruit, mainly apple scab, in North America began sometime between 1885 and 1905, following the discovery of Millardet's mixture (Bordeaux mixture) in southern France in 1885. Very little was known about insects and diseases at the time. Sharvelle et al.[4] provide a concise account of the history of spraying for control of fruit pests.

One of the first diseases to cause serious economic losses in the U.S. was fire blight, which began to devastate the apple and pear orchards of Pennsylvania and New York before the end of the 19th century. The pathogen spread rapidly to other states. Many prominent pioneers of American plant pathology worked on fire blight, notably Thomas J. Burrill at the University of Illinois, Erwin F. Smith at the U.S. Department of Agriculture, and J. C. Arthur, who worked in New York and Indiana. Burrill unraveled the mysteries of fire blight and is commonly credited with providing the first proof that bacteria can cause disease in plants. Only 2 years earlier, Robert Koch in Europe had demonstrated the role of bacteria in causing diseases of animals.

Two other diseases, peach yellows in the east and the grape plague (later known as Pierce's disease) in California, wreaked havoc by the end of the 19th century. Peach yellows appeared in New York around 1801 and reached devastating proportions there by the 1840s.

Kirtland, working in northern Ohio in 1855, published a description of plum brown rot and stated that he had known the disease for 30 years.[5] In his article, Kirtland describes his microscopic observations, one of the earliest records of the microscopical observation of diseased tissue from a fruit tree in the U.S. The fungus was known to the mycologists at the time as *Torula fructigena*, the first descriptions having been given by Persoon in 1796.[6] Prior to and for several years after Kirtland, most mention of rot of stone fruits referred to the curculio insect as the probable causal agent. As Kirtland wrote:

> The plum crop, of late years, has generally failed in northern Ohio. The result has been charged to the curculio, and in many instances correctly; but a fatal disease has been insidiously progressing among our fruit orchards which has done more injury than that insect. The effect of the two evils has not usually been discriminated one from the other. Indeed, few cultivators seem to be aware of the prevalence of any such disease.

Kirtland's observations at this period in history suggest that in the U.S. there were proponents of the germ theory of disease prior to de Bary's work on rusts and smuts published in 1853 and his other important papers on potato late blight published in the 1860s. Kirtland not only understood that brown rot was a disease, he realized it was initiated by a fungus and made an effort to identify it. In an address to the Ohio Pomological Society published in 1867,[7] he said:

> I have watched carefully the sudden and premature decay of our plum crop, at the period of
> its ripening, for the last fifteen years. From hints offered by the work of Prof. Mitchell, and
> several microscopic observations of my own, I was induced to publish an article in the "The
> Florist" of Philadelphia, in the year 1855, in which I imputed the origin of the disease to
> the Torula or some analogous species of parasite fungi. The disease still prevails among us,
> and it is sure to destroy all the plums which escape puncture by the curculio. It is, however,
> generally overlooked by our pomologists, and its effects are charged to the depredations of
> that insect. Similar disease occasionally impairs our peach and apple crops, to a less extent.
> Whenever it occurs on either of these varieties of fruit, the spurs and young wood blight or
> canker, and cease to be fruitful for several years.

The apple blight mentioned here was probably fire blight caused by *Erwinia amylovora*.
Kirtland's view of brown rot as a fungus disease was unappreciated until Peck provided
proof of pathogenicity in 1877. B. T. Galloway and E. F. Smith of the U.S.D.A. published
brief accounts of the histology of the disease on cherry fruits in 1889 and peach twig cankers
in 1891, respectively.[8,9]

IV. THE BIRTH OF MICROSCOPY AND MICROTECHNIQUE

Hans and Zaccharias Janssen invented the compound microscope about 1590. Seventy-
five years later, Robert Hooke (1635 to 1703), who in his work "Micrographia" (1665),
was the first to illustrate in detail a pathogenic microscopic fungus, probably rose rust
Phragmidium mucronatum.[10] He was not aware of the pathogenic capabilities of this fungus,
however, and the idea of spontaneous generation persisted for another 2 centuries. His
discovery of the cell, however, changed the way humankind thought about life. From the
"Micrographia":

> I took a good clear piece of cork, and with a pen-knife sharpened as keen as a razor, I cut
> a piece of it off, and thereby left its surface smooth; then examining it diligently with the
> microscope,...but that possibly, if I could use some further diligence,...I, with the same pen-
> knife, cut off from the former smooth surface an exceedingly thin piece of it: and placing it
> on a black object plate, because it was itself a white body, and casting the light on it with
> a deep plano-convex glass I could exceedingly plainly see...

Hooke's simple microscope possessed no stage. Instead, the objects were mounted on a
point attached to a pedestal at the base. Hooke was also the first to mount animal or plant
material in water or oil, although the technique did not come into general use until the 1820s.
Leeuwenhoek is credited as founding the study of microbiology. By 1683, the Dutch lens
maker had described protozoa, bacteria, and other microbes in water, oral bacteria, and
human sperm. He usually made a microscope for each specific object he wanted to view
since the objects were generally fixed to a point. For specimens that could not be mounted
to a point, he spread the specimen onto fine bits of mica that were later glued to the object
needle. Only one original Leeuwenhoek microscope remains today in Utrecht. Smith's article
contains drawings of some of these early microscopes and specimen holders and includes a
much more thorough history than that provided here.[11]

The microscopical laboratory in the modern sense did not exist in the 17th and 18th
centuries. Indeed, up until the beginning of the 19th century, the microscope was regarded
as a toy or a curiosity, rather than as a scientific instrument. Nelson mentions Pepys in 1664
paying £10 for a microscope and thinking it "a great price for a curious bauble".[12]

During the 18th century, almost all works on the microscope were written by microscope
manufacturers with great details made on their construction. Still, there was little serious
use of the microscope. Object carriers or slides were introduced near the beginning of the
18th century. In 1742, Henry Baker[13] wrote seven pages on the subject of preparation. He

suggested the use of bits of glass to cover objects and also suggested the use of glass of different colors to make some objects more distinguishable.

Historians emphasize the barrenness of the 18th century, as compared to the 17th, in the development of the microscope. With one notable exception, John Hill (1716 to 1775), this is also true of microtechnique. In his work entitled "The construction of timber, from its early growth, explained by the microscope",[14] Hill utilized techniques that had not been employed up to that time and that did not come into general use until 50 years later, and then only as rediscovered by others. He was the first to use maceration in the study of wood. He put macerated pieces of wood into alum and then spirits of wine to harden them, a procedure resembling our modern fixation methods. Hill was undoubtedly the first to use staining as an aid in the study of the microscopical anatomy of plants. He prepared an alcoholic tincture of cochineal, in which, after filtering, he placed the stems of plants. An even more advanced technique he employed was a mordanting of the tissues before developing the color. Hill used two different techniques of clearing tissues, using spirits of wine for pine tissues and turpentine for other tissues. These methods, obviously extremely valuable, were not used by others until much later. Hill apparently was a disagreeable individual and because of this had considerably less influence on his field at the time.[15] His sections were cut on a microtome (known as "cutting engines"), which were well known at the time. The first microtome in the modern sense was made by Adams around 1770.[16]

Generally, the years 1830 to 1835 saw outstanding work by the English microscopists, including the introduction of glass slides, Canada balsam, and the mounting of objects in liquids (castor oil for fungi was used in 1849).[11] In contrast to the investigators of other countries, the British microscopists maintained a great freedom and oral interchange of ideas. Thus, an individual may have devised a truly original breakthrough, and been content to communicate it only by word of mouth. Prior to the formation of the Royal Microscopical Society of London in 1839, there was no formal meeting place for microscopists, and informal verbal communications predominated. This may help explain why it is often difficult to ascribe particular discoveries to single individuals. With a few notable exceptions, the English microscopists made only a few contributions to botany that were based on their microscopical observations. However, they were largely instrumental in advancing microtechnique. The great interest in microscopy that had built upon the discoveries of the 1830s had its heyday in the period from 1850 to 1860. With the appearance of five treatises on the microscope and microtechnique, the methods that had been known to only a small group of microscopists became known worldwide and were taken up and perfected outside England.

V. THE ADVENT OF STAINING AND HISTOCHEMICAL PROCEDURES

The most important contribution of the English microscopists to microtechnique during the years 1860 to 1875 was the introduction of staining, especially for plant tissues. The German botanists did not avail themselves of the English methods, largely due to their philosophy of utmost simplicity in preparing botanical material for study. Indeed, Hugo von Mohl has been quoted as saying that the microtome was largely superfluous for scientific investigations.[11] Generally, British methods in microscopy were not used in Germany until about 1850, owing to the interests of von Mohl. Smith[11] believes that Hill should be given recognition for the first use of staining methods, whereas others believe that Theodor Hartig should be given credit. Corti is given the honor of having been the first to apply staining methods to the contents of the cell, thus giving birth to the study of cytology.[17] Gerlach (1820 to 1896), although not the first to use stains, developed methods of staining bacteria and animal tissues that came into general use.

The beginnings of histochemistry as a science have been attributed to the years 1830 through 1855, and its origins were primarily botanical. For some decades, the whole practice of histochemistry in its true sense was in the hands of botanists. Link 1807 performed the earliest microchemical reactions when he used iron sulfate for determining tannic acid in leaves. The main work in those early years was that of the French botanist Raspail, whose "Essai de Chemie Microscopique Appliquée à la Physiologie" appeared in 1830.[18] The use of the starch-iodine reaction in microscopy and the introduction of methods for protein and cell wall carbohydrates have been attributed to Raspail. The discovery of protoplasm and chemical reactions within cells by von Mohl[19] led Schleiden (1804 to 1881),[20] who worked with plant tissues, and Schwann (1810 to 1882), who worked independently of Schleiden with animal tissues, to formulate (after personal discussion and laboratory study) the "Cell Theory" in 1838. Hermann Schacht (1824 to 1864) in 1852 provided the first collection of microchemical methods for the recognition of many different plant products, including cellulose, xylogen, protein, starch, gums, and dextrines.[21] Tests for sugars and fats were not yet well developed. Smith,[11] writing in 1915, indicated that by 1850 the microchemical determination of the constituents of the plant cell was in a fairly satisfactory state. The years 1875 through 1915 saw the zoologists develop and expand the early botanical techniques. From this time through the early 20th century, botanical microtechnique developed through refinements that zoologists made in the English microscopical methods and which were then adapted by the botanists. Although the early attempts at staining were first made by microscopists or botanists on botanical material, the development of staining is due almost completely to zoologists. None of the early zoologists cite John Hill's work. From the years 1862 through 1882, of over 55 published papers on staining, only 3 were published by botanists. In discussing the development of botanical histochemistry after this period, the problem is to find out at what particular time methods devised by animal histologists were first used in botanical applications. From 1875 to 1880 the notable progress in botanical microtechnique was the adoption of staining procedures. The American microscopists are generally credited with having developed double staining methods for stems and other tissues in the late 1870s. It should be noted that many of the early phytopathological studies used a triple stain of safranin (first used by Ehrlich in 1877), gentian violet, and orange G which was first combined by Flemming in 1891.[22]

Klebs first used paraffin for embedding in 1869, while celloidin was introduced in 1879 by Duval. In botany, the first use of paraffin belongs to Francotte (1886), and it became well established in the years 1887 to 1892. Fixation is a relatively recent development, with most of the fixing mixtures having been proposed in the period from 1880 to 1890. Flemming's various mixtures, developed from 1882 to 1884, were used as standard fixatives by the early phytopathologists. The expansion of botanical microtechnique into the various subdisciplines of botany, including phytopathology, is reflected by publication of Stausburger's *Botanische Prakticum*, various editions of which appeared from 1884 through 1913.[11] Heslop-Harrison and Knox provide a good general historical perspective on plant histochemistry.[23]

VI. THE DAWN OF PLANT ANATOMY AND PATHOLOGICAL PLANT ANATOMY

Theophrastus, regarded as the father of botanical science, made many notable observations on the anatomy of plants in the 3rd century B.C. With the rise of Christianity and the concomitant "Dark Ages" in biological sciences, no significant internal studies of plants occurred until the advent of magnification. The founders of plant anatomy, Nehemiah Grew[24] (1641 to 1712) and Marcello Malpighi[25] (1628 to 1694) were stimulated by Hooke's observations; however, they left little record of their working methods. Malpighi says nothing at all about his methods, whereas Grew stated that it is important to examine a specimen from several views, including oblique, perpendicular, and transverse, "all three being requisite". It is Grew, however, who is generally regarded as the father of plant anatomy. Eames and MacDaniels' account of the history of plant anatomy illustrates these early contributions in more detail than is required here.[26]

In the late 18th century, J. J. P. Moldenhawer (1766 to 1827) introduced the technique of maceration in the study of plant tissues, thus introducing the concept of the individuality of cells. Hugo von Mohl's (1805 to 1872) attention to the living contents of the cell was the initiation point for further studies on the fundamental nature of the protoplast. Von Mohl's contributions to plant anatomy were numerous and varied, including the nature and method of formation of vessels, the structure of the epidermis and nature of the cuticle, and also the nature of lenticels and the formation of bark. The first work devoted exclusively to plant histology was that of Schacht, which appeared in 1851.[27] This work contains minute directions for the anatomical observation of different plants, and thus was of great value. Theodor Hartig (1805 to 1880) studied the anatomy of wood and phloem and discovered and named the sieve tube.[28,29] Nageli (1817 to 1891) first used the terms "xylem" and "phloem", although Theophrastus had used them in a slightly different form.

In 1877, de Bary wrote *Comparative Anatomy of the Phanerogams and Ferns*. Although lacking a comprehensive presentation of the structure of the plant body as a whole, this book was regarded as one of the most useful and usable reference books in plant anatomy during the next 50 years after its appearance.[26] The first basic treatises in pathological anatomy, written by Robert Hartig, describe the attack of *Armellaria mellea* and *Fomes annosus* on *Prunus* spp., and show how these fungi penetrate intact roots.[30,31] In contrast, Hartig also stated that many microbes cannot penetrate directly or consume intact bark, and thus many pathogens of woody plants are wound parasites. Hartig also proved that fungi cause staining and decay in timber.

VII. THE ARRIVAL OF PLANT PATHOLOGY

The unraveling of the theory of spontaneous generation can not be traced to one specific event. Several circumstances occurred over the period from the late 18th century to the time of de Bary and Pasteur. In 1729, the Italian Micheli (1679 to 1737) described the cultural experiments that he used to demonstrate that several common molds were distinct entities.[32] This work went unappreciated until the time of Spallanzani at the end of the century and Pasteur in 1864, who disproved spontaneous generation. The study of mycology developed during this period as the identities of different fungi were determined, even though they were regarded mostly as saprobes rather than parasites. The study of plant disease emerged via the examination of rust and smut diseases of cereals and potato late blight. Major contributions towards understanding the nature of rust diseases were made by Felice Fontana (1730 to 1805), Joseph Banks (1743 to 1820), T. A. Knight (1759 to 1838), and N. P. Scholer (1772 to 1851). Heteroecism in the rusts was not generally accepted until de Bary's work in 1865 to 1866.

For smuts, the works of M. Tillet (1714 to 1791) and B. Prévost are noteworthy. Prévost, whose work went unnoticed by his contemporaries, proved that smut was fungal in origin and could be controlled with copper sulfate seed treatments. His work, published in 1807, is regarded as a landmark in biology, as well as in microbiology and plant pathology.[33] In the 1830s, the Italian Agostino Bassi (1773 to 1856) demonstrated that the muscardin disease of silkworms was caused by a fungus. A few years later, David Gruby (1810 to 1898) described thrush and ringworm diseases of humans as caused by fungi. Thus, when potato late blight swept through Europe in the 1840s, it seems surprising that there was so much controversy about its etiology. Again, it was de Bary who finally established that *Phytophthora infestans* was the causal agent. Spontaneous generation was still receiving detailed discussion in books as late as 1833 by the German, Franz Unger (1800 to 1970). Berkeley's series of articles in the *Gardener's Chronicle* from 1854 to 1857 may mark the turning point for the new ideas emerging on the fungal nature of plant diseases. His work and that of de Bary and Brefeld in Germany, the Tulasne brothers in France, and Woronin in Russia influenced the approach to plant disease over the next 50 years by deepening our level of knowledge of the parasitic fungi. As the profession of plant pathology emerged with Julius Kühn's textbook in 1858, publications from Germany dominated for over 40 years. Not until the turn of the century did we see textbooks from France and England. The year 1906 marked the publication of B. M. Duggar's *Fungous Diseases of Plants* and the beginning of the dominance of English language textbooks on plant pathology in the U.S.

Thomas J. Burrill (1839 to 1916) was appointed professor of botany and horticulture at the University of Illinois in 1869.[34] He began his microscopical observations on diseases of fruits in 1871 and was the first to describe a bacterial disease of plants (1878 to 1884) when he showed that fire blight of apple and pear was caused by the bacterium now known as *Erwinia amylovora*. Credit should be given to Burrill for the first microscopical study of a bacterial fruit tree disease. Burrill was a "self-made" phytopathologist in the sense that he had no formal connection with the European school of thought. He graduated from Illinois Normal University in 1865 and travelled with Powell's first Rocky Mountain Expedition in 1867.[34] Burrill's interest in microscopy is not appreciated widely by modern phytopathologists; however, he maintained memberships in the Royal Microscopical Society, the American Society for Microscopy (President, 1885; Secretary, 1886–1889), and the American Microscopical Society (President, 1905). He taught courses in microscopy and pomology in which he provided microscopes (of both European and American origin) for his students. Two books, published around the time that Burrill was finishing his training and beginning his early research, may have been influential: *One Thousand Things for the Microscope* by M. C. Cooke in 1868,[35] and *The Microscopic Dictionary: A Guide to the Examination of the Structure and Nature of Microscopic Objects,* by Griffith and Henfrey in 1875.[36]

E. F. Smith established the study of bacteria as an important branch of phytopathology and is perhaps especially noted for his studies on crown gall.[38] It should be noted, however, that Smith was involved in many research endeavors, and his 1891 publication on brown rot cankers was one of the first histological studies of a fungus disease of fruit trees.[38]

At about the time of Burrill's classic work, Peck in New York state described brown rot as one of the more common diseases of fruits.[39] Peck was the first to demonstrate the pathogenicity of the fungus by means of inoculations. Just 2 years earlier, Sorauer in Germany (1879) had written the first monograph on diseases of fruit trees.[40]

One of the most important concepts emerging from the early studies in plant pathology is that the majority of plants are resistant to infection. Mycologists and plant pathologists before the time of de Bary appear to have been oblivious to the importance of this phenomenon until de Bary published his illuminating works. de Bary elucidated the factors responsible for susceptibility or resistance in connection with the life of facultative parasites. He demonstrated that some fungus spores germinate only when in contact with host plants, and called attention to the fact that the mycelia of some obligate parasites are confined to the immediate point of attack, while others spread widely from the point of infection over or through the host.

Resistance in plants has been investigated with reference to the morphology, genetics, cytology, and physiology of hosts. The most obvious type of host resistance is associated with some morphological or anatomical character and was earlier assumed to be the most important type.[41] For example, Cobb[42] advanced the theory that plants possessing a heavily cutinized or waxy epidermis, or a corky stem, were resistant to invasion by fungi. This idea was supported by Bolley[43] and Anderson,[44] who attempted in 1889 and 1890 to correlate resistance in cereals with certain morphological characters. This idea was supported by Sappin-Trouffy who discovered in 1896 that the mycelium of *Puccinia graminis* was restricted to the chlorophyllous parenchyma and did not develop in the collenchyma.[41] The growth of the fungus was correspondingly restricted with the amount of parenchyma being small in relation to the amount of collenchyma. J. C. Arthur confirmed this idea by his observation in 1902 that mycelia of certain rusts may be limited in their development by the anatomical nature of the tissues of *Spartina*. In contrast, Ward (1902) concluded the morphological characters of *Bromus* had little effect on rust resistance, and Cobb reported in 1892[42] that the relative hairiness of resistant wheat varieties was not sufficient to explain their resistance to rust.

In the area of fruit pathology, Valleau (1915)[45] laid the basis for studies of host morphology in pathogen resistance with his research on the brown rot fungi of plums (see Chapter 5). Some of Valleau's conclusions:

> Varieties show great difference in resistance to infection owing to the production of parenchymatous plugs which fill the stomatal cavity and to lenticels made up of corky cells through which the hyphae are unable to penetrate. Corky cells lining the stomatal cavity merely delay infection.

In addition to the obstacles to penetration given above, Valleau states that after hyphae have gained entrance to the fruit, variations in cultivar resistance are correlated with skin thickness and fruit firmness.

Higgins[46] published the first report (1914)[46] that described for a woody plant the defensive responses of foliage to fungal infection, describing the abscission layer, or cicatrice, for cherry "shot hole" leaf spot. Wiltshire,[47] was the first (1922) to describe wound wood in xylem rays and how they acted to restrict the lateral movement of *Nectria galligena* in apple wood. The lack of effective fungicide controls for many fruit diseases partially explains the plethora of anatomical and cytological studies in the 50 years following the acceptance of the germ theory of disease.

It is not clear how or from whom plant pathologists prior to the turn of the century obtained their working knowledge of botanical microtechnique. At the time, the scientific community was relatively small and most of the individuals actively involved in pathological investigations knew each other personally or through correspondence. Most could read German or had some link with the German phytopathologists. Indeed, even into the 1920s, E. C. Stakman required a year of study in Germany for new faculty members at Minnesota. Humphreys 1893 translation of Zimmerman's book on botanical microtechnique[48] was significant, as was C. J. Chamberlain's book, *Methods in Plant Histology*, first published in 1901.[49] Both of these texts undoubtedly advanced the course of phytopathological investigations in the U.S. By 1932, the latter work had been revised five times.

Within the U.S.D.A., which was established in 1862, Thomas Taylor (1820 to 1910) was appointed microscopist in 1871 to begin a systematic study of diseases of plants.[37] In that first year, Taylor, who was originally from Scotland, published illustrated reports on fungus diseases of grapes, pears, and peaches. Over the next 5 years, he published reports on fire blight, peach yellows, black knot of plum, and cranberry diseases. During this time, Taylor, regarded by his peers as a self-taught amateur in phytopathology, studied medicine at Georgetown University, receiving the M.D. in 1882. His main interest was in the area of edible and poisonous mushrooms and his *Student's Handbook* was published in 1897.

Few citations on histological methods are given in most of the early papers, although

Flemming and Haidenhain are named in association with the techniques employed. In one of Keitt's early papers on peach scab,[50] he provided a more thorough description than did many of his colleagues.

> In killing material for histological work, several standard fixing agents were employed, viz., Flemming's weak, medium, and strong fluids, picro-formal, and chrom-acetic. The stains used were Flemming's triple, Haidenhain's iron alum-haematoxylin, Durand's haematoxylin-eosin, gentian violet, and safranin. For most purposes, Flemming's medium solution and triple stain gave the best results.

(He continued, describing the action of the stain on the various tissues.)

Nixon, using paraffin-embedded material, also found Flemming's triple stain useful for observing the fire blight bacterium in his histological study of fire blight.[51] For fixing, Flemming's strong solution was used for succulent material and Benda's solution was used for cankered tissue. Nixon also provided cautions about overstaining and recommended an optimum section thickness of 7 μm. Rosen, also studying fire blight, seems to have had extensive experience in botanical microtechnique, judging from all the different methods tested, although, like the previous authors, he provided few citations for microtechnique other than those papers dealing specifically with fire blight.[52]

The early phytopathologists recognized the importance that natural plant resistance could play in the management of orchard diseases. The studies and examples cited above are by no means an exhaustive review of the literature from a historical perspective; the examples do reveal, however, the major events that shaped the evolution of the histology of tree fruit diseases. We have seen that histological studies can reveal mechanisms of pathogenesis and host resistance and thus make substantial contributions to our understanding of host/pathogen interaction. As we continue to seek new methods for controlling established diseases, and as we encounter new diseases of unknown origin, histological studies will continue to play a key role in advancing the profession of phytopathology.

VIII. SUMMARY

The history of botanical microtechnique began with the development of methods of microscopy in England from the time of Hooke and continuing into the early 1800s. The English microscopists continued to refine microscope technique through 1875, although botanical applications were based largely on the methods of the German botanists until about 1875. After this date, botanical microtechnique owes a large debt to the rise of zoological applications from which the botanists borrowed heavily. The use of the microscope in tree fruit pathology began with the observations of the early mycologists in the late 18th and early 19th centuries. Following the advent of the germ theory of disease, Robert Hartig's studies of the pathological anatomy of forest tree species, and the translation of German works on botanical microtechnique into English, fruit pathologists provided the first glimpses into the cytology, histology, and histochemistry of fruit tree diseases. In the U.S., the observations of Thomas Taylor, Beverly T. Galloway, and Erwin F. Smith of the U.S.D.A. and Thomas J. Burrill of the University of Illinois are some of the earliest anatomical accounts of tree fruit diseases. Brown rot of stone fruits is an excellent example of the long evolution of histological studies of fungal diseases of tree fruits; beginning with the mycological observations of Persoon in 1796, followed by Kirtland's early observations of the disease in 1855, succeeded by Peck's proof of pathogenicity in 1877, and culminating in the histological observations of Galloway and Smith in 1889 and 1891, respectively. The evolution of histological studies in phytobacteriology is traced from the early observations of fire blight in New York and Pennsylvania and is abbreviated, in comparison to brown rot, owing to the intellect and skills of Thomas J. Burrill.

REFERENCES

1. **Fisher, D. V. and Upshall, W. H., Eds.,** *History of Fruit Growth and Handling in the United States of America and Canada,* Regatta City Press, Kelowna, B. C., Canada, 1976, 302.
2. **Austen, R.,** A Treatise of Fruit-Trees, Shewing the Manner of Grafting, Planting, Pruning, and Ordering of Them, London, 1657, 54.
3. **Coxe, W.,** A View of the Cultivation of Fruit Trees, and the Management of Orchards and Cider; with Accurate Descriptions of the Most Estimable Varieties of Native and Foreign Apples, Pears, Peaches, Plums, and Cherries, Cultivated in the Middle States of America, Philadelphia, 1817, 253.
4. **Sharvelle, E. G., Welsh, M. F., and McIntosh, D. L.,** Diseases of fruit plants and development of control methods, in *History of Fruit Growth and Handling in the United States of America and Canada,* Fisher, D. V. and Upshall, W. H., Eds., Regatta City Press, Kelowna, B. C., Canada, 1976, 302.
5. **Kirtland, J. R.,** Premature decay of the plum, *Florist Horticult. J.,* 4, 34, 1855.
6. **Persoon, C. H.,** *Synopsis Methodica Fungorum,* Vol. 2, Gottingan, 1796.
7. **Warder, J. A.,** *American Pomology. Apples,* New York, 1867, 744.
8. **Galloway, B. T.,** Brown rot of the cherry, *Monilinia fructigena* Pers., *USDA Rep.,* 1888/1889.
9. **Smith, E. F.,** Peach blight *(Monilinia fructigena,* Persoon), *J. Mycol.,* 7, 36, 1891.
10. **Hooke, R.,** *Micrographia,* London, 1665 (original not seen, citation from Smith, 1915[11]).
11. **Smith, G. M.,** The development of botanical microtechnique, *Trans. Am. Microsc. Soc.,* 34, 71, 1915.
12. **Nelson, E. M.,** What did our forefathers see in a microscope?, *J. R. Microsc. Soc.,* 427, 1910.
13. **Baker, H.,** *The Microscope Made Easy,* 1st ed., London, 1742.
14. **Hill, J.,** The Construction of Timber, from Its Early Growth, Explained by the Microscope, Etc., London, 1770 (original not seen, citation from Smith, 1915[11]).
15. **Clark, G. and Kasten, F. H.,** *History of Staining,* 3rd ed., Williams & Wilkins, Baltimore, 1963.
16. **Adams, H.,** *Essays on the Microscope,* 2nd ed., F. Kammacher, London, 1798.
17. **Baker, J. R.,** *Cytological Technique,* 2nd ed., Methuen, London, 1945.
18. **Raspail, F. V.,** Essai de Chimie Microscopique Appliquée à la Physiologie, Paris, 1830.
19. **von Mohl, H.,** Einige Beobachtungen über die blaue Färbung der vegetabilischen Zellmembran durch Jod, *Flora,* 23, 609, 1840.
20. **Schleiden, M. J.,** Eingie Bemerkungen über den vegetabilischen Faserstoff und sein Verhältniss zum Stärkemehl, *Ann. Phys. Chem.,* 43, 391, 1838.
21. **Schacht, H.,** *Die Pflanzenzelle der innere Bau und das Leben des Gewächse,* Berlin, 1852.
22. **Flemming, W.,** Neue Beiträge zur Kenntniss der Zelle, *Arch. Mikrosk. Anat.,* 37, 685, 1891.
23. **Heslop-Harrison, J. and Knox, R. B.,** Plant histochemistry, in *Histochemistry, The Widening Horizons of its Applications in the Biomedical Sciences,* Stoward, P. J. and Polak, J. M., Eds., John Wiley & Sons, New York, 1981, 1.
24. **Grew, N.,** *The Anatomy of Plants,* London, 1682.
25. **Malpighi, M.,** *Opera Omnia,* 2 vol., 1687.
26. **Eames, A. J. and MacDaniels, L. H.,** *An Introduction to Plant Anatomy,* 1st ed., McGraw-Hill, New York, 1925, 364.
27. **Schacht, H.,** Das Mikroskop und seine Anwendung, insbesondere für Pflanzen-Anatomie und Physiologie, Berlin, 1851.
28. **Hartig, Th.,** Chlorgen, *Bot. Z.* 12, 553, 1854.
29. **Hartig, Th.,** Über die Functionen des Zellenkerns, *Bot. Z.,* 12, 574, 1854.
30. **Hartig, R.,** *Wichtige Krankheiten der Waldbäume. Beiträge zur Mykoligie und Phytopathologie für Botaniker und Forstmänner,* J. Springer, Berlin, 1874.
31. **Hartig, R.,** *Die Zersetzungserscheinungen des Holzes der Nadelholzbäume und der Eiche in Forstlicher, Botanischer und Chemischer Richtung,* J. Springer, Berlin, 1878.
32. **Micheli, P. A.,** *Nova Plantarum Genera,* Florence, 1729.
33. **Prévost, B.,** Memoir on the immediate cause of bunt or smut of wheat, translated by G. W. Keitt, *Phytopathol. Classic,* 6, 1939.
34. **Trelease, W.,** Thomas J. Burrill, portrait, *Bot. Gaz.,* 62, 153, 1916.
35. **Cooke, M. C.,** *One Thousand Things for the Microscope,* Frederick Warne and Co., London, 1869, 123 pp.
36. **Griffith, J. W. and Henfrey, A.,** *The Microscopic Dictionary — A Guide to the Examination and Investigation of the Structure and Nature of Microscopic Objects,* 3rd ed., Van Voorst Paternoster Row, London, 1975, 845 pp.
37. **True, A. C.,** A History of Agricultural Experimentation and Research in the United States, 1607—1925, *U.S. Dept. Agric./Misc. Pub.,* No. 251, 1937, 321.
38. **Smith, E. F.,** Peach blight, *Monilia fructigena* Persoon, *J. Mycol.,* 7, 36, 1891.

39. **Peck, C. H.,** Report of the botanist, *N.Y. State Mus. Nat. Hist. Annu. Rep.,* 34, 24, 1881.
40. **Sorauer, P. C. M.,** *Die Obstbaumkrankheiten,* Berlin, 1879.
41. **Reed, H. S.,** *A Short History of the Plant Sciences,* Chronica Botanica, Waltham, MA, 1942, 323.
42. **Cobb, N. A.,** Contributions to an economic knowledge of the Australian rusts, *Agric. Gaz. NSW* 3, 181, 1892.
43. **Bolley, H. L.,** Wheat rust, *Indiana Agric. Exp. Stn. Bull.,* No. 26, 1889.
44. **Anderson, H. C. L.,** Rust in wheat. Experiments and their objects, *Agric. Gaz. NSW,* 1, 81, 1890.
45. **Valleau, W. D.,** Varietal resistance of plums to brown rot, *J. Agric. Res.,* 5, 365, 1915.
46. **Higgins, B. B.,** Contribution to the life history and physiology of *Cylindrosporium* on stone fruits, *Am. J. Bot.,* 1, 145, 1914.
47. **Wiltshire, S. P.,** Studies on the apple canker fungus. II. Canker infection of apple trees through scab wounds, *Ann. Appl. Bot.,* 9, 275, 1922.
48. **Zimmerman, A.,** *Botanical Microtechnique,* translated by E. J. Humphrey, New York, 1893.
49. **Chamberlain, C. J.,** *Methods in Plant Histology,* University of Chicago Press, Chicago, 1901.
50. **Keitt, G. W.,** Peach scab and its control, *U.S. Dept. Agric. Bull.,* No. 395, 1917.
51. **Nixon, E. L.,** The migration of *Bacillus amylovorus* in apple tissue and its effect on the host cells, *Pa. State Coll. Agric. Bull.,* No. 212, 1927.
52. **Rosen, H. R.,** The life history of the fire blight pathogen, *Bacillus amylovorus,* as related to the means of overwintering and dissemination, *Univ. Ark. Agric. Exp. Stn. Bull.,* No. 244, 96, 1929.

Chapter 2

ANATOMY AND HISTOCHEMISTRY OF WOUND RESPONSES IN BARK TISSUES AND IMPLICATIONS FOR MANAGEMENT OF CANKER DISEASES

Alan R. Biggs

TABLE OF CONTENTS

I. INTRODUCTION

Many important and serious diseases of fruit trees are caused by pathogens that initiate infections at wounds caused by insects, humans, wind, hail, animals, and nutritional and physiological disorders. Research on wound responses of fruit trees is required in order to understand the processes that favor or impede the development of fungal infections in fruit, leaves, shoots, and roots. It is possible that more precise information about wound responses could lead to innovative control measures based on a better understanding of the chronology of the wound response, how wound response may be influenced by external factors, or how the wound response could be modified for improved disease control.

In trees, it is plausible to view the tissue regeneration process following wounding as being coincident with or closely allied to the defense process, given that structural responses often coincide with physiological processes that contribute to the biochemical foundation of resistant structures. In this context, the effort to distinguish between structural and physiological responses becomes unnecessary. Any distinctions between physiological processes and structural wound responses in this chapter are made merely for convenience and for the purpose of discussion. It is the goal of this chapter to discuss recent findings on the anatomical and epidemiological consequences of wounding in peach trees and to draw attention to other investigations that are relevant to host-pathogen interactions. This chapter focuses on anatomical aspects of wound response as it occurs in reaction to injuries of the living tissues of the inner bark. For literature on the wound reaction in fruits, readers are referred to other articles.[1-3]

II. ANATOMY OF BARK

Studies of the defense systems of trees in general have concentrated primarily on xylem tissues because of their economic importance to the forestry industry. Because bark tissues shield the xylem from the environment, containment of mechanical injuries and infectious microorganisms by bark tissues is critical to tree health. The integrity of normal periderm and the ability of plants to form new periderms at wounds or injuries are essential characteristics for normal plant growth and development. However, in comparison to xylem tissues, responses of periderm and other bark tissues to injury and infection are defined inadequately.

The term ''bark'' is used most often in a nontechnical context and refers to all tissues external to the vascular cambium.[4,5] Accordingly, the bark is an aggregation of tissues and organs that includes phloem and secondarily thickened tissues from the secondary plant body, as well as epidermis, cortex, and phloem derived from the primary plant body.[5] The living portion of the bark consists of the phloem and the tissues of the innermost periderm, the phellogen and phelloderm. All living tissues have been collectively termed the ''inner bark''.[6] Trockenbrodt[7] has provided an informative survey and discussion of terminology used in the bark anatomy literature.

Most pathogens of fruit trees are unable to penetrate directly the corky, suberized tissues of most outer bark tissues. These outer layers represent constitutive defenses or preformed anatomical barriers to pathogen ingress. Given that the cutinized epidermis or suberized periderms of trees are the first tissues that potential pathogens encounter,[8] and given that the majority of trees remain alive for decades (or for centuries with forest trees), these barriers apparently are very effective.

A. TISSUES OF THE INNER BARK

The living inner bark of a tree is composed of secondary phloem and periderm tissues (Figures 1 and 2). In addition, cortical tissues fulfill an important role in the bark of young

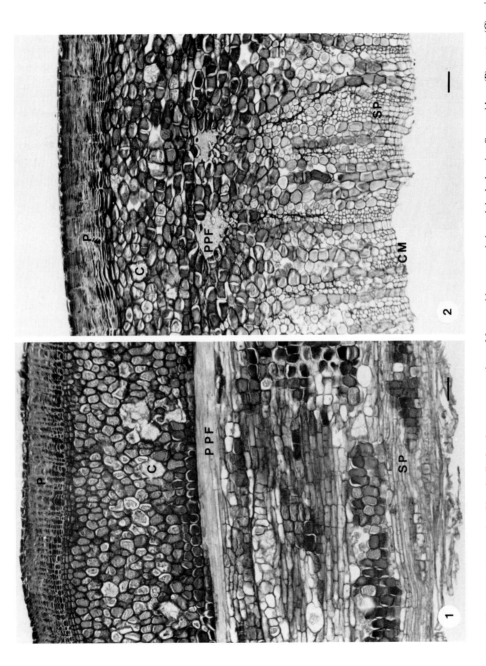

FIGURES 1 and 2. Light micrographs of longitudinal and transverse sections of 2-year-old nonwounded peach bark showing first periderm (P), cortex (C), primary phloem fibers (PPF), secondary phloem (SP), and cambial region (CM). Bar = 20 μm.

stems. Periderm, the term first used by von Mohl,[9] is a protective tissue of secondary origin which replaces the epidermis in stems and roots that have continual secondary growth. Detailed descriptions of periderm formation are available.[4,5,10] Roots, stems, and branches of gymnosperms, most dicotyledons, and a few monocotyledons develop periderm.[5] Herbaceous dicotyledons may form periderm, usually in the roots or oldest portions of the stem. In most coniferous and dicotyledonous trees, a periderm replaces the epidermis as the protective layer within the first year of growth. As trees age, sequent periderms may arise at successively greater depths, thus causing an accumulation of dead tissues on the surface of the stem or root and contributing to the formation of rhytidome on rough-barked species or simply outer bark on smooth-barked species.

The periderm consists of the following tissues, as originally described by de Bary:[11] phellogen (cork cambium), the lateral meristem which produces the periderm; the phellem (cork), the suberized protective tissue formed outwardly by the phellogen; and the phelloderm, a living parenchyma formed inwardly by the phellogen. Phellogen cells usually appear oblong in transverse and radial sections and appear polygonal or irregularly shaped when examined in tangential section.[4,5,12] Phellogen cells are characteristically thin walled, have protoplasts, are vacuolated to varying degrees, and may contain various substances, starch, and chloroplasts. The phellogen may consist of only one layer or as a zone of meristematic cells.[13] Phelloderm cells resemble cortical parenchyma cells in shape and content, although their radial arrangement makes them easily distinguished from cortical cells. Walls of phelloderm cells may be thickened and intercellular spaces may be abundant.

There are two main types of phellem cells, suberized cork cells and lignified phelloid cells. Cork cells are radially shortened, with relatively thick walls, and phelloid cells are usually thin walled and radially elongate.[14] Both cells are dead at maturity and generally lack intercellular spaces.[5] The arrangement of phellem cells in bark varies according to species.[5]

Cortical tissues are found primarily in the bark of young stems. The first periderm in the stems of most species arises in the cortex, which is ultimately shed as new periderms arise. Phloem tissues are intimately involved in the development of bark structure.[13,15-17] The inner bark of smooth-barked species consists largely of phloem tissues. In addition, patterns of phloem element deposition in conjunction with particular patterns of periderm development are responsible for the structure of ring, scale, and furrowed barks.[18] The living cells of the outer phloem give rise to deep-seated periderms.

B. NATURAL AND WOUND PERIDERMS

According to Esau,[5] natural (including first and sequent periderms) and wound periderms are basically alike in method of origin and growth. The difference between them is mainly in timing of origin and restriction of wound periderm to the place of injury. Also, wound periderm is believed to differ from natural periderms in that the former is induced by a stimulus or injury, or by factors other than those responsible for the induction of natural periderms.[4,19-21]

In peach and other angiosperms that the author has studied, wound periderms are distinct from the first periderms based on their dissimilar histochemical reactions to lignin reagents[22,23] and the formation of a ligno-suberized boundary from cells extant at the time of wounding as a prerequisite to periderm differentiation.[22,24-26] Mullick[24] provided an extensive anatomical and biochemical analysis of first, sequent, and wound periderms in gymnosperm bark. His terminology (i.e., exophylactic periderm for first and sequent periderms, and necrophylactic periderm for all wound periderms and rhytidome) is beginning to find favor among researchers.

III. ANATOMY OF WOUND RESPONSE IN BARK

A great deal is known about wound anatomy in fruit trees and there is ample, although not exhaustive, documentation in the literature on the anatomical events that lead to boundary zone and wound periderm formation in woody plants.[24,27-33]

A. LIGHT MICROSCOPY

In peach bark, the first observable response to mechanical wounding may be viewed with the light microscope as early as 24 h after wounding, although there may be cultivar differences in the timing of events and other specific features or there may be differences due to environmental effects on plant response. In injured peach bark tissues, considerable degradation of starch granules occurs within the first 12 to 24 h. By 96 h these granules will have disappeared altogether in the area of tissue undergoing dedifferentiation, although they may still be evident in the desiccated area near the wound surface and in the internal tissues some distance from the wound.[22] Cells in the incipient boundary zone undergo changes in the nucleus and the ability of the cytoplasm to take up morphological stains. In most cells of the incipient boundary zone, metaphase nuclei are easily seen and the nucleoli are prominent 48 h after wounding (Figure 3).

In *Prunus persica* and other *Prunus* spp., deposition of a polysaccharide substance has been observed with periodic acid-Schiff's (PAS) reagent in the walls of cells located in a zone about 300 μm from the wound surface.[22] Polysaccharide deposition occurred prior to the formation of a visible lignified zone. Although increased lignin can be detected with biochemical methods within the first 24 h after wounding,[34] the first signs of lignification detectable with histochemical reagents (phloroglucinol/HCl) are apparent within 72 h (Figure 4) and occur internal to the area of polysaccharide deposition (if the latter occurs). Reports of lignification as a response to wounding in tree bark are numerous.[24-26,28,31,33,35-37] The first lignified cells can be detected in areas of the wound in closest proximity to the vascular cambium. Although most cell types exhibit lignification in response to wounding, the parenchymatous cells in phloem ray tissue often are the first to stain visibly with phloroglucinol/HCl.

Suberin deposition in lignified cells occurs within 24 to 48 h after visible lignification. Results of studies on both angiosperms (including peach, apple, and sweet cherry) and gymnosperms have shown that the impermeability of the impervious layer that is formed from extant cells prior to periderm regeneration is closely related to the formation of intracellular suberin linings in cells present at the time of wounding, and which have become lignified following wounding.[22,37] Similar processes occur in wounded xylem parenchyma[38] and wounded tissues of various herbaceous and woody plant species and organs.[23]

The ligno-suberized boundary zone is most often located approximately 0.8 to 1.0 mm internal to the wound surface. Initial cells of this tissue can be detected within 4 to 7 d in wounds on actively growing trees in midsummer (Figure 5),[33,39] and usually occur in an area of the wound with closest proximity to the vascular cambium. A meristematic layer forms immediately internal to and abutting the primary ligno-suberized tissue and is usually detected 24 to 48 h after the formation of the latter tissue (Figures 6 to 8). Wound (necrophylactic) periderm may be well formed by 10 d postwounding. Complete formation of the boundary zone and new periderm around the entire wound may take up to 28 d under ideal conditions (Figures 9 and 10); however, as boundary tissues continue to form in an outward direction, new phellogen cells form immediately internal to the established boundary tissue. As phellem is produced in an outward direction, the ligno-suberized boundary is crushed and diminishes in thickness.[39]

It is important to note that the presence of tissues in various stages of wound response

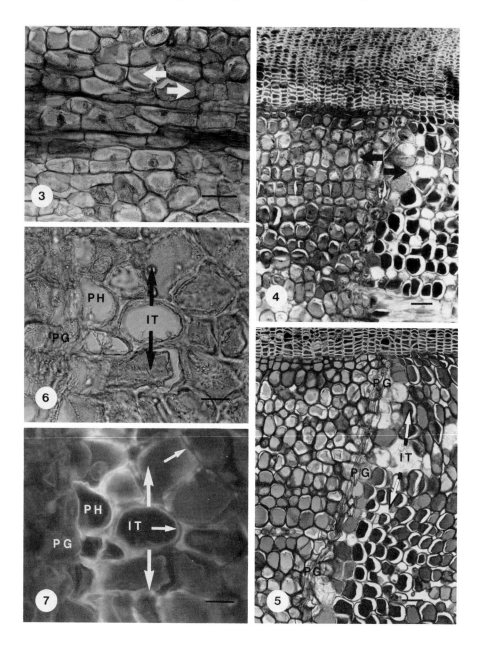

FIGURES 3 to 7.   Light micrographs of longitudinal sections through tissue undergoing dedifferentiation during generation of new phellogen after mechanical wounding of bark of healthy 2-year-old peach bark. Outside bark in all figures is oriented toward top of plate. Bar = 10 μm. **Figure 3**. Section through the secondary phloem 48 h postwounding showing mitotic activity associated with tissue redifferentiation. Cells to the left of the arrows participate in phellogen generation; cells to the right are becoming hypertrophied and are developing lignified walls and intracellular suberin linings. **Figure 4**. Section through periderm and cortex 72 h postwounding showing two distinct tissue regions in the transition zone between healthy (left) and necrotic tissue (right). Tissue to the right of the arrows shows first positive reaction with phloroglucinol/HCl, but is not yet impervious to fluid diffusion. **Figure 5**. Section of differentiated phellogen (PG) and incipient ligno-suberized impervious tissue (IT and arrows) 96 h postwounding. **Figures 6 and 7**. Bark tissue stained with phosphine, under bright field and the same under UV excitation, respectively, showing impervious tissue (IT), first phellem (PH), and newly regenerated phellogen (PG) 7 d postwounding. Note in Figure 7 the thin suberin lining of the cells in the impervious zone (small arrows). (Figures 3 through 7 from Biggs, A. R., *Can. J. Bot.*, 62, 2814, 1984. With permission.)

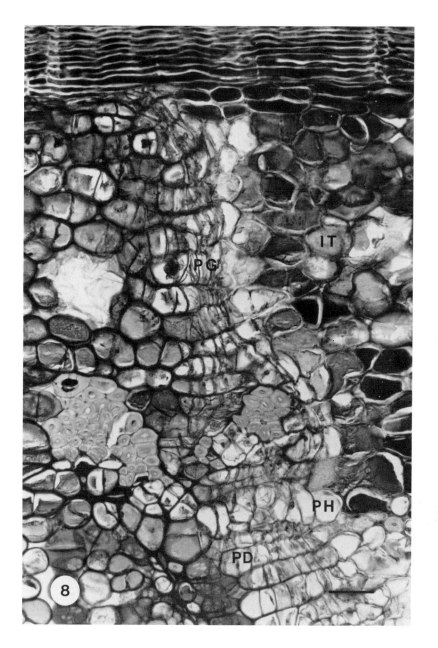

FIGURE 8. Light micrograph of stained transverse section of differentiated phellogen (PG) showing phellem (PH), phelloderm (PD), and impervious tissue (IT), 7 d postwounding. The wound is oriented toward the right side of the micrograph, healthy tissue on the left, first periderm at top. Bar = 10 μm. (From Biggs, A. R., *Can. J. Bot.*, 62, 2814, 1984. With permission.)

can be observed by sampling wounded tissues at any one time, i.e., 7 to 24 d after wounding, depending upon species, environmental conditions, and inherent regenerative capacity. Ligno-suberized cells do not form synchronously into a distinct boundary zone.[33] The cells and subsequent tissues form first between the wound surface and the vascular cambium and last in the region of the original phellogen. One can often view wound tissues ranging from no visible reaction to those exhibiting complete periderm regeneration within the same histo-logical section.[33]

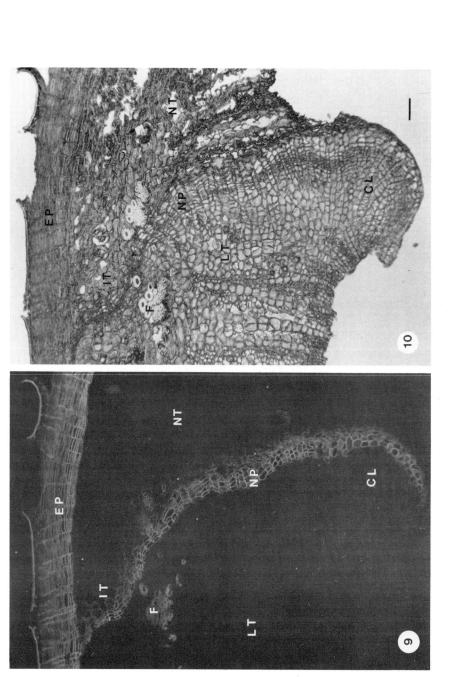

FIGURES 9 and 10.   Fluorescence and light micrographs, respectively, of transverse sections of 2-year-old peach bark at 28 d postwounding stained with phosphine GN showing necrophylactic periderm (NP) formed internal to a ligno-suberized boundary (IT). Note necrotic tissue (NT), living tissue (LT), exophylactic periderm (EP), primary phloem fibers (F), and wound callus tissue (CL). Bar = 20 μm. (From Biggs, A. R., *Phytopothology*, 76, 905, 1986. With permission.)

B. ELECTRON MICROSCOPY (TEM)

At the ultrastructural level, dramatic changes are evident within the first 24 h in cells adjacent to the wound. Generally, these changes reflect the subcellular alterations visible with light and fluorescence microscopy (described above). The most noticeable subcellular modifications are in the nucleus and include changes in nucleolar fine structure and chromatin organization, both of which suggest rRNA synthesis and active production of ribosomes. In the cytoplasm, increases occur in rough endoplasmic reticulum (ER), free ribosomes, and polysomes. Cells in the region of activity of the new phellogen show increases in the amount of cytoplasm, smooth ER, and dictyosomes.[40,41] All these changes reflect the intensified transcriptional, translational, and secretory activities of the responding cells. After 24 h, ultrastructural changes are limited to those cells either undergoing dedifferentiation to form the ligno-suberized boundary zone or those undergoing dedifferentiation to form the new phellogen and its derivatives.

Ultrastructural evidence for cell wall suberization in wounded peach bark was observed at 8 d after wounding (Figures 11 to 18). Boundary zone cell walls were completely lined on the inside with an electron-lucid material corresponding to cell wall linings with the histochemical and autofluorescence characteristics of suberin. The suberin portion of the cell walls appeared, at first, electron lucid, followed by the formation of many light and dark lammelations. The suberin lining in individual cells appeared uniform in thickness, although thickness of the lining varied from cell to cell (ca. 40 to 120 nm). Suberized cells in the boundary zone contained senescing cytoplasm with fragments of undifferentiated dense material that formed a thin, discontinuous granular deposit inside the suberin layer. The granular, electron-dense materials likely resulted from the disintegration of the cytoplasm and the various cell organelles. Between 8 and 12 d postwounding, the primary walls and the middle lamella in the boundary zone exhibited an increase in electron density, probably due to the deposition of phenolic polysaccharide material in the wall. In peach, these substances, usually referred to as gum, are produced nonspecifically in response to wounds or infections (see Chapter 11). Examination of the boundary zone with transmission electron microscopy (TEM) revealed that suberin linings were discontinuous over pit areas[41] (Figures 19 and 20). Therefore, it is likely that the impermeable nature of these primary ligno-suberized boundaries is due, to some extent, to both lignin and suberin. Vesicles of varying proportions frequently were associated with the periphery of the senescent cytoplasm (Figure 21).

Cells of the new necrophylactic phellem possess dense, granular cytoplasm with few distinct organelles (Figures 22 to 24). In mature phellem, cell contents appear as a compact mass of electron-dense amorphous material interspersed with electron-lucid deposits and dark bodies of various dimensions. Numerous vesicular elements were observed, the membrane elements appearing to be embedded randomly throughout the granular matrix. The plasmalemma typically was separated from the cell wall. Phellem cells possessed a compound middle lamella with an amorphous fine structure. This portion of the cell wall appeared red when stained with phloroglucinol/HCl. Suberin comprised the largest portion of the secondary wall and displayed fine light and dark lammelations. The thickness of phellem suberin layers (ca. 60 to 350 nm) increased with distance from the phellogen. Cell wall pits and plasmodesmata canals were not observed in the phellem. Phellem cells with intact organelles were detected infrequently and, when detected (Figure 24), were characterized by abundant mitochondria, rough ER, dictyosomes, and associated vesicles.

IV. EXTERNAL FACTORS INFLUENCING THE GENERATION OF NEW BARK TISSUES

Treatises on the role of wound healing in the resistance of plants to pathogens often

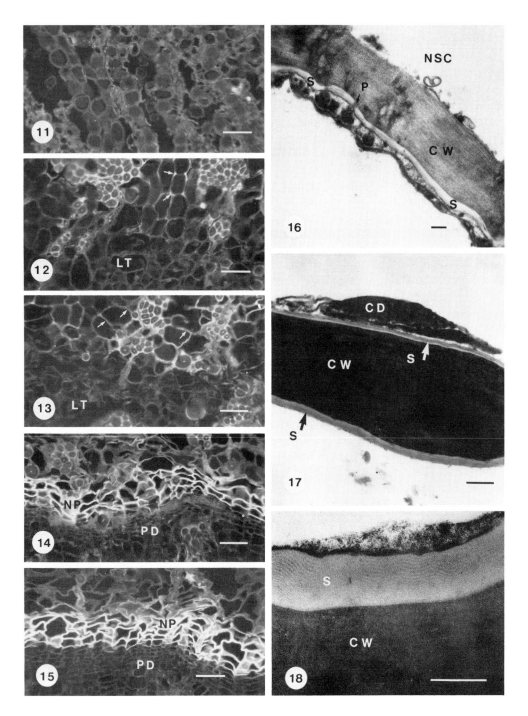

FIGURES 11 to 18.

neglect the importance of external factors as determinants of wound response rate and, indirectly, the host/pathogen interaction. Environmental factors may influence any or several of the cascade of events that are initiated following wounding (see Reference 42; Table 1), and thereby alter quality and quantity of wound-related events. Additional factors other than those discussed below probably affect the wound response either directly or indirectly, including plant nutrient status, level of herbivory and history of defoliation, carbon/nitrogen ratio as influenced by root and shoot pruning, acid rain, and air pollution.

A. TEMPERATURE

Temperature has a strong influence on the rate of wound healing in woody plant species, including fruit trees.[20,35,43-45] A significant correlation exists between temperature and rate of boundary zone and periderm regeneration in wounded bark of apple, sweet cherry, and peach.[44] Trees wounded at various times during the growing season were examined for complete formation of the primary ligno-suberized zone and new periderm. Although the tissues could be detected with a 7- to 21-d time period after wounding, degree-days (base = 0°C provided the best fit of data among the five degree-day bases examined) accumulated during the postwounding period explained over 80% of the observed variation in wound response. For peach, complete lignification and suberization were observed after 256 and 411 degee-days, respectively. Sweet cherry and apple were lignified after 212 and 192 degree-days, respectively, and suberized after 327 and 397 degree-days, respectively. Tree phenological stage did not appear to exert significant influence on the wound responses measured in this study.

In experiments to determine the formation of the primary ligno-suberized layer and phellogen following leaf abscission in peach, Biggs and Northover[43] reported that potted trees maintained in growth chambers at 7.5, 12.5 and 17.5°C showed the first indications of the primary ligno-suberized layer at 18, 9, and 6 d, respectively. Subsequent generation of phellogen and the appearance of the first phellem cells were observed at 30, 18, and 12 d, respectively. Earlier research on the influence of temperature on wound-induced cell division has shown that within limits, the time required for the first cell division is linearly and inversely related to temperature.[46] Maximum and minimum temperature limits for wound responses have not been established for any tree species, however, in potato tubers, maximum

FIGURES 11 to 18. Transverse sections of peach bark examined with UV epifluorescence illumination (Figures 11 to 15, bar = 10 μm) and TEM (Figures 16 to 18, bar = 1 μm). In Figures 12 to 15, the wound surface is approximately 800 μm above the noted cellular changes. Note the presence of phloem fibers (F) in Figures 12 to 14. **Figure 11**. Nonwounded control tissue showing ray parenchyma in the region of primary and secondary phloem. Note the lack of autofluorescence. **Figure 12**. Tissue 6 d postwounding. First traces of lignin autofluorescence are in the cell corner-middle lamella region of boundary zone cells (arrows) immediately external to living tissues (LT). **Figure 13**. Tissues 8 d postwounding. Note deposition of suberin linings (arrows) in boundary zone cells and the dedifferentiation of internal tissues in the process of forming new phellogen (LT). **Figure 14**. Tissue 12 d postwounding showing a completely differentiated necrophylactic periderm with phelloderm (PD) and two to four layers of phellem cells (NP). **Figure 15**. Tissues 14 d postwounding, showing a completely differentiated necrophylactic periderm with 3 to 5 phellem cell layers (NP) and newly differentiated phelloderm (PD). **Figures 16 to 18**. TEMs of transverse sections of peach bark boundary zone tissues. **Figure 16**. Portion of boundary zone cell 8 d postwounding with suberin lining (S) adjacent to a nonsuberized cell (NSC). Plasmodesmatal canals (P) in shared cell wall (CW) appear more electron dense than suberin. Suberin appears only slightly lamellate and is relatively electron lucid. **Figure 17**. Shared cell wall of two adjacent cells in the boundary zone 12 d postwounding. The compound wall (CW) is very electron dense. Suberin linings (S) are of uniform thickness within individual cells and vary in thickness among cells. Note the fine light and dark lamellations of the suberized wall and the cytoplasmic debris (CD) appressed to the wall of the upper cell. **Figure 18**. High magnification view of suberized cell wall in the boundary zone showing the fine light and dark lammelations characteristic of suberin. (From Biggs, A. R. and Stobbs, L. W., *Can. J. Bot.*, 64, 1606, 1986. With permission.)

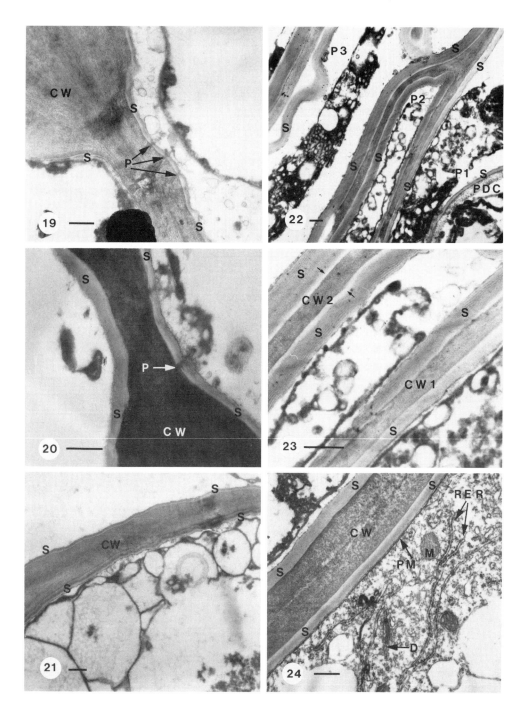

FIGURES 19 to 24.

suberization occurs at 20° to 25°C.[45,47-49] Krähmer[35] found that periderm could not be detected microscopically in leaf scars of apple at temperatures <8°C. Accordingly, fruit scars remained susceptible to infection by *Nectria galligena* for more than 4 weeks at 6°C.

When almond trees were wounded during the dormant stage, the lignin content of the bark that was determined with a thioglycolic acid assay increased, although an increase in lignin detectable with the histochemical reagent phloroglucinol/HCl was not observed.[50] Wounds became more resistant to inoculation with *Phytophthora syringae*, however, thus suggesting resistance associated with lignin at very low tissue levels or mechanisms of resistance in addition to those associated with wound repair. Wound studies during the dormant season require long observation and sampling schedules, and it is possible that epiphytic microorganisms could colonize the wound site and contribute to altered rates of pathogen colonization independent of the responses of the host.

B. EPIPHYTIC MICROBES

Biggs and Alm[51] recently studied the anatomy of wounded peach bark that was inoculated with the fungal epiphytes *Alternaria alternata*, *Trichoderma harzianum*, and *Epicoccum nigrum* alone and in combination with the peach canker pathogen *Leucostoma cincta*. Experiments were conducted at different times during the growing season in which bark tissues were sampled over a time series for light and fluorescence microscope examination to determine the influence of inoculations on the ontogeny of tissues associated with wound healing. Combined inoculations of *L. cincta* and epiphytes consistently, but not always, resulted in wound healing similar to that observed for the epiphyte alone. Inoculation of wounds with *E. nigrum* alone, or in combination with *L. cincta*, did not cause any deleterious effects on the wound responses examined in this study, whereas inoculation of wounds with *T. harzianum* resulted in delayed formation, but not prevention of tissues critical to normal wound healing. None of the epiphytic fungi were associated with the formation of wound-related tissues more quickly than was observed in the noninoculated control wounds.

FIGURES 19 to 24. TEMs of transverse sections of the boundary zone and necrophylactic periderm of wounded peach bark. Bar = 1 μm. **Figure 19**. Portion of a shared boundary zone cell wall (CW) 8 d postwounding showing a simple pit area, associated plasmodesmata (P), and lamellate suberin linings (S). **Figure 20**. High magnification view of plasmodesmatal opening through the suberized portion of the boundary zone cell wall. **Figure 21**. Shared boundary zone cell wall (CW) at 12 d postwounding with lamellate suberin linings (S) of differing thickness and the multivacuolate appearance of a senescent boundary zone cell. **Figure 22**. Portion of necrophylactic phellem showing at bottom right a portion of the lumen and shared wall of the external phellogen daughter cell (PDC), the most recently derived suberized phellem cell (P1) with thin suberized secondary walls (S), and two older phellem cells (P2 and P3) showing thick suberized secondary walls (S) and senescing cell cytoplasm. **Figure 23**. Closer view of shared necrophylactic phellem cell walls between P1 and P2 (CW1) and P2 and P3 (CW2) from Figure 22. Note the fine light and dark lamellations of the suberized secondary walls (S) and the nonlamellate electron-lucid region of the secondary wall adjacent to the primary wall (arrows). **Figure 24**. Organelle-rich cytoplasm of a phellem cell of a necrophylactic periderm sampled 14 d postwounding. Note lignified primary wall (CW), suberized secondary walls (S), plasma membrane (PM), rough endoplasmic reticulum (RER), mitochondrion (M), dictyosome (D), and associated vesicles. (From Biggs, A. R. and Stobbs, L. W., *Can. J. Bot.*, 64, 1606, 1986. With permission.)

TABLE 1
Summary of Morphological and Histochemical Changes Related to Phellogen Generation after Wounding in Peach Bark

Wound age (d)	Lignified boundary[a]	Ligno-suberized boundary[a]	Necrophylactic periderm[b]	Suberized xylem ray parenchyma[a]	Transmission (%) (lignin)[c]	Autofluorescence intensity (suberin)[d]
0	−	−	0	−	100	0.0
3	+	−	0	−	97.3	0.0
7	+	+	0	+	76.7	3.1
10	+	+	1.0	+	66.0	15.2
14	+	+	3.0	+	67.1	21.2
24	+	+	6.0	+	61.4	33.3

[a] Feature present (+) or absent (−) in the outer bark cortex in contact with the original periderm in 100% of plants examined. For suberized xylem ray parenchyma, signs indicate presence or absence in 100% of plants examined.

[b] Mean number of cells counted in transverse section at the junction of the new and original periderms (n = 20).

[c] Percent transmission at 546 nm measured over a circular area with 272 μm diameter; each measured area contained about 100 cells. Values represent mean percent transmission (n = 20) of ligno-suberized tissue or ligno-suberized tissue plus necrophylactic phellem, depending on wound age. Nonwounded bark percent transmission = 100.

[d] Autofluorescence intensity measured over a circular area with 272 μm diameter; each measured area contained about 100 cells. Values represent mean autofluorescence intensity (n = 20) of ligno-suberized tissue or ligno-suberized tissue plus necrophylactic phellem, depending on wound age. Nonwounded bark autofluorescence intensity = 0.

C. WATER STRESS

Plants under water stress are generally considered more susceptible to invasion by weak pathogens,[52] however, few studies have been able to demonstrate the mechanism by which water stress increases the susceptibility of trees to fungal invasion (see Schoenweiss[52] for a review of this literature). Water deficit influences numerous physiological processes and pathways, and wound responses are no exception. Plant water status has been shown to influence the formation of the boundary zone, affect cell division in the wound periderm, and contribute to increased susceptibility of wounds to fungal pathogens.

Biggs and Cline[53] examined the effects of irrigation treatments on the rate of boundary zone and wound periderm formation in wounds on peach limbs. No differences between irrigated and nonirrigated trees could be determined 7 to 10 d postwounding for lignin autofluorescence, the intensity of suberin autofluorescence, or in the numbers of boundary zone cells; thus, irrigation did not influence the formation of the primary ligno-suberized layer. For the cultivar 'Candor' (but not for 'Redhaven'), significant differences in suberin autofluorescence due to irrigation were measured 14 d after wounding and were related to increased numbers of suberized phellem cells in the wound periderm of irrigated trees relative to nonirrigated trees. Water stress, therefore, in the range of -0.65 to -0.80 MPa can inhibit periderm formation by diminishing the rate of cell division. In *Abies grandis*, the formation of the primary ligno-suberized layer (termed nonsuberized impervious tissue or NIT by Mullick[24]) was inhibited by water stress less than -1.5 MPa.[54] Where formation of NIT was retarded, subsequent generation of wound periderm was also thought to be slowed. Butin[31] found that susceptibility of poplar to *Cytospora chrysosperma* was increased where callusing of wounds was inhibited by increased water loss.

D. RELATIVE HUMIDITY

Relative humidity has an influence on the wound repair process in potatoes[45,48] and, presumably on that of woody species. In general, wound healing proceeds most rapidly at 20 to 25°C with relative humidities between 70 and 100% and, at 10°C, between 80 and 100%. As relative humidity approaches 100%, cell proliferation, rather than periderm formation, may occur. Relative humidities <50% may inhibit wound healing, although it is possible that as humidity declines, the depth at which the plant forms new periderm increases as desiccation of wounded tissues occurs. There are no data on relative humidity in which the studies were conducted on fruit tree tissues.

E. MISCELLANEOUS FACTORS

Other external factors may influence the expression of wound responses in plants. Suberin deposition in primary ligno-suberized tissue in wounded *Pachysandra terminalis* was affected adversely by both exposure to solar radiation and deicing salts.[55] The influence of microorganisms as irritants[56] or stimulating factors at the wound site[57] has received little attention. Studies in this area could reveal significant new information about fundamental aspects of wound responses in nature.

V. REGENERATION OF VASCULAR CAMBIUM AND WOUND CLOSURE

Given that many of the tissues involved in callus regeneration are part of the barrier zone in the process of compartmentalization, only a limited discussion is presented here. (Refer to Blanchette[58] for an excellent discussion on the anatomy of barrier zones and their role in defense against fungal pathogens.) The production of callus tissue, the differentiation of new vascular cambium within callus, and the eventual closure of the wound, usually leading to the reestablishment of vascular cambium continuity, occurs following wounds inflicted to the depth of the xylem. The amount and rate of callus production following wounding varies according to tree species and selections within species,[59,60] the size of the wound, and location of the wound on the tree.[61] The source of callus tissue also varies to some extent. In scoring experiments with poplar, silver maple, pear, and apple, callus formation is contributed mainly by living cells of vascular rays in the proximity of the cut, and, to a lesser extent, the longitudinal parenchyma of the phloem and xylem.[62] In linden, callus tissue forms from any active, newly formed cambial derivative rather than from any particular cell type.[63] Generally, there is a diversity of opinion on the source of callus in the regeneration of new bark,[19] although most researchers agree that the original vascular cambium does not contribute significantly to the formation of callus tissue. The initiation of phellogen regeneration in callus always precedes that of vascular cambium. The ventral region of growing callus, where the differentiating phellogen and vascular cambium are in close proximity, is nonlignified, nonsuberized, and is especially susceptible to disruption by pathogenic fungi.[64,65]

VI. PATHOGEN RESISTANCE IN A DYNAMIC INFECTION COURT

Following wounding, the process of establishing boundaries in the infection court between healthy tissues and the external environment confers resistance to infection in many host/pathogen interactions. Wounds generally become increasingly less susceptible to infection with age.[22,31,66-69] Resistance to infection in this manner is probably related to nonspecific plant responses leading up to and including formation of primary ligno-suberized tissues and secondary wound (or necrophylactic) periderm. The major structural components of these tissues are lignin and suberin.[70]

Upon the creation of an infection court for a wound pathogen, incidence and severity of infection will be greatest when the inoculum arrives at the infection court immediately. For wounds in which the inoculum does not arrive immediately, disease frequency and severity decline with time until the wounded tissues express resistance comparable to that of noninjured bark.[71,72] The infection court becomes a dynamic entity, its receptiveness to inoculum influenced by external events such as colonization by epiphytic organisms, and internal events such as host boundary-setting processes. Focusing on the latter, the time for complete resistance to become reestablished is dependent upon the factors discussed above as well as the pathogenic capabilities of the fungus. Histological studies with peach and *Leucostoma persoonii* showed that wounds resistant to inoculation possessed a minimum of three phellem cells in the new periderm (Table 1). At earlier stages, wounds were susceptible to the fungus, although severity of the symptoms declined beginning about 3 d after wounding.[71] The presence of periderm was critical for inhibition of the pathogen, whereas the presence of lignified and/or primary ligno-suberized tissues significantly slowed, but did not inhibit, fungal colonization (Table 2). Interestingly, none of the chemical or morphological barriers were observed at 3 d postwounding, suggesting a role in pathogen resistance for phytoalexins, lignin at levels not detectable with histochemical reagents, or other substances in the early stages of the wound reaction.[34,69,71]

The relative number of cells in periderm tissue or the thickness of the suberized layers in wounded potato tubers generally reflect the relative resistance to pathogens (i.e., wounds with thicker suberized layers are more resistant to pathogens). Generally, this is true for wound periderms in woody plants when comparisons are being made within a single genotype[30] or among different species over a time course study.[44] However, when comparing genotypes within a species, it is unlikely that numbers of wound phellem cells or the thickness of the

TABLE 2

Canker Length (millimeters + 6.0 mm) and Linear Regression Coefficients for the Relationship between Canker Length (Y) and Time (days, X) after Inoculation of Peach Bark Wounds of Varying Age with Mycelium of *Leucostoma persoonii*

Wound age (d)	Wounds infected (%)	Time postinoculation (d)				Regression coefficient
		3	7	14	21	
CK[a]	0	6.0[b]	6.0	6.0	6.0	0.388a[c]
24	0	6.2	6.0	6.0	6.2	0.395a
14	10	6.6	9.2	12.2	15.4	0 .832a
10	100	6.6	13.0	18.6	24.6	1.277b
7	100	9.2	18.4	26.6	27.4	1.589b
3	100	14.2	20.2	26.2	30.2	1.705bc
0	100	20.2	27.2	39.2	49.8	2.656c

[a]   CK, noninoculated control.
[b]   Data are means of measurements from 10 trees from 1 experiment.
[c]   Analysis, if covariance for testing homogeneity of regression coefficients was significant, F = 9.31 ($p \leq 0.01$). Letters denote significantly different regression coefficients using paired t-tests in all combinations.

suberized layer is of major importance in determining the resistance or susceptibility to wound pathogens. In peach, there was no significant correlation of these anatomical parameters with the field performance of various cultivars, canker length after inoculations, or the amount of accumulated suberin measured photometrically.[72,73]

For peach, resistance to *Leucostoma* spp. was correlated with increased rate and amount of suberin accumulation, rather than the number of suberized cells or thickness of suberized layers.[72-74] Lignin accumulation, which is thought to play an important role in many host/pathogen interactions,[75,76] appears to serve a less important role than suberin in peach bark and perhaps the bark of other tree species. Despite the abundance of correlative data presented in recent papers,[22,71-74] the precise role and importance of ligno-suberized tissue and new periderm, and their biochemical components, in resistance to fungal pathogens in trees awaits more definitive investigations.

Periderm generation is probably only one of many possible types of resistance in peach bark to *Leucostoma* spp., and most likely serves as a type of rate limiting or partial resistance. In the orchard, wound healing processes result in the progressive reduction of the relative risk of infection at wounds.[69,71] For infection of almond bark wounds by *Ceratocystis fimbriata*, 10-d-old wounds were 19 times less likely to develop cankers when compared with 2-d-old wounds.[69] In Ontario, guidelines were formulated for peach growers who elected to prune at different times during the early spring. By using historical weather data and observations on the influence of temperature on wound healing, the author[77] was able to estimate the number of days required for lignification and suberization of wounds made at various times. For example, pruning wounds made on March 15 would require up to 62 d for effective wound healing based on an estimated accumulation of 411 degree-days. In contrast, pruning on May 15 would necessitate only 27 calendar days to accumulate the 411 degree-days required for suberization. Although these estimates of wound response are only guidelines, and may vary greatly with numerous site and environmental factors, they are useful in communicating the concept of relative risk in management decisions. Indeed, the long periods of time required for wounds to develop resistance to wound pathogens may help explain why single-spray fungicide treatments have been ineffective for controlling Leucostoma canker.

It should be noted that fungi inoculated into wounds can survive in nature without causing immediate infection and/or without giving rise to immediately visible signs or symptoms. Wounded peach bark inoculated with spores of *Botryosphaeria* spp. did not develop visible macroscopic symptoms until 8 weeks after inoculation.[65] Mycelium of some forest tree pathogens can survive in wounds for periods of up to 2 years prior to development of symptoms,[78] as can the mycelium of *Valsa ceratosperma* in apple.[79] Formation of a ligno-suberized boundary zone, generation of new periderm, and callus tissue do not guarantee that wounds will not become infected, although the risk of infection is reduced greatly. Aspects of the anatomy of wound closure and microenvironment during the process of wound closure may help explain the apparent latent colonization of woody tissues by some canker-causing fungi.

VII. HERITABILITY OF THE WOUND REACTION

It is clear from the above discussion that intraspecific variation exists in peach for suberin accumulation and that this variation is related to relative susceptibility to the peach canker pathogens. Knowledge of the natural variation and the degree to which suberin accumulation rate is transmitted from parent to progeny is essential if this character is to be considered part of a breeding program for increased resistance to peach canker disease. however, there are few references in the literature on the heritability of this trait. Biggs et al.[80] initiated controlled reciprocal crosses among peach clones V68101, V68051, and New Jersey Cling (NJC) 95 during spring 1985 and parental clones and seedlings from the crosses were planted at two sites in May 1986. Wounding studies were conducted during June 1989 to quantify suberin accumulation in health phloem/cortex tissues adjacent to the wound site. Differences in suberin accumulation among the three parent clones were found to be significant, whereas all environmental sources of variation and their interactions with parent clone were not. Suberization of V68101 was significantly greater than that observed in NJC 95 and V68051, neither of which differed from each other. Suberization response in the three families involving V68101 were not significantly different from each other, and the remaining three families also were not significantly different from each other. The full-sib family heritability of 0.93 $\pm$ 0.31 reflected the significance of differences among full-sib family means. For hybrids, the individual tree heritability of 0.56 $\pm$ 0.24 indicated that approximately 55% of the phenotypic variability was accounted for by genotype. Genetic factors are sufficiently large in peach to allow for selection for increased suberization response.

Other components of nonspecific defense mechanisms in trees have been demonstrated to be under genetic control. However, it should not be assumed that the various bark and xylem resistance mechanisms are controlled by similar sets of genes. Indeed, it was demonstrated that bark and xylem wound responses in peach appeared to be independent.[64] In addition, the results of Shigo et al.[81] suggested that wound closure is a process separate from xylem compartmentalization in hybrid poplar. Lowerts and Kellison[82] demonstrated that resistance to discoloration and decay in wounded yellow poplar was under moderately strong genetic control with a family heritability estimated at 0.57. They suggested that the large amount of tree-to-tree variation and the moderately high single tree heritability estimate would allow the largest genetic gains to be made from a program of within-family selection. This observation also appears to be applicable to our findings with regard to selection for resistance to bark pathogens.

VIII. CONCLUDING REMARKS

Understanding the basis for wound responses in fruits, roots, stems, and leaves of woody plants could lead to innovative control measures for tree diseases. Research in this area should include basic studies on the molecular regulation of wound metabolism in trees, pathological studies on the role of wound responses in resistance to bark and xylem pathogens, the role of nonpathogenic microorganisms as regards properties of wound tissues, and genetic experiments to determine the heritability of favorable wound response characteristics. Well-designed time course studies with adequate controls are required in order to completely characterize wound related phenomena.

REFERENCES

1. **Skene, D. S.,** Wound healing in apple fruits: the anatomical response of Cox's Orange Pippen at different stages of development, *J. Hortic. Sci.,* 56, 145, 1981.
2. **Brown, E. A.,** Host defenses at the wound site on harvested crops, *Phytopathology,* 79, 1381, 1989.
3. **Lakshminarayana, S., Sommer, N. F., Polito, V., and Fortlage, R. J.,** Development of resistance to infection by *Botrytis cinerea* and *Penicillium expansum* in wounds of mature apple fruit, *Phytopathology,* 77, 1674, 1987.
4. **Srivastava, L. M.,** Anatomy, chemistry, and physiology of bark, *Int. Rev. For. Res.,* 1, 204, 1964.
5. **Esau, K.,** *Plant Anatomy,* 2nd ed., John Wiley & Sons, New York, 1965.
6. **Eames, A. J. and MacDaniels, L. H.,** *Introduction to Plant Anatomy,* 2nd ed., McGraw-Hill, New York, 1947, 427 pp.
7. **Trockenbrodt, M.,** Survey and discussion of the terminology used in bark anatomy, *Int. Assoc. Wood Anat. Bull.,* 11, 141, 1990.
8. **Kolattukudy, P. E. and Koller, W.,** Fungal penetration of the first line defensive barriers of plants, in *Biochemical Plant Pathology,* Callow, J. A., Ed., John Wiley & Sons, New York, 1983, 79.
9. **von Mohl, H.,** Untersuchungen über die Entwicklung des Korkes und der Borke auf der Rinde der baumartigen Dicotylen, Verm. Schr. Bot. Inh. 212, 1845.
10. **Fahn, A.,** *Plant Anatomy,* Pergamon Press, Oxford, 1967.
11. **de Bary, A.,** *Comparative Anatomy of the Vegetative Organs of the Phanerogams and Ferns,* Oxford University Press (Clarendon), London, 1884.
12. **Schneider, H.,** Ontogeny of lemon wood bark, *Am. J. Bot.,* 42, 893, 1955.
13. **Borger, G. A. and Kozlowski, T. T.,** Early periderm ontogeny in *Fraxinus pennsylvanica, Ailanthus altissima, Robinia pseudoacacia,* and *Pinus resinosa* seedlings, *Can. J. For. Res.,* 2, 135, 1972.
14. **Grozditz, G. A., Godkin, S. E., and Keith, C. T.,** The periderm of three North American conifers. I. Anatomy, *Wood. Sci. Technol.,* 16, 305, 1982.
15. **Borger, G. A. and Kozlowski, T. T.,** Effect of growth regulators and herbicides on normal and wound periderm ontogeny in *Fraxinus pennsylvanica* seedlings, *Weed Res.,* 12, 190, 1972.
16. **Borger, G. A. and Kozlowski, T. T.,** Effects of cotyledons, leaves and stem apex on early periderm development in *Fraxinus pennsylvanica* seedlings, *New Phytol.,* 71, 691, 1972.
17. **Borger, G. A. and Kozlowski, T. T.,** Wound periderm ontogeny in *Fraxinus pennsylvanica* seedlings, *New Phytol.,* 71, 709, 1972.
18. **Kozlowski, T. T.,** *The Growth and Development of Trees,* Academic Press, New york, 1971.
19. **Bloch, R.,** Wound healing in higher plants, *Bot. Rev.,* 7, 110, 1941.
20. **Bloch, R.,** Wound healing in higher plants, *Bot. Rev.,* 28, 655, 1952.
21. **Akai, S.,** Histology of defense in plants, in *Plant Pathology,* Vol. 2, Horsfall, J. and Dimond, A., Eds., Academic Press, New York, 1959, 391.
22. **Biggs, A. R.,** Boundary zone formation in peach bark in response to wounds and *Cytospora leucostoma* infection, *Can. J. Bot.,* 62, 2814, 1984.
23. **Rittinger, P. A., Biggs, A. R., and Peirson, D. R.,** Histochemistry of lignin and suberin deposition in boundary layers formed after wounding in various plant species and organs, *Can. J. Bot.,* 65, 1886, 1987.
24. **Mullick, D. B.,** The non-specific nature of defense in bark and wood during wounding, insect and pathogen attack, *Rec. Adv. Phytochem.,* 11, 395, 1977.

25. **Soo, B. V. L.**, General Occurrence of Exophylactic and Necrophylactic Periderms and Nonsuberized Impervious Tissues in Woody Plants, Ph.D. thesis, University of British Columbia, Vancouver, B. C., Canada, 1977.

26. **Biggs, A. R.**, Intracellular suberin: occurrence and detection in tree bark, *Int. Assoc. Wood Anat. Bull.*, 5, 243, 1984.

27. **Hartig, R.**, Diseases of trees, in *Diseases of Trees*, Ward, H. M., Ed., Macmillan, New York, 1894, 225.

28. **Bramble, W. C.**, Reaction of chestnut bark to invasion by *Endothia parasitica*, *Am. J. Bot.*, 23, 89, 1934.

29. **Crowdy, S. H.**, Observations on apple canker. III. The anatomy of the stem canker, *Ann. Appl. Biol.*, 36, 483, 1949.

30. **Bloomberg, W. J. and Farris, S. H.**, Cytospora canker of poplars: bark wounding in relation to canker development, *Can. J. Bot.*, 41, 303, 1963.

31. **Butin, H.**, Über den Einfluss des Wassergehaltes der Pappel auf ihre Resistenz gegenuber *Cytospora chrysosperma* (Pers.)Fr., *Phytopathol. Z.*, 24, 245, 1955.

32. **Biggs, A. R., Merrill, W., and Davis, D. D.**, Discussion: response of bark tissues to injury and infection, *Can. J. For. Res.*, 14, 351, 1984.

33. **Biggs, A. R.**, Suberized boundary zones and the chronology of wound response in tree bark *Phytopathology*, 75, 1191, 1985.

34. **Doster, M. A. and Bostock, R. M.**, Quantification of lignin formation in almond bark in response to wounding and infection by *Phytophthora* species, *Phytopathology*, 78, 473, 1988.

35. **Krähmer, H.**, Wound reactions of apple trees and their influence on infections with *Nectria galligena*, *J. Plant Dis. Protect.*, 87, 97, 1980.

36. **Biggs, A. R., Davis, D. D., and Merrill, W.**, Histology of cankers on *Populus* caused by *Cytospora chrysosperma*, *Can. J. Bot.*, 61, 563, 1983.

37. **Biggs, A. R.**, Detection of impervious tissue in tree bark with selective histochemistry and fluorescence microscopy, *Stain Technol.*, 60, 299, 1985.

38. **Biggs, A. R.**, Occurrence and location of suberin in wound reaction zones in xylem of seventeen tree species, *Phytopathology*, 77, 718, 1987.

39. **Biggs, A. R.**, Phellogen regeneration in injured peach tree bark, *Ann. Bot.*, 57, 463, 1986.

40. **Barckhausen, R.**, Ultrastructural changes in wounded plant storage tissue cells, in *Biochemistry of Wounded Plant Tissues*, Kahl, G., Ed., Walter de Gruyter, Berlin, 1978, 1.

41. **Biggs, A. R. and Stobbs, L. W.**, Fine structure of the suberized cell walls in the boundary zone and necrophylactic periderm in wounded peach bark, *Can. J. Bot.*, 64, 1606, 1986.

42. **Bostock, R. M. and Stermer, B. A.**, Perspectives on wound healing in resistance to pathogens, *Annu. Rev. Phytopathol.*, 27, 343, 1989.

43. **Biggs, A. R. and Northover, J.**, Formation of the primary protective layer and phellogen following leaf abscission in peach, *Can. J. Bot.*, 63, 1547, 1985.

44. **Biggs, A. R.**, Prediction of lignin and suberin deposition in boundary zone tissue of wounded tree bark using accumulated degree days, *J. Am. Soc. Hortic. Sci.*, 111, 757, 1986.

45. **Morris, S. C., Forbes-Smith, M. R., and Scriven, F. M.**, Determination of optimum conditions for suberization, wound periderm formation, cellular desiccation and pathogen resistance in wounded *Solanum tuberosum* tubers, *Physiol. Molec. Plant Pathol.*, 35, 177, 1989.

46. **Lipetz, J.**, Wound healing in higher plants, *Int. Rev. Cytol.*, 27, 1, 1970.

47. **Artschwager, E.**, Wound-periderm formation in the potato as affected by temperature and humidity, *J. Agric. Res.*, 35, 995, 1927.

48. **Wigginton, M. J.**, Effects of temperature, oxygen tension and relative humidity on the wound-healing process in the potato tuber, *Potato Res.*, 17, 200, 1974.

49. **Thomas, P.**, Wound-induced suberization and periderm development in potato tubers as affected by temperature and gamma irradiation, *Potato Res.*, 25, 155, 1982.

50. **Doster, M. A. and Bostock, R. M.**, Effects of low temperature on resistance of almond trees to *Phytophthora* pruning wound cankers in relation to lignin and suberin formation in wounded bark tissue, *Phytopathology*, 78, 478, 1988.

51. **Biggs, A. R. and Alm, G.**, Response of peach bark tissues to inoculation with epiphytic fungi alone and combination with *Leucostoma cincta*, *Can. J. Bot.*, 70, 186, 1992.

52. **Schoenweiss, D. F.**, The role of environmental stress in diseases of woody plants, *Plant Dis.*, 65, 308, 1981.

53. **Biggs, A. R. and Cline, R. A.**, Influence of irrigation on wound response in peach bark, *Can. J. Plant Pathol.*, 8, 405, 1986.

54. **Puritch, G. S. and Mullick, D. B.**, Effect of water stress on the rate of nonsuberized impervious tissue formation following wounding in *Abies grandis*, *J. Exp. Bot.*, 26, 903, 1975.

55. **Hudler, G. W., Neal, B. G., and Banik, M. T.**, Effects of growing conditions on wound repair and disease resistance in *Pachysandra terminalis*, *Phytopathology*, 80, 272, 1990.

56. **Kaufert, F.,** Factors influencing the formation of periderm in aspen, *Am. J. Bot.,* 24, 1936.
57. **Blanchette, R. A. and Sharon, E. M.,** *Agrobacterium tumefaciens,* a promoter of wound healing in *Betula alleghaniensis, Can. J. For. Res.,* 5, 722, 1975.
58. **Blanchette, R. A.,** Anatomical responses of xylem in injury and invasion by fungi, in *Defence Mechanisms of Woody Plants Against Fungi,* Blanchette, R. A. and Biggs, A. R., Eds., Springer-Verlag, Berlin, in press.
59. **Gallagher, P. W. and Sydnor, T. D.,** Variation in wound response among cultivars of red maple, *J. Am. Soc. Hortic. Sci.,* 108, 744, 1983.
60. **Martin, J. M. and Sydnor, T. D.,** Differences in wound closure rates in 12 tree species, *Hortic. Sci.,* 22, 442, 1987.
61. **Wensley, R. N.,** Rate of healing and its relation to canker of peach, *Can. J. Plant Sci.,* 46, 257, 1966.
62. **Soe, K.,** Anatomical studies of bark regeneration following scoring, *J. Arnold Arbor. Harv. Univ.,* 40, 260, 1959.
63. **Barker, W. G.,** A contribution to the concept of wound repair in woody stems, *Can. J. Bot.,* 32, 486, 1954.
64. **Biggs, A. R.,** Comparative anatomy and host response of two peach cultivars inoculated with *Leucostoma cincta* and *L. persoonii, Phytopathology,* 76, 905, 1986.
65. **Biggs, A. R. and Britton, K. O.,** Presymptom histopathology of peach trees inoculated with *Botryosphaeria obtusa* and *Botryosphaeria dothidea, Phytopathology,* 78, 1109, 1988.
66. **Cline, M. N. and Neely, D.,** Wound-healing processes in geranium cuttings in relationship to basal stem rot caused by *Pythium ultimum, Plant Dis.,* 67, 636, 1983.
67. **Russin, J. S. and Shain, L.,** Initiation and development of cankers caused by virulent and cytoplasmic hypovirulent isolates of the chestnut blight fungus, *Can. J. Bot.,* 62, 2660, 1984.
68. **Riffle, J. W. and Peterson, G. W.,** Thyronectria canker of honeylocust: influence of temperature and wound age on disease development, *Phytopathology,* 76, 313, 1986.
69. **Bostock, R. M. and Middleton, G. E.,** Relationship of wound periderm formation in resistance to *Ceratocystis fimbriata* in almond bark, *Phytopathology,* 77, 1174, 1987.
70. **Kolattukudy, P. E.,** Biochemistry and function of cutin and suberin, *Can. J. Bot.,* 62, 2918, 1984.
71. **Biggs, A. R.,** Wound age and infection of peach bark by *Cytospora leucostoma, Can. J. Bot.,* 64, 2319, 1986.
72. **Biggs, A. R.,** Temporal changes in the infection court following wounding of peach bark and their association with cultivar variation in infection by *Leucostoma persoonii, Phytopathology,* 79, 627, 1989.
73. **Biggs, A. R. and Miles, N. W.,** Association of suberin formation in noninoculated wounds with susceptibility to *Leucostoma cincta* and *L. persoonii* in various peach cultivars, *Phytopathology,* 78, 1070, 1988.
74. **Biggs, A. R. and Miles, N. W.,** Suberin deposition as a measure of wound response in peach bark, *HortScience,* 20, 902, 1985.
75. **Craft, C. C. and Audia, V.,** Phenolic substances associated with wound barrier formation in vegetables, *Bot. Gaz.,* 123, 211, 1962.
76. **Vance, C. P., Kirk, T. K., and Sherwood, R. T.,** Lignification as a mechanism of disease resistance, *Annu. Rev. Phytopathol.,* 18, 259, 1980.
77. **Biggs, A. R.,** Integrated control of Leucostoma canker of peach in Ontario, *Plant Dis.,* 73, 869, 1989.
78. **Ostry, M. E. and Anderson, N. A.,** Infection of trembling aspen by *Hypoxylon mammatum* through cicada oviposition wounds, *Phytopathology,* 73, 1092, 1983.
79. **Sakuma, T.,** Studies on apple Valsa canker. IV. Seasonal differences in the susceptibility of the host plant, *Bull. Fruit Tree Res. Stn. C, Jpn.,* 10, 76, 1983.
80. **Biggs, A. R., Miles, N. W., and Bell, R. L.,** Heritability of suberin accumulation in wounded peach bark, *Phytopathology,* 82, 83, 1992.
81. **Shigo, A. L., Shortle, W. C., and Garrett, P. W.,** Genetic control suggested in compartmentalization of discolored wood associated with tree wounds, *For. Sci.,* 23, 179, 1977.
82. **Lowerts, G. A. and Kellison, R. C.,** Genetically controlled resistance to discoloration and decay in wounded trees of yellow poplar, *Silvae Genet.,* 30, 98, 1981.

Chapter 3

THE CYTOLOGY AND HISTOLOGY OF APPLE SCAB

Christopher M. Becker

TABLE OF CONTENTS

0-8493-2939-6/93/$0.00 + $.50

© 1993 by CRC Press, Inc.

I. INTRODUCTION

Apple scab is caused by *Venturia inaequalis* (Cke.) Wint. (anamorph = *Spilocaea pomi* Fr.), and the disease is of economic importance in most apple-growing regions of the world.[1] Lesions can occur on developing leaves, fruit, twigs, sepals (Figure 1), petioles, petals, and bud scales. The symptoms are visible as distinct, circular lesions or, less commonly, as diffuse lesions colonizing entire tissues. Scab lesions that develop on the fruit during the growing season or in storage reduce the value of fruit, and severe infections can cause premature fruit drop. The disease is managed primarily with multiple applications of fungicides that prevent or eradicate infections. Apple cultivars immune to the pathogen are available, also.

Both ascospores and conidia of *V. inaequalis* can initiate scab on susceptible *Malus* tissue. Ascospores, the major source of primary inoculum, are produced within pseudothecia that develop in the diseased foliage overwintering on the orchard floor. Conidia produced from the primary infections supply secondary inoculum for repeating cycles that can occur through the remainder of the season. Conidia can function as initial inoculum when lesions overwinter on 1-year-old twigs, and in rare cases, when conidia overwinter within buds.[2,3]

II. HISTOLOGY AND CYTOLOGY OF PSEUDOTHECIA, ASCOSPORES, AND CONIDIA

A. Pseudothecia

V. inaequalis is heterothallic, and the formation of pseudothecia and ascospores results from mutual recognition, plasmogamy, and karyogamy between hyphae of opposite (+ / −) mating types. The formation of pseudothecia occurs only in leaves and begins after leaf fall. Individual lesions are of a single mating type, so hyphae from that lesion must cross with hyphae from a lesion of the opposite mating type. After leaves fall to the ground, hyphae that typically grow subcuticularly now grow between walls of adjacent epidermal cells and extend intercellularly into the palisade and mesophyll layers. The degree of leaf disintegration may affect which tissues are colonized: *V. inaequalis* colonizes the mesophyll region when little disintegration occurs, whereas colonization is limited to the palisade region when leaf disintegration is extensive. In abscised leaves, intracellular colonization of host cells occurs frequently. The hyphae observed in the disintegrating leaves are multiseptate, with uninucleate cells.[4]

The formation of pseudothecia has been examined in naturally infected leaves and in culture.[4-8] In as little as 3 days after leaf fall, subcuticular hyphae of *V. inaequalis* begin to ramify intercellularly between subjacent epidermal cells.[4] Pseudothecial initials are visible 2 weeks later in either mesophyll or palisade tissue[6] and consist of either two branches of the same filament coiled together or one branch spiraled around itself to form a compact, coiled mass.[4,5,7] The coil becomes increasingly intricate as additional branches from the original hyphae enter the coil. Within about a month after leaf fall, the tightly packed coil enlarges

FIGURE 1. Symptoms of apple scab on leaves and fruit (A), shoots (B), and sepals (C). (Magnification: A × 2/5; B × 4; C × 3.)

to a diameter of ∼ 50 to 60 μm (Figure 2). The outer three or four layers of pseudoparenchymatous cells thicken and darken to form the pseudothecial wall (10 to 18 μm thick).[4,6]

During the late fall, the developing pseudothecium appears multicellular and, in the center, relatively large, dark-staining cells differentiate into the ascogonium. Further enlargement and differentiation of the pseudothecium is halted or slowed during the winter, but continues following a period of dormancy and an increased accumulation of degree days in the spring. Lack of moisture at leaf fall inhibits the formation of pseudothecial initials, and moisture lacking during the spring slows or delays the maturation and release of ascospores.[6,8,9]

As maturation resumes in the spring, large cells differentiate into an ascogonium. A tongue-like trichogyne then expands outward from the ascogonium; it is nonseptate, binucleate, and frequently extends beyond the periphery of the coil (Figure 3).[7] Swollen-tipped hyphae that appear to be from the opposite mating type function as antheridia and congregate

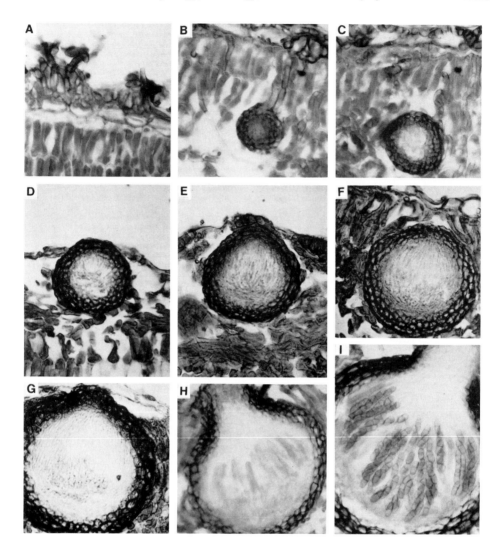

FIGURE 2. Maturation of pseudothecium of *Venturia inaequalis* in apple leaves. A, Subcuticular stroma; B, pseudothecial initial showing coiling of the hyphae; C, formation of the ascogonium from the initial; D, pseudo-paraphyses beginning to appear in the lumen of pseudothecium, concurrently as the ascogonium disappears; E, pseudoparaphyses fill the lumen and the pseudothecium has increased in diameter; F, pseudothecium that has increased in diameter; G, appearance of asci within the pseudothecium; H, asci have formed but the contents appear nondifferentiated; I, asci with pigmented ascospores. (All figures × 300.) (From James, J. R. and Sutton, T. B., *Phytopathology*, 72, 1073, 1982. With permission.)

around the portion of the trichogyne that is outside the coil. The antheridia and the trichogyne fuse, and a pore develops between the two structures. One or more nuclei from the antheridia have been observed to migrate towards the trichogyne, but actual plasmogamy and karyogamy have not been observed.[5] Sections through the trichogyne following the presumed fertilization show cells that are enlarged, more densely stained, and more septate than they are before fertilization. Also, prior to plasmogamy, the cells of the trichogyne contain a pair of nuclei, while afterwards, four nuclei are visible. The nuclei are presumed to migrate into the centrally located ascogonium, which consists of only a few, rather large cells. The ascogenous hyphae then branch and form numerous uninucleate or binucleate cells.[4,5,7,10] Croziers form

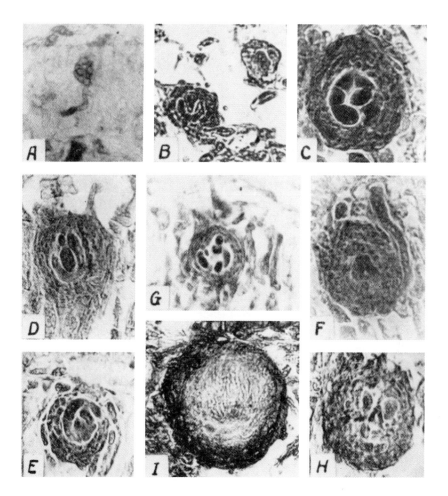

FIGURE 3. Development of ascogonium within pseudothecia of *V. inaequalis.* A, Pseudothecial initials originating from coil of hyphae (×382); B, formation of ascogonium from ascogonial initial (×535); C and D, ascogonia
curved with one to several septations (×535); E, the trichogyne is formed by the elongation from one end of the
ascogonium and pushes through the wall of the pseudothecium; F, pseudothecium with an antheridium nearby
(×125); G, pseudothecium with long trichogyne (×535); H, ascogonium beginning to disappear, as hyphae
differentiate within the lumen of the pseudothecium (×522); I, remnants of the ascogonium are apparent within
the base of the psuedothecium, and the remainder of the lumen is filled with hyphae. From this stage, asci will
begin to differentiate from the ascogenous hyphae (×352). (From Wilson, E. E., *Phytopathology,* 18, 375, 1928.
With permission.)

from the tips of the ascogenous hyphae that appear hooked or club-shaped,[4,7,11,12] elongated,[5]
or without any characteristic shape.[4] The tip of the binucleate, elongated hypha bends over
and develops septations to form a crozier that has a uninucleate terminal cell and a subtending
binucleate, penultimate cell. Within the subtending cell, the nuclei fuse and that cell enlarges
to become the young ascus.[10,11]

Observations of the nuclei within the developing ascus reveal that the first two of four
divisions are meiotic, resulting in reduced numbers of chromosomes. The third and forth
divisions are mitotic; the third division creating eight ascospores, while the fourth division
forms two unequal-sized cells.[10-12] Seven chromosomes have been observed during meiosis
and mitosis (Figure 4).[10,12]

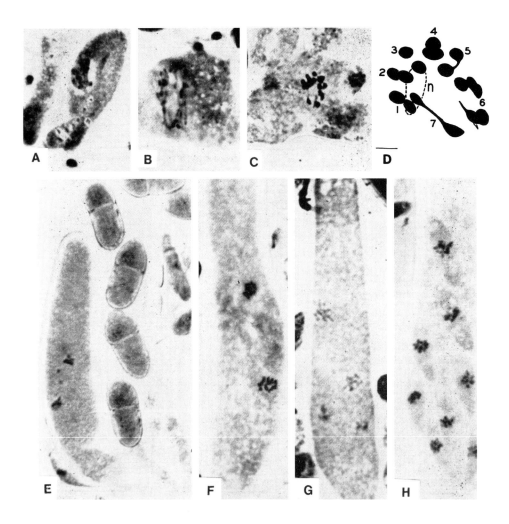

FIGURE 4. Stages in the development of the ascus of *V. inaequalis*. A, Young ascus at synapsis with paired chromosomes; B, pachytene nucleus with chromosomes at maximum extension; C, late metaphase I with seven bivalent chromosomes; D, hand-drawn version (from Figure C) of the 7 bivalents and the remains of the nucleolus; E, anaphase I with two lagging chromosomes (plus an ascus with 4:4 segregation of normal and abnormal ascospores); F, metaphase II; G, metaphase III; H, metaphase IV in uninucleate ascospores; Figures A to G ~ ×1800; Figure H ~ ×8000. (From Day, P. R., Boone, D. M., and Keitt, G. W., *Am. J. Bot,* 43, 835, 1956. With permission.)

Maturing asci and uninucleate, filiform paraphyses begin to fill the entire lumen of the pseudothecium concurrently as the ascogonium appears to disintegrate.[10] Mature pseudothecia (120 to 225 μm in diameter) develop wholly within the leaf (Figure 2), with only the ostiole erupting through the cuticle. Setae may or may not circumscribe the ostiole.[13,14] Pseudothecia are commonly produced beneath only the periphery of definite, marginate lesions, but beneath any portion of diffuse-type lesions.[4] The walls of asci are bitunicate, consisting of an inner and an outer wall layer, the endoascus and ectoascus, respectively. Immature asci are filled with hyaline periplasmic fluid. The developing ascospores appear hyaline when first visible, and olive or brownish at maturity (Figure 5). The maturing ascospores replace the periplasmic fluid, and the ascal walls become thinner. The mature asci are 60 to 70 μm long × 7 to 12 μm wide, and 80 to 140 asci develop per pseudothecium.[13,15] During intervals of wetting, the tip of an ascus elongates into the ostiole and

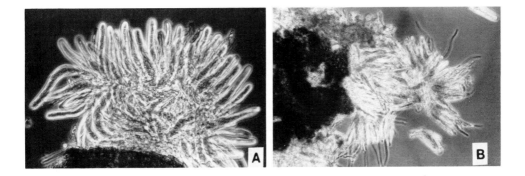

FIGURE 5. Pseudothecia of *V. inaequalis* that were teased from overwintered apple leaves in the spring and crushed in water between a coverslip and a microscope slide. A, asci with mostly pigmented ascospores (= morphologically mature) ×243; B, ascospores have been ejected from most asci, and the endoasci have been extruded beyond the ectoascus (×190).

protrudes about 40 μm above it.[16] The tip of the protruding ascus ruptures, and eight ascospores are ejected rapidly into the atmosphere. During ascospore release, the endoascus extrudes beyond the ectoascus. Pseudothecia that have been teased from leaves during the spring and squashed on a microscope slide show immature, mature, and discharged asci and a range of maturity of ascospores (Figure 5).

The negatively geotropic pseudothecia develop the ostiole towards the upper leaf surface, regardless of the surface initially colonized by the fungus.[5,17-19] Observations of pseudothecia with ostioles on both leaf surfaces result from leaves that turn over during the winter (Figure 6).[17,20,21]

B. MISCELLANEOUS OBSERVATIONS

1. Inhibition of Maturation of Pseudothecia by *Athelia bombacina*

An application of the basidiomycete *Athelia bombacina* Pers. to scab-infected apple leaves at leaf fall inhibits the springtime production of *V. inaequalis* ascospores. Sections of treated leaves that were sampled over time reveal that both subepidermal growth of hyphae and formation of pseudothecial initials of *V. inaequalis* are normal. However, pseudothecia fail to develop to full size, which results in the absence or incomplete formation of asci and ascospores (Figure 7). Immunocytochemical observations, using a polyclonal antibody preparation that was sufficiently specific to *A. bombacina*, reveal that *A. bombacina* grows endophytically and epiphytically. Also, there is no particular spatial association between *A. bombacina* and either hyphae or pseudothecia of *V. inaequalis*. The treated leaves appear considerably more degraded than do the controls; however, the mode of action for inhibiting the maturation of *V. inaequalis* pseudothecia remains unclear. Pectinases are known to be produced by *A. bombacina* and are, therefore, suspected to be an important component in the interaction. Their role, however, has not been documented.[14]

2. Pseudothecia of Coinhabiting Fungi in Overwintered Apple Leaves

In New Zealand, pseudothecia of *V. asperata* Samules and Sivanesan have been detected in overwintered apple leaves that also are colonized by *V. inaequalis*.[22] In Utah, *Mycosphaerella tassiana* (de Not.) Johans. overwinters in apple leaves with and without pseudothecia of *M. tassiana*.[23] The pseudothecia and ascospores of both coinhabitants are somewhat similar in size and shape to *V. inaequalis,* and their presence can result in misidentification of the scab fungus or its stage of maturity. Proper identification of *V. inaequalis* and the

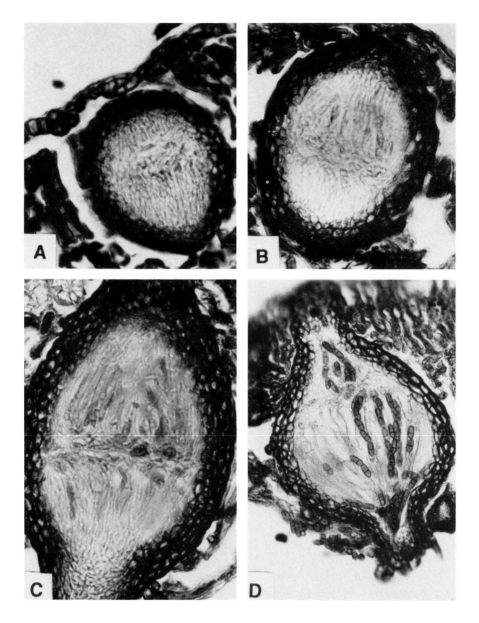

FIGURE 6.   Stages in the development of biostiolate pseudothecia of *V. inaequalis*. A, Pseudoparaphyses developing from either end of the immature pseudothecium; B, formation of asci in the locule of the pseudothecium; C, ascospores beginning to differentiate within the pseudothecium; D, pigmented ascospores available for discharge through ostioles that protrude through both leaf surfaces. Figure A to C ×400; Figure D ×250. (From James, J. R., Sutton, T. B., and Grand, L. F., *Mycologia*, 73, 564, 1981. Copyright 1981, The New York Botanical Garden. With permission.)

coinhabitant requires teasing pseudothecia from the leaf and microscopically viewing the crushed contents. In the ascospores of *V. asperata*, the lower of the two, unequal-sized cells is the smaller, tapered cell, in contrast with the opposite arrangement in *V. inaequalis*. Additionally, *V. asperata* produces *Fusicladium*-like conidia singly or as short chains from successively produced growing points of conidiophores.[22] *M. tassiana* is differentiated from *V. inaequalis* by the presence of saccate asci, mature ascospores that remain hyaline, a *Cladosporium herbarum* (Pers.) Link anamorph, and pseudothecia not completely immersed

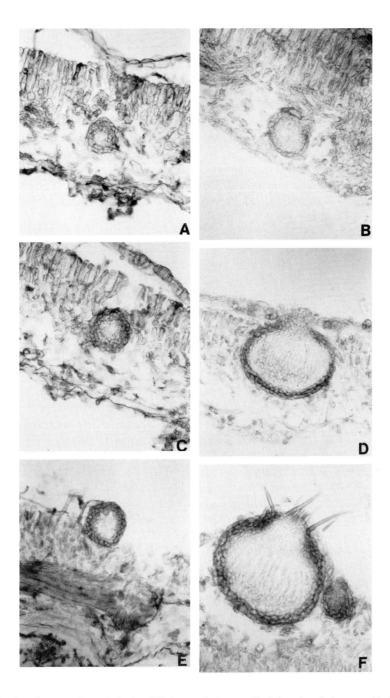

FIGURE 7. Development of pseudothecia of *V. inaequalis* in naturally infected apple leaves, with (A, C, E) and without (B, D, F) colonization by *Athelia bombacina*. A and B, collected February 1, 1987, immature pseudothecia; C and D, collected April 1, 1987, pseudoparaphyses and ascal initials visible within pseudothecia; E and F, collected May 1, 1987, near full-sized pseudothecium with asci visible, but ascospores not differentiated. All figures × 205. (From Young, C. S. and Andrews, J. H., *Phytopathology,* 80, 536, 1990. With permission.)

in the leaf.[23] Neither the anamorph nor the teliomorph stages of either *V. asperata* or *M. tassiana* is pathogenic.

C. ASCOSPORES

Ascospores are olive or brown, measuring 13 to 18 × 6 to 8 μm, with a septum in the upper third. The apex of the smaller cell is tapered, whereas that of the lower cell is rounded.[13] A single pore, 1 to 2 μm in diameter, is positioned centrally in the septum.[24] In nongerminated ascospores, the cells are uninucleate, and the cytoplasm contains small vacuoles, lipid bodies, mitochondria, endoplasmic reticulum, free ribosomes, and membrane complexes (Figure 8).[24,25] In addition, spherical bodies with a convoluted or tubular outer layer have been observed (Figure 9).[24] The smooth wall of ascospores is tripartic with (1) and electron-dense outer layer that appears porous, (2) an electron-dense, solid middle layer, and (3) a thicker electron-lucent inner wall.[25]

D. CONIDIA AND ANNELLOPHORES

Ultrastructural details of conidia and conidiophores (= annellophores) have been provided by several transmission electron microscopy (TEM) studies (Figures 10 to 13).[26-28] The annellophores erupt singularly through the cuticle form subcuticular mycelium or as a fascicle of several annellophores.[26,29] Annellophores erupt through the cuticle from the uppermost cells of the fungal stroma.[30,31] Each annellophore has a bilayered cell wall at the apex (Figure 11). However, after the production of several conidia the wall of the conidiogenous cell becomes trilayered or multilayered. The cytoplasm within annellophores contains lipids, vacuoles, and mitochondria.[26,27]

Conidiogenesis is blastic-annelidic:[32] conidia are "blown out" from the apex of the annellophore, while the wall of the conidium remains continuous with the wall of the annellophore (Figures 11 and 12).[27] Thus, the cytoplasm of the annellophore becomes incorporated into the conidium. When conidia appear to be full-sized, a percurrent proliferation develops at the apex of the annellophore that lengthens the conidiophore and results in the formation of a distinct annellation (Figure 12). A relatively thick septum then forms between the mature conidium and the annellophore. The septum splits schizolytically, so that both the base of the succeeded conidium and the apex of the annellophore have complete bilayered cell walls (Figure 13).[27] Concentric bodies, appearing as finely radiate, pinwheel-like inclusions (~270 nm in diameter), have been viewed close to the nucleus of annellophores; their function is unknown, and they do not migrate into the conidium.[26,33]

Conidia are typically one-celled, obclavate, and have a truncate base. A basal frill remains attached at the outer edge of the truncate base, apparently as remnants of the outer wall layer following the split from the annellophore (Figure 12).[26,27] The wall of the conidium consists of an outer, granular, electron-dense layer, and an inner layer that is electron transparent. A single nucleus occurs in each conidium, and the cytoplasm contains numerous lipid and protein bodies and small vacuoles.[24,26-28] Two-celled conidia are common.

III. HISTOLOGY AND CYTOLOGY OF GERMINATION AND PENETRATION

A. GERMINATION AND FORMATION OF APPRESSORIA

Conidia and ascospores form at least one germ tube in the presence of free water; the germinated propagule is referred to as a germling (Figure 14). Multiple germ tubes can develop from a single spore.[15,34] A single-pored septum separates the spore from the germ tube and separates each cell in the germling (Figure 15).[28,31] Woronin bodies are juxtaposed near the septal pore. After a germling forms one or more cells, an appressorium is differ-

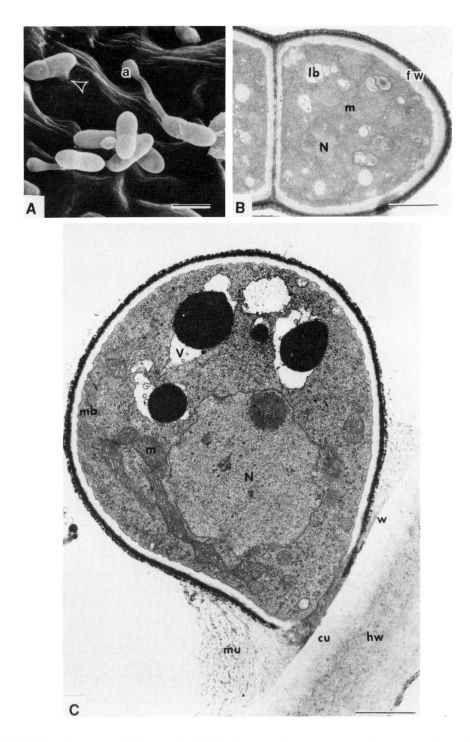

FIGURE 8. Ascospores of *V. inaequalis*. A, SEM of germinated and nongerminated ascospores on the surface of an apple leaf; scale bar = 5.0 μm, B, TEM of nongerminated ascospore; scale bar = 1.0 μm, C, TEM sections through an ascospore 5 h after inoculation, showing organization of the organelles within the cytoplasm and the formation of mucilage around the penetration site; scale bar = 1.0 μm. (From Smereka, K. M., MacHardy, W. E., and Kausch, A. P., *Can. J. Bot.* 65, 2549, 1987. With permission.)

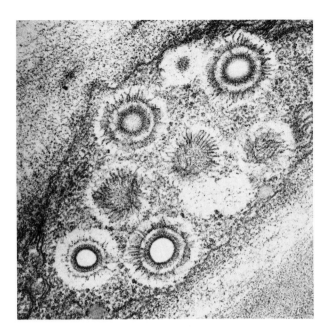

FIGURE 9.   Concentric bodies that appear as pinwheel inclusions in cytoplasm of *V. inaequalis* ($\times 18,750$). (Courtesy of A. L. Granett, American Cyanamid, Princeton, NJ.)

entiated from the apex of the terminal cell, and a septum is formed subtending the appressorium. The cell wall of appressoria appears bilayered and continuous with the germinating hypha.[28,31] The cell wall of appressoria from conidia and ascospores consists of a granular, electron-dense outer layer and a granular, somewhat electron-transparent, inner layer.[25,35] In appressoria formed recently from conidia, the cytoplasm contains vacuoles of different sizes with or without electron-dense inclusions, a nucleus, cytoplasmic vesicles, mitochondria, concentric bodies, microtubules, glycogen granules, lipid droplets, and some endoplasmic reticulum (Figures 16 and 17),[28,35] whereas, the cytoplasm in appressoria from ascospores appears granular with mitochondria.[25]

A mucilaginous substance has been observed covering an entire appressorium or circumventing only the region closest to the leaf surface (Figures 16 and 17).[25,31,35-38] The sheath of mucilage often extends for some distance away from the appressorium. The mucilage becomes visible concurrently with the apparent disintegration of the outer layer of the appressorial wall.[25] A portion of the outer electron-dense wall appears granular or particulate in nature, and near the leaf surface it appears continuous with the mucilage that circumvents the appressorium.[35] The mucilage has been conjectured to aid attachment to the host.[28,39] Alternatively, it may represent extracellular secretions involved in host penetration.[28]

Tropic responses are unknown for directional growth by germlings of *V. inaequalis*, and host-mediated signals have not been linked with the differentiation of appressoria from the terminal cells. However, on foliage, higher percentages of appressoria develop in the slight depressions at anticlinal walls and over veins (Figure 16).[15,28,31,36,38-40]

B. PENETRATION

Several detailed studies of pathogenesis by *V. inaequalis* have shown that an "infection sac" (Figures 16, 18, and 19) forms within appressoria, and a penetration pore develops in the appressorial wall subjacent to that sac. The host cuticle then appears to be dissolved

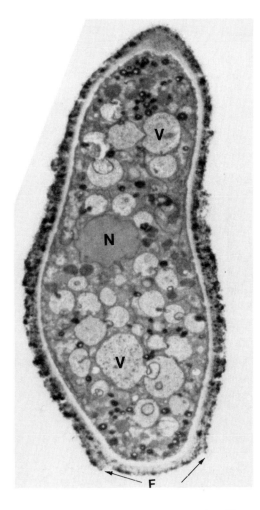

FIGURE 10. TEM section of conidium of *V. inaequalis* (×9000), N, nucleus; V, vacuole; F, basal frills. (From Hammill, T. R., *Trans. Br. Mycol. Soc.*, 60, 65, 1973. With permission.)

enzymatically, followed by extension of a penetration peg across the breached cuticle (Figures 19 to 21). In leaves, hyphae establish an association between the cuticle and epidermal cells and ramify away from the initial breach point in the cuticle in a radiate pattern (Figure 22). Penetration pores and "infection sacs" developed also within spores that had formed only short or almost nonexistent extensions of the spore wall rather than germ tubes.[25,31,39]

1. Penetration Pore

Single, translucent pores develop in the wall of appressoria (or extensions of the spore wall) in the region that adpresses to the cuticle of the host (Figure 16). Light- and electron-microscopy show that single, centrally positioned pores are ∼2 μm in diameter.[25,28,35,39] TEM sections of appressoria from conidia show that the pore forms where the inner wall layer tapers off and disappears. Around the pore, the outer wall layer may thicken slightly, but at the rim it tapers off.[25,35] The thickened region is probably generated by changes that occur with the fungal plasma membrane over the penetration pore during the formation of the infection sac. Subjacent to the developing infection sac, the fungal plasmalemma is invaginated and folded in on itself at the border of the pore. The resulting apposed plasma membrane circumscribes the pore and resembles intracellular junctional structures described

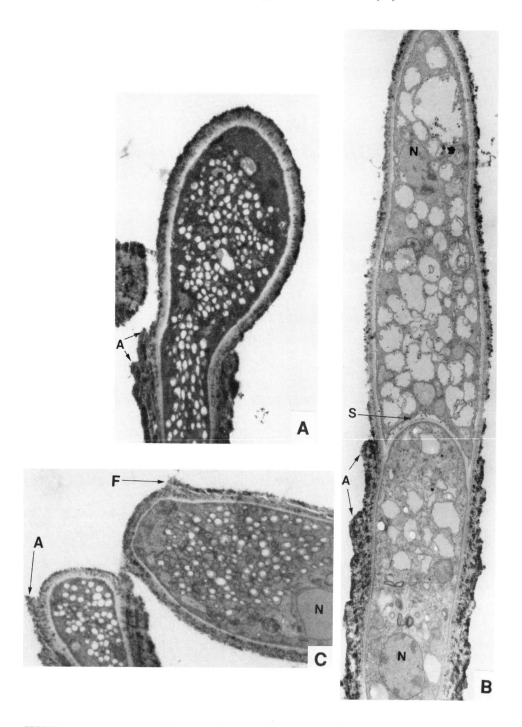

FIGURE 11.   Conidiogenesis of *V. inaequalis*. A, Proliferation of annellophore and blowing out of conidium initial (×8500); A, annellations are remnants from conidium that was previously released; B, recently formed conidium delimited from annellophore by a septum (S) (×9500); C, secessation of conidium from annellophore (×8500), F, frill; and A, annellations. (From Hammill, T. R., *Trans. Br. Mycol. Soc.*, 60, 65, 1973. With permission.)

FIGURE 12. Annellophore wall of *V. inaequalis,* showing that each annellation (A) results from a distinct layer of wall material ($\times$43,500). (From Hammill, T. R., *Trans. Br. Mycol. Soc.* 60, 69, 1973. With permission.)

in invertebrates (Figure 23). Within the thickened region, the junctional structures contain a 17 nm-wide intermembrane space with 5 nm-wide septa at 2.5 nm intervals providing a ladder-like appearance. The junctions appear to anchor the infection sac and provide structural rigidity during penetration of the cuticle.[41]

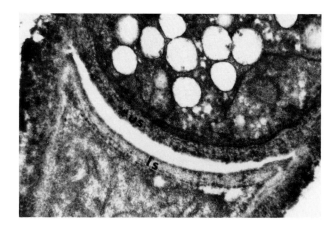

FIGURE 13.   Schizolytic separation between conidium of *V. inaequalis* and annellophore (×25,800). (From Cortlett, M., Chong, J., and Kokko, E. G., *Can. J. Microbiol.*, 22, 1144, 1976. With permission.)

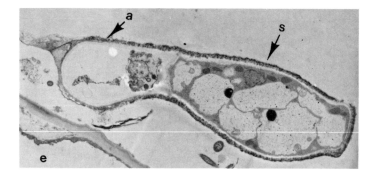

FIGURE 14.   Conidium of *V. inaequalis* with an appressorium that is prepared to penetrate the cuticle of a susceptible cultivar (×2,500): a, appressorium; s, spore; note the presence of a host-formed fibrillar deposit of cell wall material on the inner surface of the epidermal cell (e). (Courtesy of L. M. Yepes-Martinez, Cornell University, New York State Agricultural Experiment Station, Geneva.)

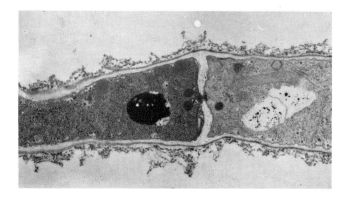

FIGURE 15.   Transverse TEM section through germling of *V. inaequalis* showing a septal pore, with Woronin bodies nearby (×7200). (From Yepes-Martinez, L. M., Development of Techniques for Selection of Disease-Resistant Apple Germplasm in Vitro, M.S. thesis, Cornell University, Ithaca, NY, 1987, chap. 2.)

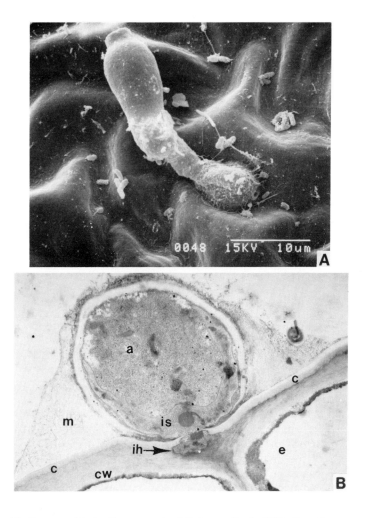

FIGURE 16. Mucilage associated with appressoria of *V. inaequalis*. A, SEM of conidium on apple leaf with appressorium covered with mucilage (arrow) (scale bar = 10 μm) and B, TEM section through appressorium at the penetration site; the bilayer walled appressorium (a) is covered by a mucilaginous sheath (m). The infection sac (is) is in contact with the host cuticle (c) at the penetration site; an infection hypha (ih) is seen below the breached cuticle (×6000); cw = cell wall; e = epidermal cell. (A, unpublished photograph; B, from Yepes-Martinez, L. M., Development of Techniques for Selection of Disease-Resistant Apple Germplasm in Vitro, M.S. thesis, Cornell University, Ithaca, NY, 1987, chap. 2.)

2. Infection Sac

A cup-shaped open-ended "infection sac" is first visualized within appressoria (or within spores with extended walls) as a single-layered membrane that presses against the developing pore (Figures 18 and 19).[25,28,35] No membranes are observed around the portion of the sac that opens into the center of the appressorium. A second membrane, separate from the appressorial wall, is formed at the pore, so that together with the cup-shaped membrane, a double membrane appears to be the base of the infection sac.[25,35]

3. Dissolving of the Cuticle

Subjacent to the pore and infection sac, the host cuticle appears thinner than in bordering regions.[28,30] The thinner cuticle contains electron-transparent platelets that are dispersed within an electron-dense material and fibrils that are darker than in adjoining areas.[28] Beneath

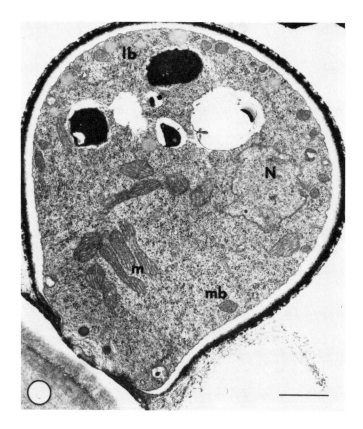

FIGURE 17. TEM section through appressorium of *V. inaequalis* and a nearly complete pore; scale bar = 1.0 μ. (From Smereka, K. M., MacHardy, W. E., and Kautsch, A. P., *Can. J. Bot.*, 65, 2549, 1987. With permission.)

the penetration pore, regions of the epicuticular wax and the cuticle stain darker and appear to be dissolved enzymatically by secreted fungal hydrolases; *V. inaequalis* appears to simply eat its way slowly through the cuticle until it arrives between the cuticle and the epidermal cell wall (Figures 20 and 21).[25,31,36,42] A cutinolytic enzyme has been isolated and characterized from *V. inaequalis* that is capable of and likely responsible for dissolving the cuticle.[42] Small quantities of the enzyme are released by conidia during the earliest stage of germination, and cutin monomers induce de novo cutinase production by germinating conidia and mycelium; therefore, the role of the cutinase appears to be two-fold. A constant release of small quantities of cutinase appears to be involved in an important first step in pathogenesis: cuticle sensing. During that process, cutin monomers are released from the long-chain, cutin polymers and the monomers induce a more prolific production of cutinase for dissolving the cuticle and facilitating penetration by *V. inaequalis*.[43] Applications of an active cutinase inhibitor prevent the establishment of a subcuticular stroma by *V. inaequalis*, which further verifies that a cutinase is produced by the fungus and is necessary for penetration of the cuticle. The recently characterized cutinase was probably responsible for the apparent dissolution of the cuticle prior to penetration as shown by TEM sections through the penetration site.[25,28,31,36] The esterase that has been observed to be active within appressoria is probably the cutinase that was described recently.[44]

4. Breaching the Cuticle

The leading edge of the infection sac expands into the dissolved region of the cuticle, with substantial differentiation and cytological changes occurring in both the fungus and especially, the host. As the fungal plasmalemma is invaginated above the penetration pore, electron-dense inclusions accumulate in host tissues near the penetration site.[25,28] TEM sections reveal that the inclusions release their contents at the leading edge of the fungal membrane and dissipate below. As the cuticle becomes sufficiently dissolved, the upper membrane of the double membrane of the infection sac then broadens below the pore; it becomes a primary hypha that radiates between the cuticle and epidermal cells (Figures 20 and 21).[25,28,35] Once the primary hypha develops, the infection sac is no longer detected. After penetration, the infection hypha flattens out along the host wall.[39] An inner, often convoluted wall forms that is electron transparent and appears continuous between the appressorium and the developing primary hypha.[28] Electron-dense, crystal-like deposits are visible along this second wall in the appressorium, but only in susceptible hosts; on hosts with hypersensitive reactions, the deposits are on the membranes of the infection hypha. These deposits have sharp irregular borders; however, they are restricted to the penetration site and are not found prior to the formation of the second wall. In a susceptible host the hyphae ramify between the cuticle and the epidermis away from that point of entry in a spoke-like pattern.[28,30,31,39,45] Depending on the susceptibility of the host and environmental conditions, sporulating lesions develop within 8 to 15 d.

IV. HISTOLOGY AND CYTOLOGY OF HOST-PATHOGEN INTERACTIONS IN FOLIAR LESIONS

A. INTERACTION ON A SUSCEPTIBLE HOST
1. The Fungus
Scab lesions on apple foliage typically have a velvety, olive, green, or gray surface with margins that can range from distinct to indeterminate. The velvety appearance is due to the production of conidia on the annellophores all over the lesion. Following penetration of the cuticle, the primary hypha remains one cell thick as it branches and radiates from the site of penetration (Figure 22).[34] Growth and expansion of mycelium can be hindered by junctures of epidermal cells. One row of epidermal cells may be covered by the stroma, while neighboring host cells remain free.[30] A stroma of hyphae is formed that is typically one cell layer thick, but it consists of two to five layers of smaller diameter hyphae in regions where annellophores erupt through the cuticle (Figure 24).[26,39] Annellophores erupt singly or clustered in groups of two to five, and they sporulate abundantly.[46] The cytoplasm of the subcuticular hyphae is mostly lipid and glycogen, but also includes ribosomes, mitochondria, Golgi bodies, vesicle-like inclusions, endoplasmic reticulum, and ground cytoplasm.[26] Vacuoles are observed with inclusions that are either spherical and electron opaque, or membranous and appearing as invaginations of the tonoplast. In addition, multivesicular bodies are observed in the cytoplasm (Figures 25 and 26). Ribosomes are associated with the endoplasmic reticulum, the nuclear membrane, or clustered into groups resembling polysomes.[47] Mitochondria are elongated, with a dense matrix and platelike cristae. Hyphae usually contain a single nucleus, but multiple nuclei have been observed.[11,28] Nuclei contain discrete nucleoli and a finely granular nucleoplasm. Within the subcuticular hyphae, Woronin bodies are juxtaposed to septa and as in germlings, they are common near septal pores.[28,31,48] The plasmalemma is adpressed closely to the hyphal wall. Deposits of glycogen-like rosettes are present in the cytoplasm beginning 10 d after inoculation of the host.[48]

2. The Host
The cuticle above the hyphae of *V. inaequalis* is stretched and conformed to the advancing hyphae without tearing.[26] Hyphal cells are typically adpressed to the upper epidermal wall with minor gaps evident only at anticlinal walls. However, between the subcuticular mycelium and either the cuticle or the epidermal cells, gaps develop that have ranged from

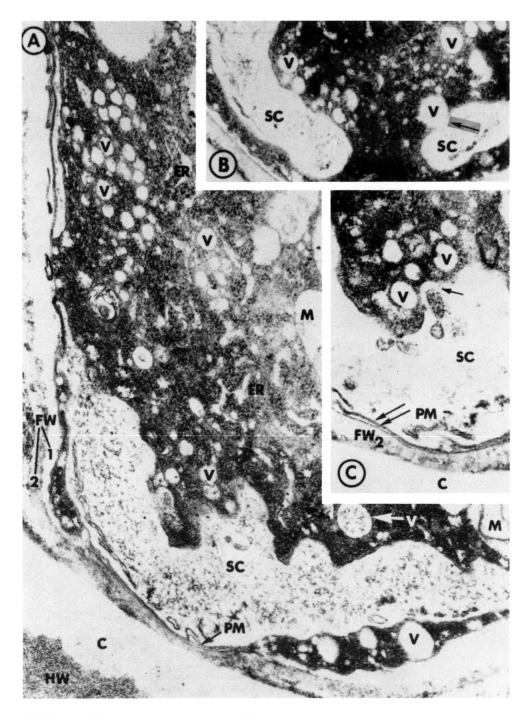

FIGURE 18. TEM section through appressorium of *V. inaequalis*. A, The membrane bound infection sac (SC) is adpressed to the developing penetration pore. The inner wall layer (FW1) of the appressorium is tapered off at the apex of the sac, and the outer layer (FW2) has contacted the host cuticle (C). Numerous vesicles (V) and endoplasmic reticulum are visible above the infection sac ($\times 37,800$); B, TEM section through the infection sac (SC) within appressoria showing vesicles (V) are fused with the infection sac ($\times 40,500$); C, TEM section of the infection sac (SC) showing that the plasma membrane (PM) and the infection sac membrane appear to form a double membrane at the outer fungal wall (FW2). Vesicles appear to fuse with the infection sac (single arrow) ($\times 46,800$). (From Maeda, K. M., An Ultrastructural Study of *Venturia inaequalis* (Cke.) Wint. Infection of *Malus* Hosts, M.S. thesis, Purdue University, Lafayette, IN, 1970, 112 pp.)

insignificant to nearly the thickness of the hyphae.[25,26,28,31] The walls of the subcuticular hyphae are moderately electron dense, granular, and interspersed with electron-transparent zones. Fungal walls often appear closely interlocked with or gradually blending into the cell walls of the more densely stained host epidermis (Figure 26).[26] Interlocking patterns are conjectured to be cell surfaces degraded by extracellular enzymes produced by the subcuticular mycelium.[26] TEM sections show that subjacent to the fungal stroma, a translucent zone can develop within the epidermal cell wall (Figure 27). A cellulase[49] and polygalacturonase[50] have been detected in liquid cultures of *V. inaequalis* that are candidates of fungal products that are capable of creating the localized degradation. The polygalacturonase has been isolated and characterized from cultures of *V. inaequalis* that were amended with pectin or ^{14}C-labeled cell wall extracts. Additionally, a ragged interface is observed above and below the subcuticular hyphae in some sections, indicating that physical tearing or separation may occur as the fungus grows between the cuticle and the epidermal wall (Figure 28).[26]

For a period of several days or until macroscopic lesions are visible, the epidermal cells subjacent to subcuticular hyphae remain intact and relatively unaffected (Figure 29).[28,30,39,45,51,52] However, the epidermal wall may become altered, minor changes may occur within the organelles, or after sporulation the entire cell may collapse.[26,39,45] Following contact with hyphae of *V. inaequalis*, fibrils darken within the epidermal wall, and electron-dense fibrils aggregate between cells and on the inner surface of the epidermal cell walls (Figure 30).[28] Additionally, cells that contain aggregations of fibrils develop osmiophilic droplets in the vacuoles and along the tonoplast. The aggregation of fibril-like deposits, especially within the cell, appear to be physical barriers that the host forms to combat the ensuing infection.[28,31,51] A host-formed physical barrier has been observed as ''an extension of the palisade layer'' that could curl the leaf.[30] In some epidermal cells subjacent to stroma, the chloroplasts have larger plastoglobuli and longer thylakoids than those in noninoculated controls.[51] Individual cells are collapsed or plasmolyzed within sporulating lesions of crab apple cv 'Hopa,'[26] and 21 d after 'Delicious' leaves have been inoculated with conidia.[45]

The palisade and mesophyll cells, which are randomly dispersed, with abundant spacing between cells, appear relatively unaffected in several studies (Figure 26).[26,28,34,39,45,51] However, chloroplasts within the palisade and mesophyll cells can lose their integrity and organization and coalesce into larger disorganized masses. As the chloroplasts become disorganized, the cytoplasm becomes increasingly vacuolated, and many cells collapse.[26,39]

3. Interaction on Tissue Culture Plantlets

Following inoculation of conidia of *V. inaequalis* onto tissue culture plants growing inside jars, the conidia germinate and produce branched germlings and numerous appressoria. The cuticle is penetrated, but only a rudimentary primary hypha forms. Secondary hyphae are essentially absent, and sporulation has not been observed.[31]

B. INTERACTION ON A RESISTANT HOST

Genes at six loci are believed to be involved in resistance to *V. inaequalis*.[53] The V*f* gene from the clone *M. floribunda* 821 and the V*m* gene from *M. micromalus* induced a pin-point, hypersensitive response around the subcuticular hyphae. Hybrids with the V*f* gene express a range of symptoms that have been categorized as follows: class 1, a pin-point reaction with no sporulation, which is considered a hypersensitive response; class 2, chlorotic lesions without sporulation; class 3a(=M), limited hyphal growth, usually no sporulation; class 3b, few restricted sporulating lesions; class 4, abundant sporulation that is equated with field susceptibility and was described in the previous section.[53] Summaries of the histological examinations of the host-pathogen interactions will be presented according to the above-mentioned classes.

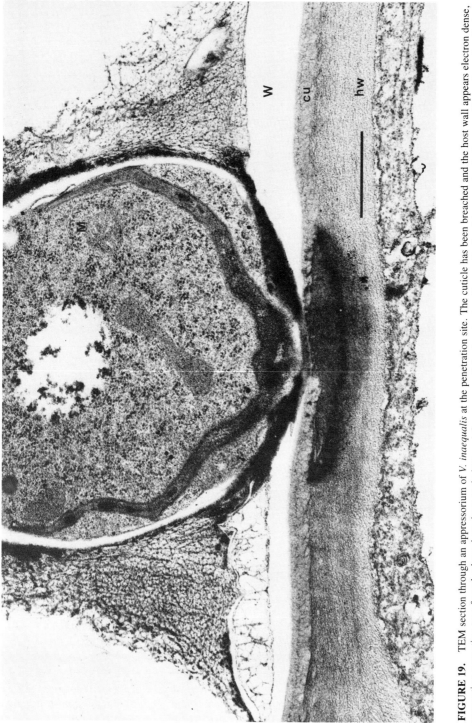

FIGURE 19. TEM section through an appressorium of *V. inaequalis* at the penetration site. The cuticle has been breached and the host wall appears electron dense, suggesting a reaction to a fungal released product; scale bar = 1.0 μm; cu, cuticle; hw, host wall; m, mitochondria; w, epicuticular wax. (From Smereka, K. M., MacHardy, W. E., and Kautsch, A. P., *Can. J. Bot.*, 65, 2549, 1987. With permission.)

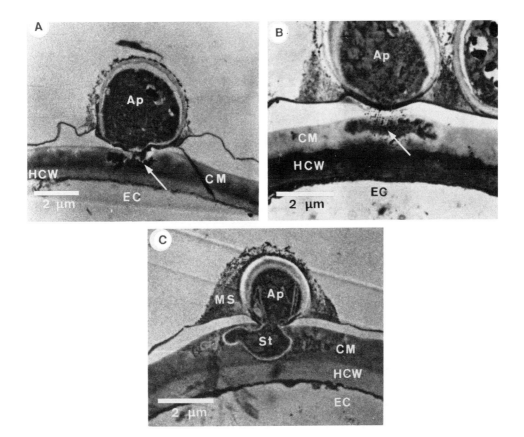

FIGURE 20. The cuticle of apple leaves appears to dissolve subjacent to appressoria of *V. inaequalis*. A and B, change in the density of the cuticle subjacent to the penetration pore, suggesting enzymatic degradation; C, primary hypha has grown into the cuticle to near the host epidermal wall; Ap, appressoria; CM, cuticular membrane; HCW, host cell wall; EC, epidermal cell; MS, mucilaginous sheath; St, primary hypha, All scale bars = 2 μm. (From Valsangiacomo, C. and Gessler, C., *Phytopathology*, 78, 1066, 1988. With permission.)

1. Hypersensitive, Pin-Point Symptoms; Class 1

In cultivars that are genetically programmed for the hypersensitive reaction, the epidermal cells react to the hyphae of *V. inaequalis* by the vacuoles accumulating osmiophilic droplets and the nuclei migrating towards the hyphae.[28,45] Within the epidermal cells the endoplasmic reticulum is abundant, with nearby vesicles that fuse with the plasmalemma. The cellular structures of the endoplasmic reticulum are replaced by a heterogenous material and sometimes become necrotic.[51] Spaces develop between the cytoplasm and the cell wall and contain membrane remnants and granular materials (Figure 30); these structures resemble the wall apposition materials that are observed in the susceptible host.[28,31] Another heterogenous substance is formed additionally in the space between the epidermal and palisade cells. Following necrosis of the cytoplasm and the folding of the anticlinal walls, the epidermal cells eventually collapse. Four to 20 epidermal cells typically collapse and form a depression 100 to 500 μm in diameter (Figure 31). The hypersensitive response occurs 48 to 72 hours after inoculation, regardless of whether the host is in the field, the greenhouse, or as tissue culture plants within jars.[31] In the orchard or greenhouse, *V. inaequalis* forms a primary stroma of no more than four or five cells. However, following the collapse of the subjacent epidermal cells, all hyphal cells become necrotic or collapse (Figure 32–34).[28,39,45,52] No primary hyphae are detected in tissue culture-grown plants. In leaves, the subjacent palisade and mesophyll cells are slightly disorganized and discolored, with amorphous, osmiophilic

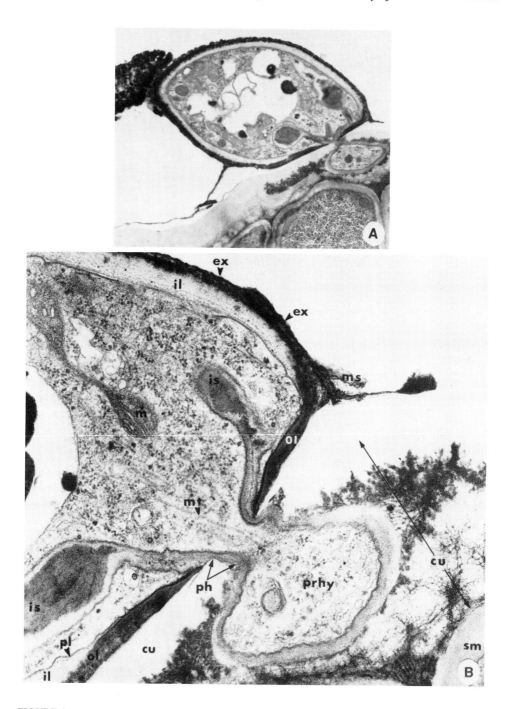

FIGURE 21. Expansion of primary hyphae of *V. inaequalis* following the disintegration of the cuticle. A and B are serial sections through the site of the breached cuticle Figure A ($\times$ 8640); Figure B ($\times$ 34,200); ap, appressorium, cu, cuticle; ex, external particulate layer; il, inner wall layer; is, infection sac; m, mitochondria; mt, microbutules; ol outer wall layer; ph, penetration hyphae; pl plasmalemma; prhy, primary hypha; sm, subcuticular mycelium. (From Cortlett, M. and Chong, J., *Can. J. Bot.*, 55, 5, 1977. With permission.)

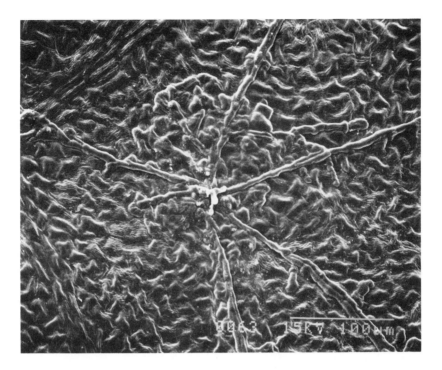

FIGURE 22. SEM of hyphae of *V. inaequalis* radiating subcuticularly from site of penetration on a susceptible apple leaf; scale bar = 100 μm. (Photograph courtesy of L. M. Yepes-Martinez, Cornell University, New York State Agricultural Experiment Station, Geneva.)

droplets in the cytoplasm.[28,51,52] Chloroplasts have an increased number of grana, thylakoids, plastoglobuli, plastoribosomes, and stroma density, with a decrease in the number and size of starch grains. Palisade cells are close together, the intercellular spaces are reduced and often filled with homogenous deposits, and the vacuoles contain polymorphic granules. Palisade cells are collapsed in some interactions. Sometimes, epidermal cells are observed that remain turgid beneath health hyphae; these are often over leaf veins.[28]

In an extremely resistant cultivar (*M. sieboldii*), following penetration by *V. inaequalis* tiny granules are formed near the host cell walls and the cytoplasm becomes slightly discolored. Thirty days after penetration, there is no microscopic or macroscopic evidence of hyphae ramifying from the inoculation site.[45]

2. Chlorotic Flecks and No Sporulation; Class 2

On leaves in this class, only a small proportion of the propagules penetrate the cuticle and form lesions. Lesions that form appear chlorotic or necrotic;[45] subcuticular hyphae seldom expand more than the length of several epidermal cells.[52] Annellophores are altogether absent, or they abort after some initial differentiation. Hence, conidia are never produced. Subjacent to the hyphae, the epidermal cells are observed to partially plasmolyze or collapse, or they contain disorganized organelles (Figure 33 and 34). Beneath the lesion, the palisade cells are slightly smaller, compared to noninfected areas, and they contain smaller vacuoles that contain atypically shaped granules. Additionally, the epidermal and palisade cells can become discolored and occasionally fill with yellowish droplets. Between the fungal cell wall and the epidermal cell wall, as well as in the lumen of the epidermal cell, there is an accumulation of an electron-dense fibrillar material. Also, amorphous deposits are formed along the anticlinal wall junctions of the epidermal cells. Ultimately, epidermal cells collapse, but not

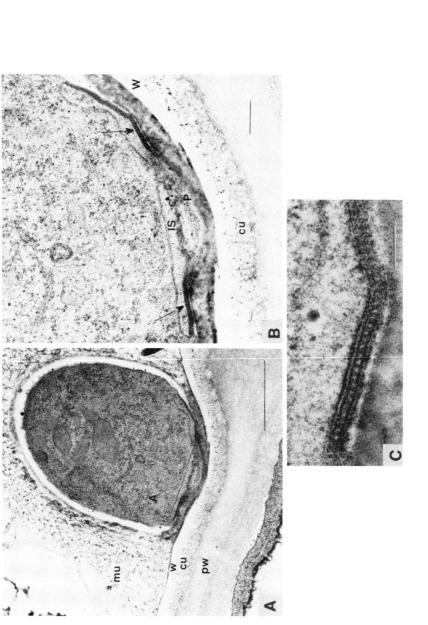

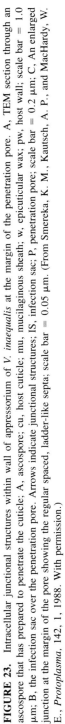

FIGURE 23. Intracellular junctional structures within wall of appressorium of *V. inaequalis* at the margin of the penetration pore. A, TEM section through an ascospore that has prepared to penetrate the cuticle; A, ascospore; cu, host cuticle; mu, mucilaginous sheath; w, epicuticular wax; pw, host wall; scale bar = 1.0 μm; B, the infection sac over the penetration pore. Arrows indicate junctional structures; IS, infection sac; P, penetration pore; scale bar = 0.2 μm; C, An enlarged junction at the margin of the pore showing the regular spaced, ladder-like septa; scale bar = 0.05 μm. (From Smereka, K. M., Kautsch, A. P., and MacHardy, W. E., *Protoplasma*, 142, 1, 1988. With permission.)

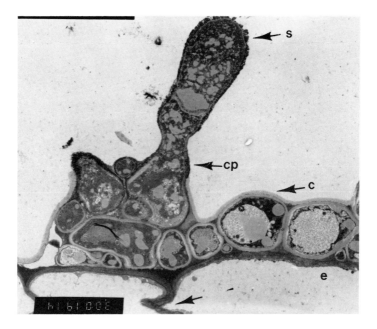

FIGURE 24. TEM section through subcuticular mycelium of *V. inaequalis* including an annellophore, a conidium, and susceptible leaf tissue. The fungal stroma (st) is between the cuticle (c) and the epidermal cell (e) and is usually one cell thick; except it is several cell layers thick where annellophores (cp) erupt through the cuticle. The conidium (s) is attached to the annellophore. Note the anticlinal wall of the epidermal cells has started to collapse subjacent to the stoma (arrow) ($\times$2300). (Courtesy of L. M. Yepes-Martinez, Cornell University, New York Agricultural Experiment Station, Geneva.)

as many or as rapidly as in the hypersensitive reactions. The mesophyll cells are unaffected.[52] After 16 days, hyphae of *V. inaequalis* have a few vacuolated cells, with large lipid droplets and mitochondria, and the cytoplasm appears dense. Eventually dead and collapsed hyphae are visible over the discolored and collapsed epidermal cells,[39] yet, hyphae appear viable over epidermal cells that appear healthy or that contain slightly disorganized organelles. In one reported class 2 interaction, the epidermal cells formed yellow droplets and turned brown, and the hyphae remained healthy. In this interaction, the palisade cells remained intact and unaffected.[45]

3. Intermediate Interactions; Class 3a

In the intermediate (3a) reaction, *V. inaequalis* produces a stroma that ranges from the sparse production of a few radiate hyphae to the formation of extensive secondary stromata (Figure 35). Annellophores are produced frequently, but only from the center of the lesion; they are typically aborted, and rarely produce conidia. Over time, the hyphae become necrotic in the center of the lesion, the structural integrity of the epidermal walls is broken down, and the cytoplasm disintegrates. Below the necrotic epidermal cells, the upper layers of palisade cells are strikingly affected. Either the upper layer of the palisade cells collapses, or each cell forms a cross wall internally, so that the portion of the cell that is adjacent to the necrotic cell becomes walled off. When the upper palisade layer becomes necrotic and plasmolyzed, cells in the second layer of the palisade region form the transverse walls. Below the necrotic portions of the lesion the mesophyll cells are unaffected, or they appear slightly hypertrophied with dark globules near the intercellular spaces.[52]

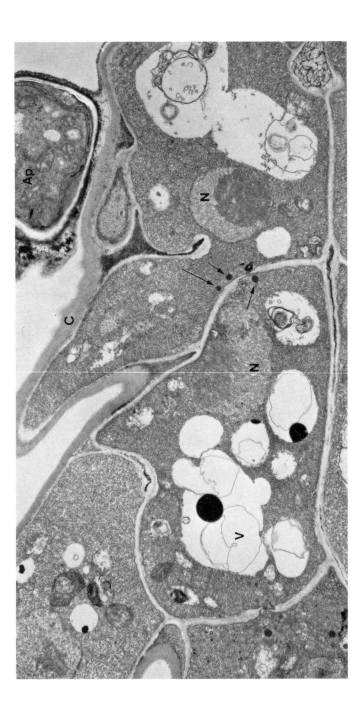

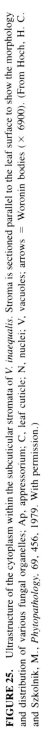

FIGURE 25. Ultrastructure of the cytoplasm within the subcuticular stromata of *V. inaequalis*. Stroma is sectioned parallel to the leaf surface to show the morphology and distribution of various fungal organelles; Ap, appressorium; C, leaf cuticle; N, nuclei; V, vacuoles; arrows = Woronin bodies (× 6900). (From Hoch, H. C. and Szkolnik, M., *Phytopathology*, 69, 456, 1979. With permission.)

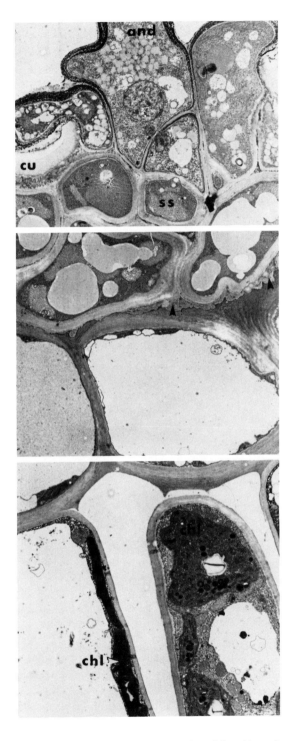

FIGURE 26. Ultrastructure of subcuticular stroma of *V. inaequalis* and the subjacent host epidermal and palisade cells. Portions of lower epidermal cell wall appear to be degraded (arrowheads) (×4200); and, annellophore; chl, chloroplasts; cu, cuticle; ss, subcuticular stroma. (From Cortlett, M., Chong, J., and Kokko, E. G., *Can. J. Microbiol.*, 22, 1144, 1976. With permission.)

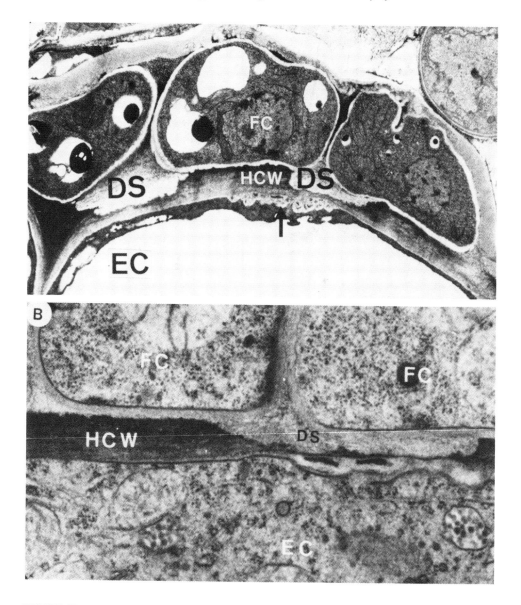

FIGURE 27. Degradation of epidermal cell wall by subcuticular stroma of *V. inaequalis*. A and B, subjacent to the fungal stroma (FC) degradation spots (DS) appear as electron-translucent sections of the periclinal wall (HCW) of the epidermal cell. The degradation is suspected to be due to polygalacturonase secreted by the fungal stroma; A, tissues were prepared by chemical fixation; B, tissues were prepared by high pressure freeze-drying technique. (From Valsangiacomo, C., Aspects of Pathogenesis and Host Resistance in the Interaction between *Venturia inaequalis* and Apple Leaves, Ph.D. thesis, Swiss Federal Institute of Technology, Zurich, 1990.)

4. Restricted Lesion and Some Sporulation; Class 3b

The lesions with class 3b appear somewhat smaller-sized, compared to a susceptible reaction. The subcuticular hyphae thicken irregularly and branch abundantly. Annellophores are produced, but there are considerably fewer of them, compared to the susceptible interaction. Annellophores are formed only from a stroma that has two to three layers of hyphae. They erupt through the cuticle and either sporulate profusely or fail to develop completely. Depending on the cultivar, the epidermal cells react to the subcuticular hyphae with one of two contrasting responses: they appear relatively unaffected, or they collapse and become

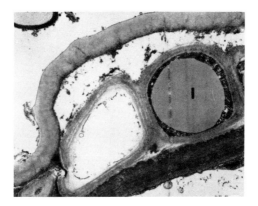

FIGURE 28. Separation of host tissues from above and below the subcuticular stroma of *V. inaequalis*. Gaps suggest that tearing or physical separation may occur during growth by the fungus. ($\times$6350); l, lipid body. (From Cortlett, M., Chong, J., and Kokko, E. G., *Can. J. Microbiol.* 22, 1144, 1976. With permission.)

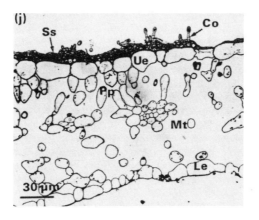

FIGURE 29. Interaction between *V. inaequalis* and host cells within a susceptible cultivar. Despite the abundant subcuticular stoma (Ss) and numerous annellophores (Co), the subjacent leaf tissues are only slightly modified; scale bar = 30 μm. Pp, palisade parenchyma; Mt, mesophyll tissue; Ue, upper epidermis; Le lower epidermis. (From Chevalier, M., Lespinasse, Y., and Renaudin, S., *Plant Pathology,* 40, 249, 1991. With permission.)

necrotic. In the latter response, the epidermal cells in the center of the lesion are destroyed by the time symptoms become macroscopic. The palisade cells appear relatively unaffected.[52] Over the disintegrated epidermal cells the fungal hyphae subsequently collapse. At the margins of the necrotic epidermal cells, sporulation is abundant.

Nusbaum and Keitt[39] described a reaction that is somewhat similar to the class 3b interaction, except that only one layer of the fungal stroma is formed and sporulation is absent. The primary hyphae form a weakly branching and feebly growing stroma that is restricted to 6 mm in diameter. The fungal cells are sparsely septate and devoid of contents except for inconspicuous nuclei. The subjacent host tissues remain unaffected.

A wound type periderm layer has not been observed in leaves infected with *V. inaequalis*.[54] However, cross walls have been observed to form within cells at the interface between collapsed and healthy tissues, so that a physical barrier appears to form at the margin of the infection (Figure 36).[39]

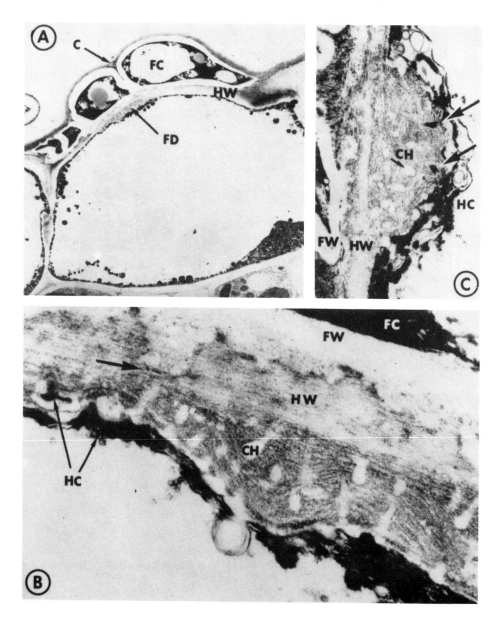

FIGURE 30. Ultrastructure of apposition-like reactions in host epidermal cells subjacent to hyphae of *V. inaequalis*. A, Subjacent to subcuticular stoma (EC) the periclinal epidermal wall (HW) has stained less densely, compared to in regions without the presence of fungal stroma. A fibrillar deposit (FD) appears to have formed on the inner surface of the host wall ($\times 2550$); B, enlargement of Figure A shows the fibrillar deposit and irregular staining of the host wall that is subjacent to the fungal cell wall (FW). A darkened and uneven border delimits the irregularly stained wall region and almost intercepts the fibrillar deposit in one region (arrow). Electron-transparent channels (CH) are apparent, and plasma membrane with material resembling host cytoplasm (HC) is in the channel to the left ($\times 59,500$); C, fibrillar deposit on the inner wall of infected tissue appears dome-shaped with channels. Two channels near the inner margin of the deposit (arrows) are filled with material that has stained like the host cytoplasm ($\times 22,100$). (From Maeda, K. M., An Ultrastructural Study of *Venturia inaequalis* (Cke.) Wint. Infection of *Malus* Hosts, M.S. thesis, Purdue University, Lafayette, IN, 1970, 112 pp.)

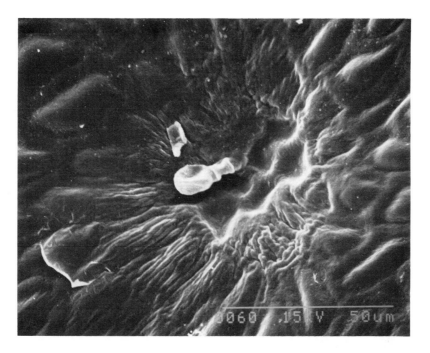

FIGURE 31. Collapse of the epidermal cells following penetration by *V. inaequalis* on apple leaf. (Scale bar = 50 μm.) (Courtesy of L. M. Yepes-Martinez, Cornell University, New York State Agricultural Experiment Station, Geneva, 1987, chap. 2.)

C. ONTOGENIC RESISTANCE OF FOLIAGE

V. inaequalis establishes an extensive subcuticular mycelium on the youngest leaf of a susceptible cultivar. However, as the age of the leaf increases, penetration of the cuticle occurs as on young leaves, but a decreasing proportion of the propagules forms a stroma, and stroma formation becomes restricted to a small area. On mature foliage, mycelium extends from the infection site, but it is considerably less than on younger leaves, and sporulation is rare.[30,34,55,56] Following the inoculation of all the leaves on potted trees, distinct, circular scab lesions were observed to develop on the youngest leaves after 9 to 11 d, but diffuse lesions formed on the abaxial surface only after 39 or 55 d.[9] The reduced activity by the fungus on the older foliage of susceptible cultivars has been described as ontogenic resistance.[34] Because host cells remain intact subjacent to the penetration site, ontogenic resistance is speculated to be controlled by a similar mechanism that governs the class 3b interactions associated with the *Vf* gene.[34] No trend has been observed in the ability of crude enzyme extracts from *V. inaequalis* to degrade cell wall preparations from leaves of increasing ages and from susceptible and resistant cultivars. Therefore, because *in vitro* degradability of cell walls could not be correlated with expression of resistance, the role of the polygalacturonase detected in the crude enzyme extracts from *V. inaequalis* remains uncertain.[50]

D. HOST/PATHOGEN INTERACTION IN FOLIAGE TREATED WITH FUNGICIDES

Following postinfection applications of phenylmercuric acetate (PMA), fenarimol, benomyl, or dodine to *V. inaequalis* on foliage, changes within the hyphae range from no apparent effects to necrosis.[48] The effects are not uniform within stroma that received the same treatments. After an application of PMA 50 h after inoculation, the protoplasm of the hyphae is necrotic with unrecognizable structural components (Figure 37). When the ap-

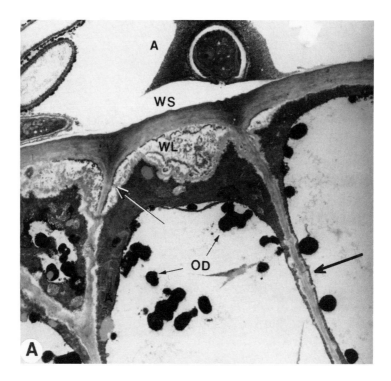

FIGURE 32. Hypersensitive reaction in apple leaves subjacent to penetration by *V. inaequalis*. A, Subjacent to the appressorium (A) and mucilaginous sheath (SH) lesions of amorphous material (WL) appears within the upper portions of the epidermal cell walls. Plasmodesmata (arrows) are suspected as links between epidermal cells. Osmiophilic droplets (OD) are present in the epidermal cell vacuoles and adjacent to the tonoplast. The white space (WS) is proably an artifact created in preparation (×7000); B, following the penetration by the infection hyphae (IH), epidermal cells appear necrotic (X) with coagulated cytoplasm. Immediately beneath the infection hyphae, lesions appear on the epidermal wall and osmiophilic droplets concentrate within the vacuole (×7,000); C, subjacent to the fungal cells (FC and FW), several epidermal cells collapse (X), which probably are the cause of the visible "pitting" of the leaf. In addition to the epidermal cells, osmiophilic droplets can form in palisade cells (×4000). (From Maeda, K. M., An Ultrastructural Study of *Venturia inaequalis* (Cke.) Wint. Infection of *Malus* Hosts, M.S. thesis, Purdue University, Lafayette, IN, 1970, 112 pp.)

plication of PMA is 100 h after inoculation, a gradation of effects is visible in individual hyphal cells. Septa separate the drastically different effects of PMA; some cells are entirely necrotic, others have localized moribund areas, and some are less affected or have only deformed organelles. The fungal organelles appear swollen, especially the mitochondria, cisternae of the endoplasmic reticulum, and the nuclear envelope. Glycogen-like rosettes develop in the cytoplasm and within vacuoles and can comprise 20 to 40% of the area within hyphae. Within 3 weeks after an application of PMA, chlorotic flecks develop on leaves, but sporulation is absent. Even though only 10% of the hyphal cells collapse or appear necrotic and the fungus could be successfully reisolated from the flecks, PMA appears to keep *V. inaequalis* from advancing within lesions.

Following an application of fenarimol to *V. inaequalis* on leaves, a progressive necrosis is viewed within the protoplasm along fungal profiles. Distortions within treated hyphae appear initially as membrane-confined, necrotic, or electron-opaque areas within the cytoplasm. Entire cells and hyphae become necrotic, which correlates with the inability to reisolate *V. inaequalis* from the leaf tissue (Figure 38). No glycogen-like rosettes are formed, indicating that the cells become necrotic shortly after treatment.[48]

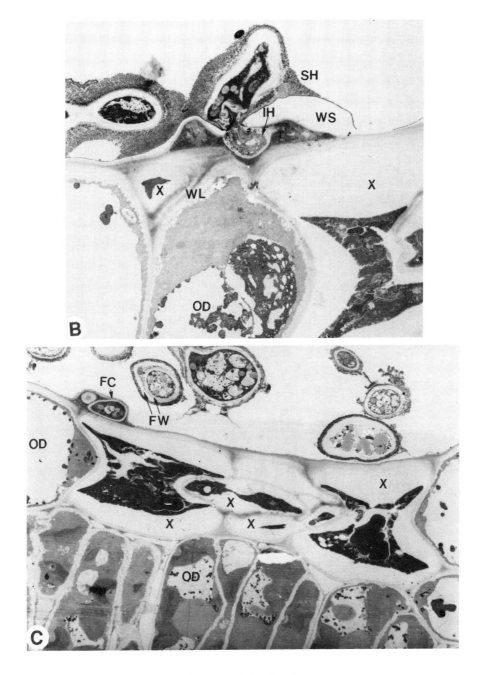

FIGURE 32 (continued).

Single applications of benomyl or dodine, 50 or 100 h after inoculation, result in less pronounced effects and less necrosis of hyphae of *V. inaequalis*, compared to PMA or fenarimol (Figure 38). Benomyl or dodine inhibit further development of subcuticular hyphae. Within the cytoplasm, the endoplasmic reticulum cisternae are dilated, the mitochondria are swollen, and there are scattered plate-like and glycogen-like deposits. Sporulation is inhibited.[48]

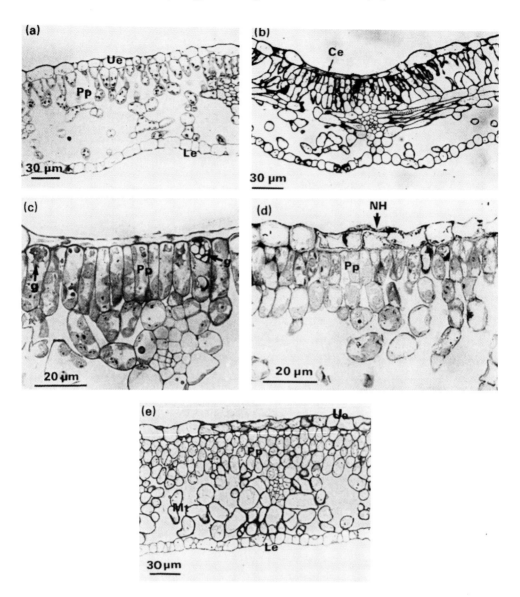

FIGURE 33. Cross-sections through apple scab lesions on leaves with class 1 and 2 interactions following penetration by *V. inaequalis*. A, Noninoculated control leaf; Le, lower epidermis; Pp, palisade tissue; Ue, upper epidermis; B, class 1 interaction: the upper epidermis of the leaf collapses (Ce). The palisade tissue has a high cellular density; some cells are transversely partitioned. The subcuticular stroma is limited; C, class 1 interaction: the intercellular spaces between the upper epidermis and palisade cells are occupied by homogeneous deposits. The vacuoles contain polymorphic granulations (g); D, class 2 interaction: the subcuticular stoma is necrotic (NH). The epidermal and palisade cells show significant modifications; E, class 2 interaction: no fungal stroma is visible within this section. The number of layers of palisade cells increases and cell dimensions decrease. Numerous cells of the mesophyll tissue (Mt) are hypertrophied. (From Chevalier, M., Lespinasse, Y., and Renaudin, S., *Plant Pathology*, 40, 249, 1991. With permission.)

Following multiple applications of benomyl in an orchard with abundant apple scab inoculum, *V. inaequalis* was still capable of forming subcuticular hyphae, but fewer lesions were able to form compared to untreated leaves. In the resulting lesions the cell walls of the subcuticular hyphae are indistinct, either swollen or gelatinized, and the contents of the hyphae are clumped or disintegrated. Most of the relatively few annellophores that erupt

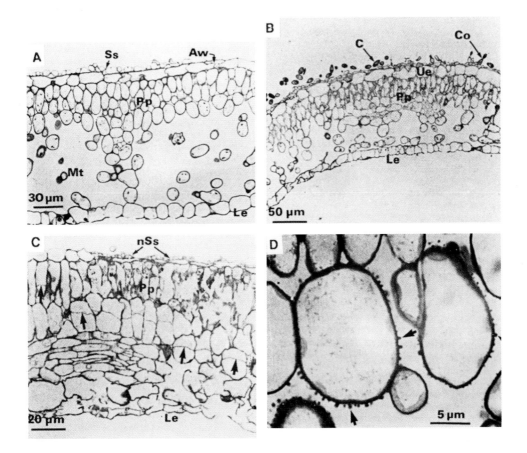

FIGURE 34.   Cross-sections through apple scab lesions on leaves with class 3a interaction following penetration by *V. inaequalis*. A, Class 3a interaction: the subcuticular stroma (Ss) in uniseriate. The upper epidermis is partially collapsed, with buckled anticlinal walls (Aw); Pp, palisade tissue; Mt, mesophyll tissue; Le, lower epidermis; B, class 3a interaction: the stroma is visible with limited production of conidia (C) from annellophores (Co). The leaf tissues show few modifications; C, class 3a interaction: in regions with host tissues that appear necrotic, the upper epidermis and the first layer of the palisade tissue are destroyed; the second-layer cells are transversely partitioned (arrows); D, Class 3a interaction: some cells of the mesophyll tissues are hypertrophied and exude numerous globules (arrows). (From Chevalier, M., Lespinasse, Y., and Renaudin, S., *Plant Pathology*, 40, 249, 1991. With permission.)

through the cuticle collapse, and only a few remain cylindrical and produce two or more conidia. Many of the annellophores appear ampulliform, as if the differentiation of the primary conidium has ceased.[57]

On leaves treated with bitertinol, conidia of *V. inaequalis* germinate and form appressoria, as on untreated leaves, except that the germ tubes are shorter. Bitertinol allows the initial formation of a subcuticular stroma; however, hyphae ramify only a short distance from the penetration site and no annellophores are produced. Following an application of bitertinol 3 d after conidia are inoculated onto leaves, subcuticular growth is halted, and even though annellophores are produced, sporulation is absent.[29]

E. INFECTION BY BIOCHEMICAL MUTANTS OF *VENTURIA INAEQUALIS*

Pathogenic and nonpathogenic mutants of *V. inaequalis* have been selected that require a single amino acid for growth on artificial media. When inoculated onto leaves, the patho-

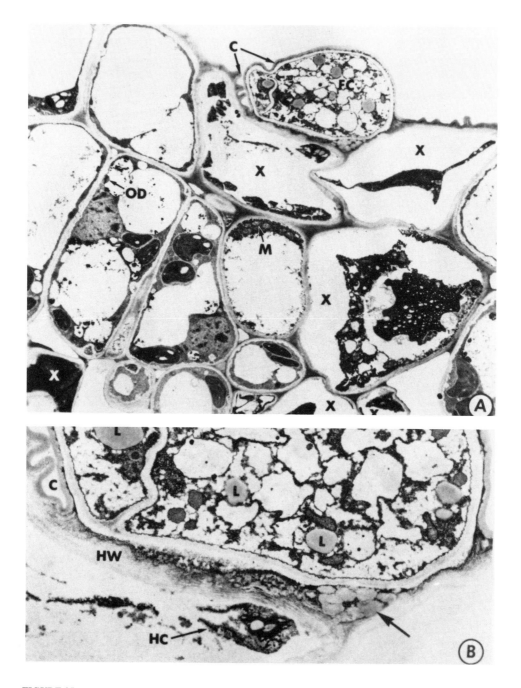

FIGURE 35. Ultrastructure of the interaction between *V. inaequalis* and the host tissues within a resistant cultivar. A, Subjacent to the healthy appearing fungal cell, the epidermal, palisade, and mesophyll cells (X) have begun to disintegrate and collapse. Extensive vacuolation is visible in palisade cells and osmiophillic droplets (OD) accumulate along the tonoplast. An aggregation of mitochondria (M) is apparent in the palisade cell adjacent to the disorganized epidermal cell; C, cuticle, (×4,000); B, the cuticle is uplifted and distorted between the subcuticular hyphae and the host wall. The host wall (HW) stains irregularly adjacent to the hyphae. Amorphous droplets (arrow) with the same density as the lipid droplets (L) are visible at the anticlinal junctions of the epidermal cells. Larger vacuoles and lipid bodies are visible in the hypha (×16,000). (From Maeda, K. M., An Ultrastructural Study of *Venturia inaequalis* (Cke.) Wint. Infection of *Malus* Hosts, M.S. thesis, Purdue University, Lafayette, IN, 1970, 112 pp.)

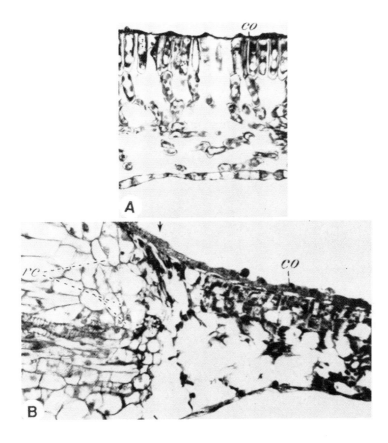

FIGURE 36. Cross-section through apple scab lesions with mild (A) and severe (B) necrosis of subjacent host cells. A, lesion on apple leaf that is similar to class 2 interaction; fungal hyphae and epidermal cells collapse 3 weeks after inoculation. Palisade and mesophyll tissues are relatively unaffected ($\times$330); B, severe collapse of epidermal, palisade, and mesophyll cells subjacent to the hyphae. The stroma may often sporulate early, but declines and becomes necrotic over the senescent host cells. A row of rhomboid cells (rc) forms internal, transverse septa between the totally collapsed cells (right) and the impoverished tissues (left). Fungal hyphae are collapsed to the right of the arrow and intact to the left of it ($\times$330). (From Nusbaum, C. J. and Keitt, G. W., *J. Agric. Res.*, 56, 610, 1938. With permission.)

genic mutants establish a subcuticular stroma and sporulate without an additional source of their required substances.[58] The nonpathogenic mutants germinate and develop subcuticular mycelia as do wild type isolates. However, histological observations of these isolates on leaves show that the formation of subcuticular stromata and sporulation is enhanced following the applications of the nutrient required to overcome auxotrophy, except for the mutant that required adenine (Figure 39).

V. HISTOLOGY OF APPLE SCAB LESIONS ON FRUIT

Germlings of *V. inaequalis* form appressoria on apple fruit and penetrate the cuticle as they do on leaves.[30] In mature fruit, hyphae appear to grow horizontally for some distance within striations of the cuticle, which appear to have been laid down in layers.[30] *V. inaequalis* can penetrate the cuticle of pear fruit, but hyphae fail to become established.[30] Scab lesions on fruit are typically gray or brownish, with variously sized cracks. The cracks radiate from the center of larger or older lesions and may split the lesion into two or more unequal

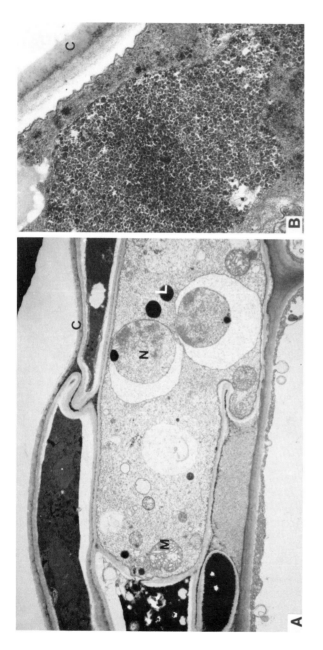

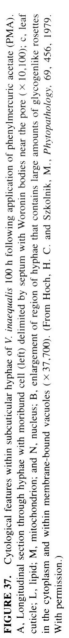

FIGURE 37. Cytological features within subcuticular hyphae of *V. inaequalis* 100 h following application of phenylmercuric acetate (PMA). A, Longitudinal section through hyphae with moribund cell (left) delimited by septum with Woronin bodies near the pore (× 10,100); c, leaf cuticle; L, lipid; M, mitochondrion; and N, nucleus; B, enlargement of region of hyphae that contains large amounts of glycogenlike rosettes in the cytoplasm and within membrane-bound vacuoles (×37,700). (From Hoch, H. C. and Szkolnik, M., *Phytopathology*, 69, 456, 1979. With permission.)

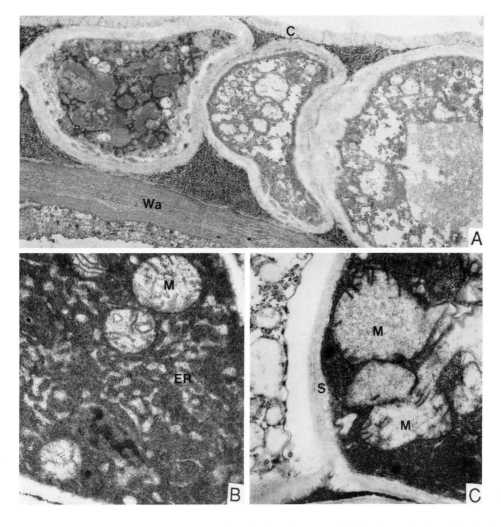

FIGURE 38. Cytological features within subcuticular hyphae of *V. inaequalis* following application of fenarimol (A), dodine (B), and benomyl (C). A, After an application of fenarimol, the cytoplasm within the fungal stroma appears disrupted or necrotic ($\times 27,400$); C, leaf cuticle; Wa, periclinal wall of a leaf epidermal cell; B, following an application of dodine, the endoplasmic reticulum cisternae (ER) appear dilated, and the mitochondria appear swollen within the fungal cytoplasm ($\times 30,000$); C, following an application of benomyl, a septum delimits cells that appear healthy and slightly disrupted with swollen mitochondria ($\times 23,300$). (From Hoch, H. C. and Szkolnik, M., *Phytopathology*, 69, 456, 1979. With permission.)

portions. The margins are usually distinct; however, an interrupted pattern of narrow, light and dark bands may occur. Lesions 3 to 6 mm in diameter also develop on pedicels. Relatively young lesions appear velvety due to prolific sporulation of *V. inaequalis* all over the lesion. Histological observations of the lesions on fruit and pedicels show that a cushion of parenchymatous-like hyphal cells is initially formed at the surface of the epidermal cells, and that layer produces the annellophores.[30,59,60] Subjacent to the hyphae, a thin or thick layer of cork-like cells is produced by the host (Figure 40).[59,60] Annellophores become visible above the cuticle after the immature conidiophore either dissolves its way upward through cuticle in a manner similar to that observed during the initial penetration of the cuticle, or by a pattern of growth that results in the annellophores "bursting through the cuticle".[30] In older lesions, sporulation is rare or may occur only near the margins. Sections through

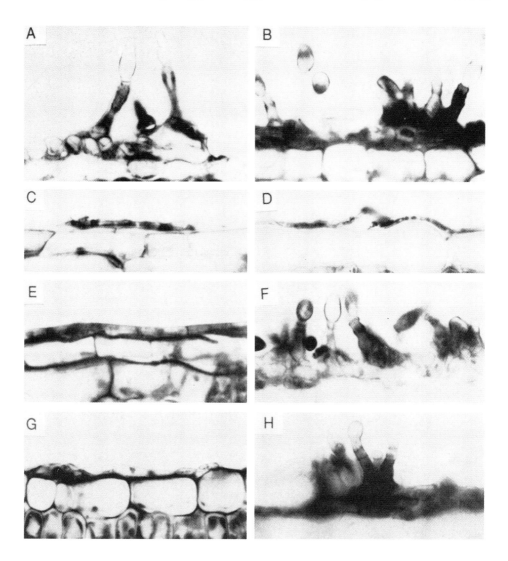

FIGURE 39. Colonization of apple leaves by biochemical mutants of *V. inaequalis* without (A, C, E, G) and with (B, D, F, H) applications of the substance required to overcome auxotrophy. Subcuticular stroma is produced by all mutants without nutrient applications, but stroma development and sporulation is enhanced by the application of the required nutrient, except for the adenine requiring mutant. A and B, mutants that require choline; C and D, adenine; E and F, arginine; and G and H, methionine. (All figures ×875.) (From Kline, D. M., Boone, D. M., and Keitt, G. W., *Am. J. Bot.*, 51, 634, 1964. With permission.)

mature lesions show that the cushion of hyphal cells is often lacking, annellophores are rare, and portions of the cork layer are often sloughed off late in the season. Additionally, the cuticle is often observed to be lacking in older lesions, or a cuticle-like material appears to be deposited on the lesion surface. Subjacent to the cushion of hyphae, regardless of whether it remains intact, three or four rows of cork-like host cells become darkly stained. They appear crushed, disintegrated, and devoid of contents.[59-61] Below the damaged cork layers are five to eight or more rows of slightly crushed but intact host cells;[60] there can be 12 to 14 rows in older lesions.[54] These cells form a layer of tightly packed and suberized cells.

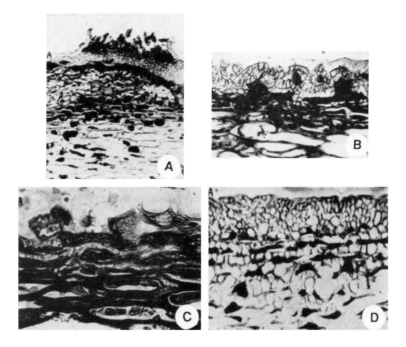

FIGURE 40. Infection of fruit by *V. inaequalis*. A, section through lesion on fruit that developed in the orchard. A, Layer of cork cells has developed subjacent to the sporulating stroma (×400); B, after a period of storage, mycelium radiates subcuticularly beyond the margin of a summer lesion; C and D, in lesions that develop on fruit in storage, hyphae grow mostly intercellularly among the cortical cells without cork formation (C) (×780); extensive colonization occurs within the cuticular region and intercellularly within the cortex devastates the cortical cells (D) (×615). (From Walker, E. A., *Peninsula Hortic. Soc., 54th Trans.,* 1940, 84. With permission.)

They appear to be pushed out of a phellogen layer and function as a protective periderm layer.[54,60] The periderm layer inhibits further penetration by hyphae of *V. inaequalis*, but it has no effect on lateral proliferation by hyphae.[54]

Scab can develop on fruit in storage as (1) an enlargement of existing lesions with additional sporulation beyond the margin, (2) small, blackish lesions that are smooth and devoid of sporulation, and (3) small, brown or black lesions that are rough with an occasional cluster of annellophores that erupts through the surface.[60,62,63] In lesions that sporulate at the margins after a period of storage, the host epidermal cells are somewhat crushed and completely surrounded by hyphae of *V. inaequalis*. Hyphae are detected intercellularly one or two cortical cells beneath the epidermis (Figure 40). The formation of periderm is lacking in this region. In the smooth, nonsporulating lesions, hyphae ramify subcuticularly and extend intercellularly eight or ten cells beneath the epidermis. Subjacent to an abundance of hyphae, host cells are often crushed and necrotic. Annellophores are never observed. Rough lesions are only slightly larger than the smooth type, but there are two distinctive characteristics that separate them from smooth lesions: annellophores frequently erupt through the cuticle, and a layer of suberized host cells is formed that limits the hyphae to the upper two layers of cells.[60,63]

On fruit pedicels, scab lesions are formed that sporulate profusely all over lesions. *V. inaequalis* forms a subcuticular stroma that is similar to that on leaves; however, a periderm layer is formed that consists of several layers of phellem cells. The infected pedicel can be deformed, weakened, or twisted 15 to 30 degrees.[54]

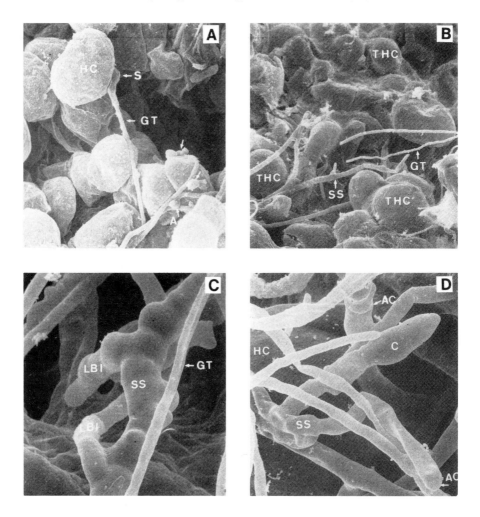

FIGURE 41.   Colonization of apple callus tissue by *V. inaequalis*. A, conidia (S) germinate and form long germ tubes (GT) with appressoria (A) that attach to host callus cells (HC); B, secondary stroma (SS) growing intercellularly among turgid host cells (THC); C, secondary stroma with lateral branch initials (LBI); D, secondary stroma form annellophores (AC) and conidia (C). (From Beech, I. and Gessler, C., *J. Phytopathology,* 116, 317, 1986. With permission.)

VI. HISTOLOGY OF SCAB INFECTION OF APPLE CALLUS TISSUE

In one study, hyphae of *V. inaequalis* were unable to grow on young cultures of callus cells, but they grew intercellularly on senescing callus cells.[64] However, in a more recent study, *V. inaequalis* was reported to germinate and form appressoria on the surface of apple calli as well as on young leaves, and a histological study describes the host/pathogen interaction (Figures 41 and 42).[65] A primary hypha is formed on the surface of the callus and subsequently broadens into stolon-like hyphae and secondary stromata with lateral branches. However, unlike on leaves, the secondary stromata grows intercellularly among the callus cells. On the surface of the callus tissue, annellophores form typical conidia. Between the host cells and fungal hyphae, the extracellular space is filled with an extracellular

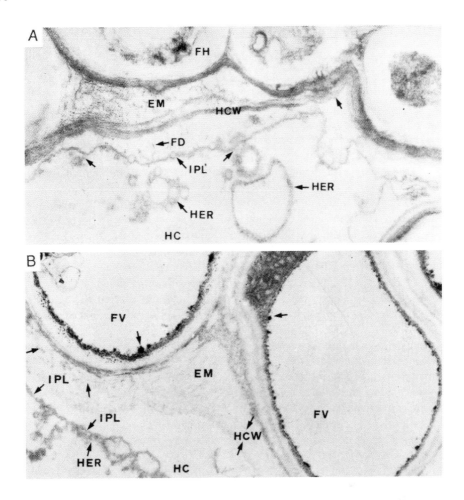

FIGURE 42. Ultrastructure of apple callus cells colonized by *V. inaequalis*. A, interaction between fungal hyphae (FH) and host callus cells (HC) of the susceptible cultivar Golden Delicious. The extracellular space between the hyphae and the callus cell walls (HCW) is filled with a fibrillar matrix (EM). The host cell wall adjacent to the hyphae stains irregularly, becoming electron dense (arrow). The host plasmalemma (IPL) is invaginated and highly vesiculated. The space between the invaginated plasmalemma of the host and the cell wall contains fibrillar deposits (FD). The complexes of host endoplasmic reticulum (HER) appear continuous with the invaginated host plasmalemma (arrows); B, the host cell wall appears to be chemically decomposed (arrows) adjacent to the hyphae. The fungal cell contains a central vacuole (FV) with irregular electron-dense deposits (arrows). (From Beech, I. and Gessler, C., *J. Phytopathology*, 116, 317, 1986. With permission.)

matrix that consists of small electron-dense fibrillar deposits. After 11 days, the host cell walls adjacent to the fungal hyphae stain irregularly, suggesting alteration by a secretion product from the fungus. The host plasmalemma in cells opposite the hyphae becomes invaginated and is often highly vesiculated. Portions of the host endoplasmic reticulum are adpressed to the invaginated plasmalemma and appear to form organized complexes of large and small cisternae. The zone between the host cell wall and the invaginated host plasmalemma is either electron lucent and devoid of contents or contains a densely staining fibrillar material. The latter material is presumed to be precursors to wound-healing wall appositions. The ultrastructural interactions between *V. inaequalis* and host callus appear similar both for calli produced from cultivars susceptible and resistant to the scab.[65]

VII. HISTOLOGY OF APPLE SCAB LESIONS ON SHOOTS

Spores of *V. inaequalis* penetrate the cuticle of green, unwounded apple shoots as they do that on leaves; however, appressoria may be absent or poorly developed.[2] Hyphae become established subcuticularly, but after about a week the epidermal cells collapse and become invaded and occupied by the hyphae. Infections viewed in the summer on green shoots of old trees show that the initial subcuticular stroma develops into a cushion of hyphae (Figures 43 and 44).[66-68] In the center of the lesion, the cuticle ruptures and partially peels back. An abundance of annellophores is formed throughout the lesion, and they sporulate profusely.[66-70] The presence of the stroma appears to interfere with the normal formation of phellogen and stimulates the formation of a new region of cork subjacent to the infection. In lesions sectioned in midsummer, the cork layer is detected subjacent to all lesions, and it appears to have joined the normal phellogen. The thickness of the cork layer and the depth that it reaches vary considerably. The periderm usually is formed five to eight cortical cells below the surface; however, occasionally it is formed immediately adjacent to the surface.[66,67] As summer progresses, the cork layer thickens and becomes convoluted, concomitant with decreased sporulation. By autumn, portions of the fungal stroma slough off, probably as a direct result of the thickened periderm and conidia rarely are detected on the lesions.[66-69] However, hyphae can ramify laterally and breach the periderm layer, so that the lesion expands into healthy tissue. During the following spring, no conidia are evident on the lesions, and the hyphae that breach the periderm laterally fail to erupt though the bark.[66-68] Lesions on the shoots of young trees develop like those on older trees. However, the cork cambium fails to thicken during the summer, so a larger stroma develops with the production of abundant conidia throughout the summer. Conidia production diminishes during the autumn but is reactivated the following spring. During the second season, the host often forms several layers of cork cambium in attempts to isolate the fungal stroma (Figure 44). However, some stromata remain viable in those pustules during the subsequent winter and even produce a few conidia during the third season. Finally, during the third season the cork cambium thickens sufficiently to isolate the fungal hyphae.[68]

On eight cultivars with a range of susceptibility to scab, lesions develop on shoots only on cultivars that support sporulating lesions on the foliage. Observations have not been conducted, however, to determine if hypersensitive reactions occur on shoots. Following inoculations of shoot tips, lesions develop mostly on the shoot tissue immediately above and below the youngest leaf that was unfolded at the time of the inoculation.[69]

VIII. HISTOLOGY OF SCAB INFECTIONS OF ETIOLATED APPLE HYPOCOTYLS

Etiolated apple hypocotyls have been used to study the host/pathogen interactions between *V. inaequalis* and apple cultivars with varying levels of resistance. The lack of host pigments in etiolated apple hypocotyls permits observations of the growth of *V. inaequalis* *in vivo* and of the subsequent response by the host tissues. Cultivars that respond to penetration of the foliage with either the susceptible or the pit-type reaction had a similar response on the hypocotyl. On a susceptible hypocotyl, *V. inaequalis* forms an extensive subcuticular stroma that is initially one cell in depth, but it thickens to several cell layers (Figure 45). The stroma supports abundant sporulation, beginning after 8 days. Granulation of the cytoplasm and formation of yellow-orange globules occur in host cells near the well-developed stromata. Responses by the host are less extensive in the susceptible reaction.[71] A comparison between the responses by the host cells in the hypocotyl to those in foliage following stroma formation reveals several interesting facts. On the hypocotyls, the epidermal walls stain more irregularly subjacent to the stromata, and the organelles are more dispersed. Additionally, the hyphae of *V. inaequalis* grow between epidermal cells on the hypocotyls, and a layer of hyphae becomes established at the walls of the cortical cells. The lumen of the epidermal cells can be penetrated by hyphae of *V. inaequalis* and the penetrated cells are devoid of recognizable organelles.[28] The cortical cells remain turgid. Thus, the host/pathogen interaction within hypocotyl tissues more closely resembles interactions occurring in green shoots than those in the foliage.

On cultivars that form a pit-type hypersensitive interaction on the foliage, *V. inaequalis* only forms a subcuticular stroma that is limited to one layer of up to five stromatic cells (Figure 46). The cytoplasm of epidermal cells appears granular or occasionally is filled with small ''bubbles'' 2 or 3 d following inoculation. The epidermal cells appear discolored as yellow-orange globules develop within cells. Eventually, extreme granulation occurs, the globules dissipate, and the cells in the infected area turn brown. Forty-eight to 96 h after inoculation, red-brown streaks are visible macroscopically due to coalesced, necrosing host cells. Sporulation is absent. The formation of the yellow-orange globules is suspected to be due to accumulation of phenolic compounds present in apple leaves, e.g., phloridzin. In fact, the oxidation of phenolic compounds has been suggested to inhibit the development of *V. inaequalis* in hypersensitive or resistant reactions. However, stroma formation occurs in the absence of browning, browning occurs after macroscopic symptoms are visible, and no changes have been detected in the level of phloridzin or phloretin at the site of the hypersensitive response until 24 h after inoculation.[71] These observations indicate that the oxidation of phenolic compounds probably does not inhibit growth and sporulation of *V. inaequalis in vivo*.

IX. HISTOLOGY OF APPLE SCAB LESIONS ON BUD SCALES

Apple scab lesions have been detected on the abaxial and adaxial surfaces of the outermost scales of apple buds.[61,66,67,72,73] Lesions on the exposed, abaxial surface are suspected to be initiated early in the summer when the bud scale tissues are morphologically similar to that of young foliage. A subcuticular stroma appears to be formed during mid-fall, and it consists of a cushion of hyphal cells 20 to 200 μm thick (Figure 47).[72] Annellophores appear to be formed with or without a significant subcuticular stroma.[61,72] The stroma is usually subcuticular, but hyphae ''deep in the host tissue'' have been observed, especially at the base of the scales. During the late winter, the outer host cells appear to be pushed up or ruptured by the fungal stroma. Concurrently, a dense mass of annellophores and conidia erupts from the cushion of hyphae. Sporulating lesions appear as circular or elliptical pustules and are usually less than one millimeter in diameter.[67] Lesions on the adaxial surface appear strikingly similar to infections on the foliage; hyphae of *V. inaequalis* appear to grow subcuticularly, from which annellophores erupt through the cuticle singly or in clusters (Figure 48).[67] Individual annellophores with more than 20 annellations are observed frequently. Additionally, hyphae of *V. inaequalis* are observed growing above the cuticle. This observation is in contrast to observations of fungal growth patterns on foliar lesions and indicates that the microclimate (e.g., high relative humidity) between the bud scales is conducive to this pattern of growth.

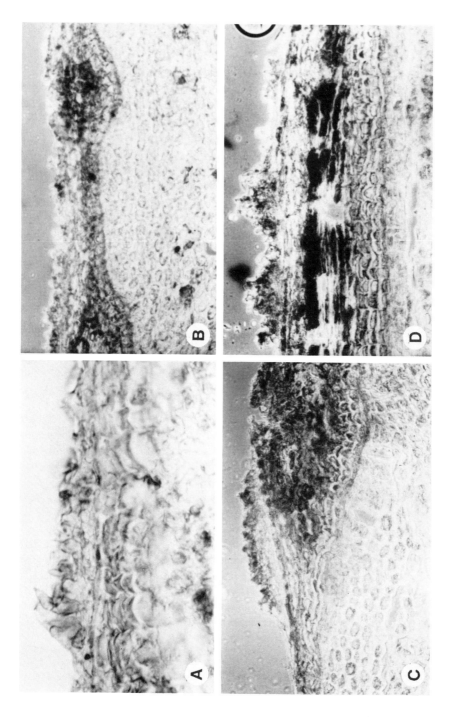

FIGURE 43.

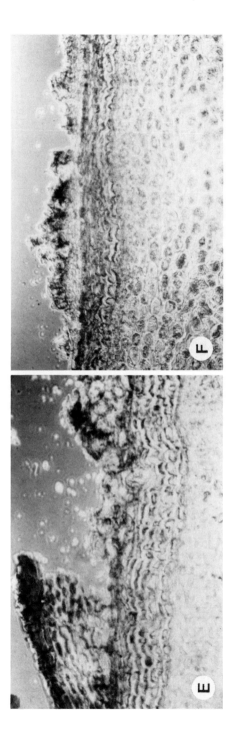

FIGURE 43. Sections through scab lesions on apple shoots caused by *V. inaequalis*. Sections in early summer reveal conidia and annellophores typical of the anamorph *Spilocaea pomi* (A), and the formation of a periderm layer of cork-like cells four to six cell layers beneath the fungal stroma (B). Within a lesion sectioned in the fall, sporulation is only evident at the margin of the lesion (C). The most common lesion type observed in the fall has disorganized, and disrupted host tissues above the periderm layer (D), with over 50% sloughed off (E). Some lesions that were sectioned in the fall continued to sporulate, and had periderm layer formed immediately subjacent to the stroma (F) on the surface of the shoot tissue (A ×1320; B ×530; C ×680; D ×680; E ×680; F ×530.) (From Becker, C. M., Overwintering of the Anamorph of *Venturia inaequalis* (*Spilocaea pomi*) in Apple Buds and the Viability of Conidia as Affected by Discontinuous Wetting, Ph.D. thesis, Cornell University, Ithaca, NY, 1990.)

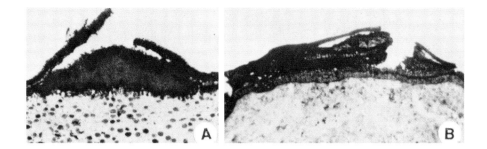

FIGURE 44.   Sections through scab lesions on shoots during the second and third years after *V. inaequalis* initiated the lesion. A, sporulation of conidia during the spring from 1-year-old lesion (×70); B, sporulation of conidia during the spring from 2-year old lesion. Several periderm layers have been formed by the host in attempts to wall off the pathogen (×22). (From Swinburne, T. R., *Plant Pathology,* 14, 23, 1965. Reproduced with permission of Her Britannic Majesty's Stationery Office.)

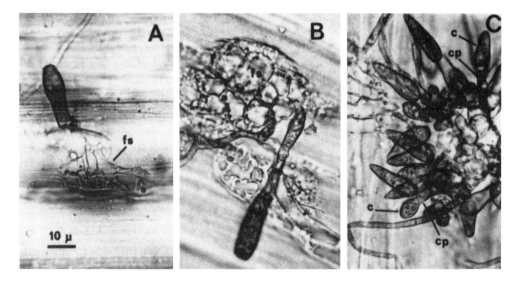

FIGURE 45.   Susceptible interaction between *V. inaequalis* and etiolated hypocotyl tissues. A, subcuticular stroma (fs) forms from the appressorium and is initially one-cell layer thick; B, the stroma expands to several cell layers in thickness; C, prolific production of conidia (c) from annellophores (cp); scale bar = 10 μm for all photographs. (From Nicholson, R. L. et al., *Phytopathology,* 63, 365, 1973. With permission.)

X. SUMMARY AND CONCLUDING REMARKS

The abundant histological and cytological studies of *V. inaequalis* have provided a detailed account of how this fungal pathogen breaches the cuticle and survives within the leaf. High quality observations of the interactions between *V. inaequalis* and shoot and fruit tissues are lacking. Since *V. inaequalis* grown subepidermally in those tissues, such studies would provide insight into the mechanisms of resistance. Observations of *V. inaequalis*-infected flower petals and sepals are also lacking. Histological studies can provide information about these sites as additional sources for production of secondary inoculum, especially on sepals, which can be the first tissues exposed to inoculum in the spring[9] and remain attached to the fruit until harvest. Histological observations of diffuse, late season lesions on the under surface of leaves will show not only the availability of inoculum late

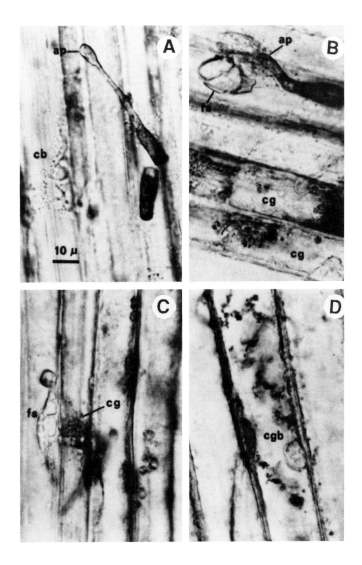

FIGURE 46. Hypersensitive interaction between *V. inaequalis* and etiolated hypocotyls. A, Germling and appressorium (ap) formation appears typical as on a susceptible host. Cytoplasmic bubbling (cb) is sometimes observed prior to cytoplasmic granulation; B, cytoplasmic granulation (cg) within the host epidermal cells and limited extension away from the site of penetration by the fungal stromata; C, cytoplasmic granulation at the interface of two epidermal cells directly beneath a fungal stroma; D, cytoplasmic granulation becomes extensive and browning of host tissues (cgb) occurs. Fungal stroma is essentially absent. Scale bar = 10 μm for all photographs. (From Nicholson, R. C. et al., *Phytopathology*, 63, 364, 1973. With permission.)

in the season, but will show the distribution of subcuticular stromata available for penetrating through the leaf and producing pseudothecia. Both SEM and TEM have proven to be effective tools for examining the above scenarios. However, there are new and exciting tools available that provide specific information about biochemical interactions at the cellular level: immunocytological techniques and fluorescence and confocal microscopy. These tools are becoming more commonplace, and they will undoubtedly be coupled with evaluations of structural interactions between the host and pathogen (e.g., apposition formation). Additionally, future research and development of options for managing apple scab, such as target-site specific fungicides and potential biocontrol microorganisms, will likely utilize these tools.

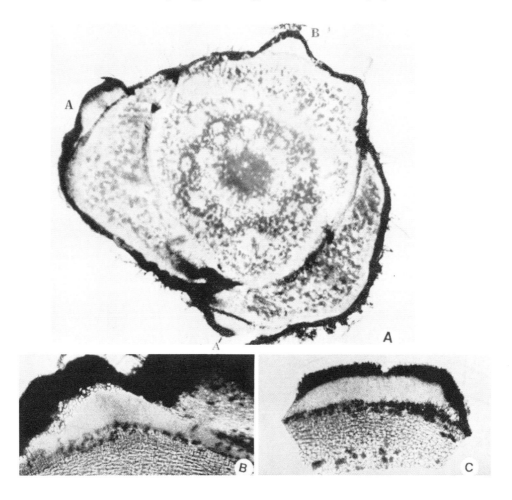

FIGURE 47. Lesions on dormant apple bud scales caused by *V. inaequalis*. A, Sectioned bud in late winter with scab lesions on abaxial surface at positions A, A′, and B′ (×34); B, in late winter, an extensive cushion on mycelium growing within the cortical tissues. The stromata appears to be expanding to the right (×64); C, in early spring, the cushion of light colored mycelium appears to produce an abundance of annellophores and conidia on the surface (×64). (From McKay, R., *Sci. Proc. R. Dublin Soc. NS,* 21, 623, 1938. With permission.)

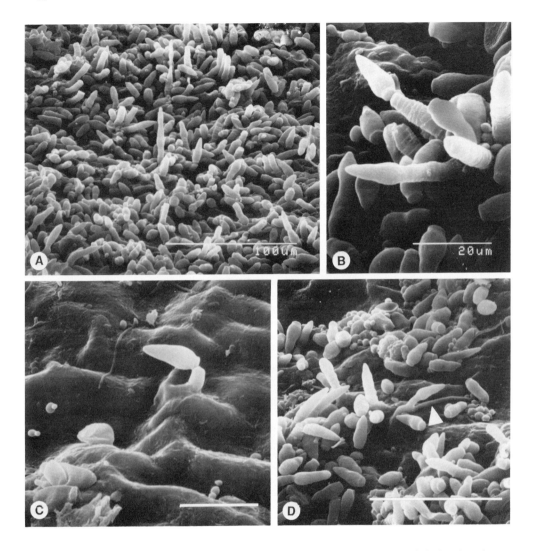

FIGURE 48. SEM of lesion caused by *V. inaequalis* on the adaxial surface of an apple bud scale, prior to budbreak in the spring. A, Abundant production of annellophores and conidia within center of lesion; scale bar = 100 μm; B, cluster of annellophores with mature conidia showing numerous annellations per conidiophore; scale bar = 20 μm; C, at the margin of the lesion, swollen ridges are probably the result of subcuticular hyphae, especially since an annellophore has erupted from one such ridge; scale bar = 20 μm; D, hypha growing along the surface of the host (arrow), which is atypical compared with foliar infections; scale bar = 50 μm. (From C. M. Becker, Ph.D. thesis, Cornell University, Ithaca, 1990.)

REFERENCES

1. **Sivanesan, A. and Waller, J. M.,** *Venturia inaequalis, C. M. I. Descriptions of Pathogenic Fungi and Bacteria,* 401, Commonwealth Mycological Institute, Kew, England, 1974.

2. **Marsh, R. W. and Walker, M. M.,** The scab fungus *Venturia inaequalis* on apple shoots, *J. Pom. Hortic. Sci.,* 9, 171, 1932.

3. **Becker, C. M., Burr, T. J., and Smith, C. A.,** Overwintering of conidia of *Venturia inaequalis* in apple buds in New York orchards. *Plant Disease,* 76, 121, 1992.

4. **Wilson, E. E.,** Studies of the ascigerous stage of *Venturia inaequalis* (Cke.) Wint. in relation to certain factors of the environment, *Phytopathology,* 18, 375, 1928.

5. **Frey, C. N.,** The cytology and physiology of *Venturia inaequalis* (Cooke) Winter. *Wisc. Acad. Sci. Arts, and Letters,* 21, 303, 1924.

6. **James, J. R. and Sutton, T. B.,** Environmental factors influencing pseudothecial development and ascospore maturation of *Venturia inaequalis, Phytopathology,* 72, 1073, 1982.

7. **Killian, K.,** Über die sexualitat von *Venturia inaequalis* (Cooke) Ad., *Z. Boton.,* 9, 353, 1971.

8. **Louw, A. J.,** Studies on the influence of environmental factors on the overwintering and epiphytology of apple scab [*Venturia inaequalis* (Cke.) Wint,] in the winter-rainfall area of the Cape Province, *South Afr. Dep. Agric. Sci. Bull.,* 310, 1951.

9. **Keitt, G. W. and Jones, L. K.,** Studies of the epidemiology and control of apple scab. *Wis. Agric. Exp. Stn. Res. Bull.,* 73, 1926.

10. **Julien, J. B.,** Cytological studies of *Venturia inaequalis, Can. J. Bot.,* 36, 607, 1958.

11. **Backus, E. J. and Keitt, G. W.,** Some nuclear phenomena in *Venturia inaequalis, Bull. Torrey Bot. Club,* 67, 765, 1940.

12. **Day, P. R., Boone, D. M., and Keitt, G. W.,** *Venturia inaequalis* (Cke.) Wint. XI. The chromosome number, *Am. J. Bot.,* 43, 835, 1956.

13. **Sivanesan, A.,** The taxonomy and pathology of *Venturia* species, *Bibl. Mycologica,* Band 59, J. Cramer, Vaduz, 1977, 139 pp.

14. **Young, C. S. and Andrews, J. H.,** Inhibition of pseudothecial development of *Venturia inaequalis* by the basidiomycete *Athelia bombacina* in apple leaf liter, *Phytopathology,* 80, 536, 1990.

15. **Becker, C. M.,** personal observations.

16. **Brook, P. J.,** The ascospore production season of *Venturia inaequalis* (Cke.) Wint., the apple black spot fungus. *N. Z. J. of Agric. Res.,* 9, 1064, 1966.

17. **Gadoury, D. M. and MacHardy, W. E.,** Negative geotrophism in *Venturia inaequalis, Phytopathology,* 75, 856, 1985.

18. **Jehle, R. A. and Hunter, H. A.,** Observations on the discharge of ascospores of *Venturia inaequalis* in Maryland, *Phytopathology,* 18, 943, 1928.

19. **Wallace, E.,** Scab disease of apples, *Cornell Univ. Exp. Stn. Bull.* 335, 38 pp, 1913.

20. **James, J. R., Sutton, T. B., and Grand, L. F.,** Unusual ascocarp formation in *Venturia inaequalis, Mycologia,* 73, 564, 1981.

21. **Jeger, M. J.,** Overwintering of *Venturia inaequalis* in relation to lesion intensity on leaf surfaces, and leaf surface exposed, *Trans. Br. Mycol. Soc.,* 83, 495, 1984.

22. **Samuels, G. J. and Sivanesan, A.,** *Venturia asperata* sp. nov. and its *Fusicladium* state on apple leaves, *N. Z. J. of Bot.,* 13, 645, 1975.

23. **Becker, C. M., Shotwell, K. M., Thomson, S. V., and Burr, T. J.,** Differentiating between *Mycosphaerella tassiana* and the apple scab pathogen *Venturia inaequalis* on overwintering apple leaves, *Phytopathology* (Abstr.), 80, 1042, 1990.

24. **Granett, A. L.,** Ultrastructure of the ascospores and conidia of *Venturia inaequalis, Proc. Electronmicroscopy Soc. Am.,* 31, 466, 1973.

25. **Smerka, K. M., MacHardy, W. E., and Kausch, A. P.,** Cellular differentiation in *Venturia inaequalis* ascospores during germination and penetration of apple leaves, *Can. J. Bot.,* 65, 2549, 1987.

26. **Corlett, M., Chong, J., and Kokko, E. G.,** Ultrastructure of the *Spilocaea* state of *Venturia inaequalis* in vivo, *Can. J. Microbiol.,* 22, 1144, 1976.

27. **Hammill, T. M.,** Fine structures of annellophores. IV. *Spilocaea pomi, Trans. Br. Mycol. Soc.,* 60, 65, 1973.

28. **Maeda, K. M.,** An Ultrastructural Study of *Venturia inaequalis* (Cke.) Wint. Infection of *Malus Hosts,* M.S. thesis, Purdue University, Lafayette, IN, 1970, 112 pp.

29. **Paul, H. V.,** Biology of *Venturia inaequalis* (Cooke) Winter, the pathogen of apple scab, *Pfanzenschutz-Nach.,* 34, 60, 1981.

30. **Wiltshire, S. P.,** Infection and immunity studies on the apple and pear scab fungi (*Venturia inaequalis* and *V. pirina*), *Ann. App. Biol.*, 1, 335, 1915.

31. **Yepes-Martinez, L. M.,** Development of Techniques for Selection of Disease-Resistant Apple Germplasm in Vitro, M.S. thesis, Cornell University, Ithaca, NY, 1987, chapter 2.

32. **Kendrick, B.,** Conclusions and recommendations, in *Taxonomy of Fungi Imperfecti,* Kendrick, B., Ed., University of Toronto, Toronto, 1971, 253.

33. **Granett, A. L.,** Ultrastructural studies of concentric bodies in the ascomycetous fungus *Venturia inaequalis, Can. J. Bot.,* 52, 2137, 1974.

34. **Gessler, C. and Strumm, D.,** Infection and stroma formation by *Venturia inaequalis* on apple leaves with different degrees of susceptibility to scab, *Phytopath. Z.,* 110, 119, 1984.

35. **Corlett, M. and Chong, J.,** Ultrastructure of the appressorium of *Spilocaea pomi, Can. J. Bot.,* 55, 5, 1977.

36. **Alderhold, R.,** Die Fusicladien unserer Obstbaume. I, *Landw. Jahrb.,* 25, 875, 1896.

37. **Valsangiacomo, C. and Gessler, C.,** Role of the cuticular membrane in ontogenic and VF-resistance of apple leaves against *Venturia inaequalis, Phytopathology,* 78, 1066, 1988.

38. **Voges, E.,** Die Bekampfung des *Fusicladium, Zietschr. Pflanzenkr.,* 20, 385, 1910.

39. **Nusbaum, C. J. and Keitt, G. W.,** A cytological study of host-parasite relations of *Venturia inaequalis* on apple leaves, *J. Agric. Res.,* 56, 595, 1938.

40. **Busgen, M.,** Über einige Eigenschaften der keimlinge parasitische Pilze, *Bot. Zeit.,* 53, 1893.

41. **Smereka, K. M., Kausch, A. P., and MacHardy, W. E.,** Intracellular junctional structures in germinating ascospores of *Venturia inaequalis, Protoplasma,* 142, 1, 1988.

42. **Koller, W. and Parker, D. M.,** Purification and characterization of cutinase from *Venturia inaequalis, Phytopathology,* 79, 278, 1989.

43. **Koller, W., Parker, D. M., and Becker, C. M.,** Role of cutinase in the penetration of apple leaves by *Venturia inaequalis, Phytopathology,* 81, 1375, 1991.

44. **Nicholson, R. L., Kuc, J., and Williams, E. B.,** Histochemical demonstration of transitory esterase activity in *Venturia inaequalis, Phytopathology,* 62, 1242, 1972.

45. **Enochs, N. J.,** Histological Studies on the Development of *Venturia inaequalis* (Cke.) Wint. in Susceptible and Resistant Selections of *Malus,* M.S. thesis, Purdue University, Lafayette, IN, 1964, 58 pp.

46. **Locci, R. and Biasiach, M.,** Scanning electron microscopy of the invasion of leaf tissues by the apple scab fungus, *Instituto Di Path. Veg, (Univ. Di Milano),* p. 15, 1971.

47. **Dahmen, H. and Hobot, J. A.,** Ultrastructural analysis of *Erysiphe graminis* haustoria and subcuticular stroma of *Venturia inaequalis* using cryosubstitution, *Protoplasma,* 131, 92, 1986.

48. **Hoch, H. C. and Szkolnik, M.,** Viability of *Venturia inaequalis* in chlorotic flecks resulting from fungicide application to infected *Malus* leaves, *Phytopathology,* 69, 456, 1979.

49. **Wagner, K., Hitz-Germann, L., Seng, J. M., and Gessler, C.,** Cellulytic ability of the scab fungus, *Venturia inaequalis, J. Phytopathology,* 123, 217, 1988.

50. **Valsangiacomo, C.,** Aspects of Pathogenesis and Host Resistance in the Interaction between *Venturia inaequalis* and Apple Leaves, Ph.D.thesis, Swiss Federal Institute of Technology, Zurich, 1990, 56 pp.

51. **Chelvalier, M. and Lespinasse, Y.,** La resistance du pommier (*Malus* × *domestica* Borkh.) a *Venturia inaequalis* (Cke.) Wint. Etude histologique et ultrastructurale du symptome de resistance en piqure d'epingle, *C. R. Acad. Sci., Paris,* t. 312, Serie III, 117, 1991.

52. **Chevalier, M., Lespinasse, Y., and Renaudin, S.,** A microscopic study of the different classes of symptoms coded by the V*f* gene in apple for resistance to scab *(Venturia inaequalis), Plant Pathology,* 40, 249, 1991.

53. **Williams, E. B. and Kuc, J.,** Resistance in *Malus* to *Venturia inaequalis, Annu. Rev. Phytopathol.,* 7, 223, 1969.

54. **Drake, C. R. and Shear, G. M.,** The periderm, a factor limiting apple scab development, *Phytopathology* (Abstr.), 61, 890, 1971.

55. **Johnstone, K. H.,** Observations on the varietal resistance of the apple to scab (*Venturia inaequalis,* Aderh.) with special reference to its physiological aspects, *J. Pom. and Hortic. Sci.,* 9, 30, 1931.

56. **Schwabe, W. F. S.,** Changes in scab susceptibility of apple leaves as influenced by age, *Phytophylactica,* 11, 53, 1979.

57. **Corlett, M. and Ross, R. G.,** Morphology of *Spilocaea pomio* on untreated and benomyl-treated McIntosh apple leaves, *Can. J. Plant Pathol.,* 1, 79, 1979.

58. **Kline, D. M., Boone, D. M., and Keitt, G. W.,** *Venturia inaequalis.* XV. Histology of infections by biochemical mutants, *Am. J. Bot.,* 51, 634, 1961.

59. **McAlpine, D.,** The fungus causing "black spot" of the apple and pear, *J. Dept. Agric., Victoria,* 1, 703, 1902.

60. **Walker, E. A.,** Histological studies on storage scab of apples, *Peninsula Hortic. Soc. 54th Trans.* 1940, 84.

61. **Barakat, F. M. and El-Shehedi, A. A.,** Pathological and anatomical studies on apples infected by *Venturia inaequalis, Agric. Res. Rev.,* 60, 245, 1982.
62. **Tomerlin, J R. and Jones, A. L.,** Development of apple scab on fruit in the orchard and during cold storage, *Plant Dis.,* 67, 147, 1983.
63. **Bratley, C. O.,** Incidence and development of apple scab fungus on fruit during the late summer and while in storage, *U. S. Dept. Agric. Tech. Bull.,* No. 536, 1937.
64. **Saad, A. T.,** The Culture of Apple Callus Tissue and Its Use in Studies on Pathogenicity of *Venturia inaequalis,* Ph.D. thesis, University of Wisconsin, Madison, 1965, 111 pp.
65. **Beech, I. and Gessler, C.,** Interaction between *Venturia inaequalis* and apple callus tissue cultures: an electron microscopic study, *J. Phytopathology,* 116, 315, 1986.
66. **Becker, C. M.,** Overwintering of the Anamorph of *Venturia inaequalis (Spilocaea pomi)* in Apple Buds, and the Viability of Conidia as Affected by Discontinuous Wetting, Ph.D. thesis, Cornell University, Ithaca, 1990, 86 pp.
67. **Becker, C. M. and Burr, T. J.,** Apple scab lesions, caused by *Venturia inaequalis,* on shoots in New York; histology and enumeration of inoculum, *Phytopathology* (Abstr.), 80, 117, 1990.
68. **Swinburne, T. R.,** Apple scab infection of the young wood of Brambley's Seedling trees in Northern Ireland, *Plant Pathology,* 14, 23, 1965.
69. **Jeger, M. J. and Alston, F. H.,** Resistance in apple to shoot infection by *Venturia inaequalis, Ann. Appl. Biol.,* 108, 387, 1986.
70. **Salmon, E. S.,** Apple scab or black rot, *Gard. Chron.,* 40, 21, 1906.
71. **Nicholson, R. L., Van Scoyoc, S., Williams, E. B., and Kuc, J.,** Etiolated apple hypocotyls: a useful host tissue in apple scab research, *Phytopathology,* 63, 363, 1973.
72. **McKay, R.,** Conidia from infected budscales and adjacent wood as a main source of primary infection with the apple scab fungus *Venturia inaequalis* (Cooke) Wint., *Sci. Proc. Royal Dublin Soc. NS,* 21, 623, 1938.
73. **Salmon, E. S. and Ware, W. M.,** A new fact in the life-history of the apple scab fungus, *Gard. Chron.,* 89, 437, 1931.

Chapter 4

THE CYTOLOGY AND HISTOLOGY OF PEAR RUST

Mitsuru Kohno and Hitoshi Kunoh

TABLE OF CONTENTS

0-8493-2939-6/93/$0.00 + $.50

I. INTRODUCTION

The rust fungi (class Basidiomycetes, order Uredinales) parasitize most green plants and have worldwide distribution, ranging from tropical to arctic regions, from desert to marsh, and from valley bottoms to mountain tops.[1] Rust diseases frequently cause serious problems for fruit cultivation. *Gymnosporangium fuscum* DC. (*G. sabinae* Dicks. ex Wint.) is the most widely distributed rust fungus found on pear (*Pyrus communis*), and pear rust has been reported to occur over much of the temperate world, including Europe, North America, North Africa, and Asia Minor.[2] In Japan, *G. asiaticum* Miyabe ex Yamada (*G. haraeanum* Syd.) is the rust pathogen found on Japanese pear (*Pyrus serotina* var. *culta*). The disease, Japanese pear rust, is a considerable problem, since cultivars of *P. serotina* var. *culta* are more popular in Japan than those of *P. communis*. Almost all known cultivars of Japanese pear are susceptible to *G. asiaticum,* but cultivars of *P. communis* are resistant.[3] *G. shiraianum* Hara (*G. juniperi* Link) also occurs on Japanese pear, but only in restricted areas of the country. This chapter focuses mainly on the disease caused by *G. asiaticum.*

The rust fungi have a complex life cycle that may produce up to five different spore stages (basidiospores, pycniospores, aeciospores, urediospores, and teliospores), some of which, depending on the species, form on two different host species obligatorily (heteroecious rusts). Other rusts may pass their entire life cycle on one host species (autoecious rusts).[4] The rusts that exhibit all five spore stages are termed macrocyclic. Such stages may be restricted to a single host, or may infect alternate hosts. Another type of life cycle is termed demicyclic, and deletes the uredial stage from an otherwise macrocyclic form.[5] The genus, *Gymnosporangium,* is a typical demicyclic rust.

G. asiaticum forms telia on some cultivars of *Juniperus chinensis* (*J. chinensis, J. chinensis* var. *procumbens, J. chinensis* var. *pyramidabilis,* etc.), while pycnia and aecia are produced on *Chaenomeles lagenaria, Ch. sinensis,* and *Cydonia oblonga* in addition to most cultivars of Japanese pear.[3] *G. shiraianum* differs from *G. asiaticum* in that the former fungus forms telia on *Juniperus rigida, J. conferta, J. communis,* and *J. sibirica.*

G. asiaticum infects leaves, petioles, and fruits of Japanese pear. Photosynthetic activity is remarkably reduced in heavily infected leaves. Leaves on which a number of aecia are formed often fall off. Infected fruits are frequently deformed. Thus, this disease reduces the commercial value and yield of pear and brings serious economic consequences to growers. The rust does not seem to cause appreciable damage to any of its alternate hosts.

II. OVERALL FEATURES OF THE DISEASE ON HOST TREES

A. SYMPTOMS AND SIGNS ON JAPANESE PEAR AND THE ALTERNATE HOST JUNIPER

In the spring (usually middle to late April in the central parts of Japan), several slightly yellow blotches that are covered with honeydew appear on the upper surfaces of young leaves of pear (Figure 1; refer to Figures 1 to 6 legend for a list of the abbreviations found in the figures). The dew dries 7 to 10 d later, leaving numerous dark dots in the blotches (Figure 2). These dots correspond to the ostioles of pycina. In middle to late June, swellings begin to appear on the lower surfaces of pear leaves directly below the pycnia, and are followed by the emergence of aecial horns (Figures 3 and 4). The aecial part is very hard.

In early to mid-March, brown, scab-like structures suddenly appear on leaves and twigs of juniperous trees. Until then, any symptom and sign cannot be detected on any parts of this host. Subsequently, several yellow to orange horns appear on the scab-like structures (Figure 5). Initially, telial horns are only a few millimeters in width. During or after a period of rain in late March, the scab-like structures and horns absorb water and swell, becoming

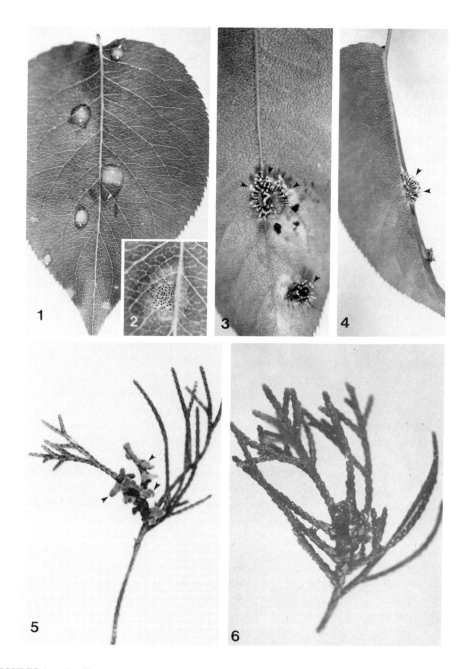

FIGURES 1 to 6. **Figure 1.** The honeydew secreted from the pycnia which have been produced on the upper surface of a pear leaf. **Figure 2.** The ostioles (dark dots) of the pycnia which are visible after the honeydew dries. **Figure 3.** The aecial horns (arrowheads) that have emerged from the lower surface of a pear leaf. **Figure 4.** A side view of the aecial horns (arrowheads). **Figure 5.** The telial horns (arrowheads) which have appeared on twigs of juniper. **Figure 6.** The gelatinous telial horns which have swollen by absorbing water. (Figure abbreviations ai: aeciospore initial, ap: appressorium, as: aeciospore, C: chloroplast, CU: host cuticle, CW: host cell wall, cw: fungal cell wall, EP: host epidermal cell, ep: encapsulation, er: endoplasmic reticulum, gt: germ tube, h: hypha, ha: haustorium, ic: intercalary cell, ih: intracellular hypha, il: inner layer of cell wall, inh: intercellular hypha, l: lipid body, m: mitochondrion, ml: middle layer of cell wall, N: host nucleus, n: fungal nucleus, Nd: microneedle, nl: nucleolus, nm: nuclear membrane, ol: outer layer of cell wall, os: ostiole, pa: paraphyses, pe: peridium, ped: pedicel, pi: pycnial initial, PP: palisade parenchyma tissue, ps: pycniospore, pw: pycnial wall, py: pycnium, se: septum, sp: pycniosporophore, SP: spongy parenchyma tissue, spo: sporidium, st: sterigma, STO: stoma, ts: teliospore, V: host vacuole, v: fungal vacuole.

gelatinous (Figure 6). The sequence of events is initiated by rainfall, but is apparently governed by temperature. The optimum temperatures for swelling of horns are 12.5° to 20°C, and the ideal temperature for the teliospore germination is 20°C.[3] This optimum range of temperature is consistent with that for the development of young leaves of Japanese pear. Although conspicuous galls are formed on junipers infected by *G. juniperi-virginianae*,[1,6] only slightly swollen galls appear on twigs of juniper infected by *G. asiaticum*.

B. GENERAL VIEW OF THE PATHOGEN LIFE CYCLE

In the early spring (late March to early April in the central parts of Japan), young pear leaves become infected by airborne basidiospores, which are disseminated form the telial horns on infected junipers in the vicinity. These basidiospores have a thin wall, and infection of the pear leaves can therefore take place only if the juniperous hosts are nearby, within a radius of approximately 1 km. Within 10 to 15 d after infection has taken place, pycnia appear in yellow blotches on the upper surface of the pear leaves. Fruits, twigs, and petioles of pear may also become infected. In early summer swellings begin to appear on the lower surfaces of pear leaves, particularly on the yellow leaf blotches directly below the pycnia. On these swellings aecial horns eventually develop which mature and begin to release their aeciospores in early summer. These aeciospores, like the basidiospores, become airborne after they are disseminated and infect only juniper. Aeciospores very often land on pear leaves and produce a long germ tube, but never form an appressorium, and thus never attempt penetration.[7] In contrast to basidiospores, aeciospores having a thick cell wall are relatively resistant to temperature and moisture changes, allowing them to be carried by wind currents for hundreds of kilometers without loss of viability.[2]

By early summer, usually late June to early July, almost all aeciospores are disseminated, and aecial horns are dried up. Although Borno and van der Kamp[8] reported that junipers were infected by aeciospores of *G. fuscum* soon after spore deposition, it is still unknown at what point infection by airborne aeciospores takes place on the alternate hosts of *G. asiaticum*. The symptom and sign remain hidden in the junipers and cannot be identified until the following spring after infection has occurred. It is during this initial period that the rust of juniper escapes detection. The number of months required for the appearance of the first telia on junipers after infection by *G. fuscum* has been debated. Ziller[2] reported that 18 months were required before the first telia appeared on the infected junipers, while Borno and van der Kamp[8] emphasized that only 6 months were required. In the case of *G. asiaticum*, the time required for the development of mature telia on junipers is unknown. In the spring, 1 to 2 years after the initial infection, the first crop of telial horns develops on infected juniper twigs. The rust fungus is then established in the juniper twigs and will produce a new crop of telial horns from the same twig canker in the spring for the following 4 to 5 years, without apparent damage to its host.[9] With the infection of pear by basidiospores released from the first crop of telial horns on juniper, the life cycle of the rust is completed.

III. CYTOLOGY OF THE PATHOGEN LIFE CYCLE

A. TELIAL STAGE

Telia are sori that contain teliospores, the basidium-producing spores of the life cycle.[4,10]

1. Teliospore

A mature teliospore is ellipsoid with slightly tapered ends. Most are about 30 to 60 μm in length and about 20 to 30 μm in diameter. Each is divided by a transverse septum into an apical and a basal cell (Figure 7). A long, hyaline, cylindrical pedicel is attached to the basal cell of the spore (Figure 8). Within a few hours after teliospores encounter water, the germ pore regions near the septum begin to swell, giving rise to globose germ tube initials (Figure 8).

Teliospores embedded in telial horns are composed of two cells, each of which is binucleate (Figure 9 and 10). The binucleate condition possibly represents a stage prior to the fusion of two haploid nuclei. The cytoplasm is packed with small lipid bodies surrounded by a dense cytoplasmic substratum that contains mitochondria and ribosomes. Mature teliospores of rust fungi are difficult to prepare for transmission electron microscopy (TEM), probably due to several problems encountered with infiltration of fixatives, washing buffers, and resins, as pointed out by Mims et al.[11]

2. Germination of Teliospores

The first indication of germination, distinguishable by light microscopy, is the formation of globular protrusions through the germ pores (Figure 8). Figures 11 through 16 demonstrate changes occurring in the cell wall of germinating teliospores before the first sign of germination can be confirmed by light microscopy.[12]

The teliospores have a thick cell wall composed of several layers (Figure 11). The thin outer layer is electron dense and is continuous around the entire two-celled spore. Each cell of the spore is covered by two to three middle and innermost layers. The first sign of germination is an increase in thickness of the innermost electron-lucent layer at the point of germ tube emergence (Figure 12). This increase in thickness of the innermost layer results in an apparent reduction in the thickness of middle layers. The outer layer splits and the components of the middle layers are released through this split (Figures 13 and 14). Subsequently, all the wall layers break up, apparently due to the protrusion of cytoplasm into the germ tube or promycelium (Figure 15). A strong pressure of the outgrowing cytoplasm seems to bring about the initiation of germ tube emergence. The fibrous material around the emerging germ tube is slightly electron dense and apparently continuous with the innermost layer of the spore. Only a part of this layer serves as the wall of emerging germ tube (Figure 16). The overall feature of changes of cell walls during germination is basically similar to that of *G. juniperi-virginianae*[11] and *G. clavipes*.[13]

B. BASIDIOSPORE STAGE
1. Basidiospore Formation

Within a few hours after teliospores encounter water at favorable temperatures, they begin to germinate, forming a promycelium which is a basidium of this fungus. The entire cytoplasm of the teliospore, including a nucleus, enters the promycelium (Figure 17), and a transverse septum is then formed in the middle of the promycelium to separate the empty teliospore from the promycelium. The first nuclear division, probably mitotic, occurs to give rise to two daughter nuclei (Figure 18). The subsequent septum formation divides the promycelium into two cells, each containing a single nucleus (Figure 19). The two nuclei undergo the second division simultaneously (Figure 20). Two septa are laid down, resulting in a four-celled promycelium with a single nucleus in each cell (Figure 21). A sterigma on which one basidiospore (= sporidium) will be formed emerges from each cell. Shortly after the basidiospore forms, the nucleus migrates from the promycelium cell into the basidiospore (Figure 22). The formation of basidiospores does not simultaneously occur in each of the four cells of the promycelium. The nucleus of each spore may divide mitotically, rendering the mature spore binucleate (Figure 23), as is the case in *G. juniperi-virginianae*.[14] Mitosis, however, does not seem to occur in every spore. Of 517 basidiospores observed by Kohno et al.,[15] 47.8% had a single nucleus, 34.6% had two nuclei, and the nuclear number was not determined in 17.6%. Such nuclei appeared elongate, suggesting possible nuclear overlapping. The two nuclei in basidiospores may fuse to give rise to a single large nucleus

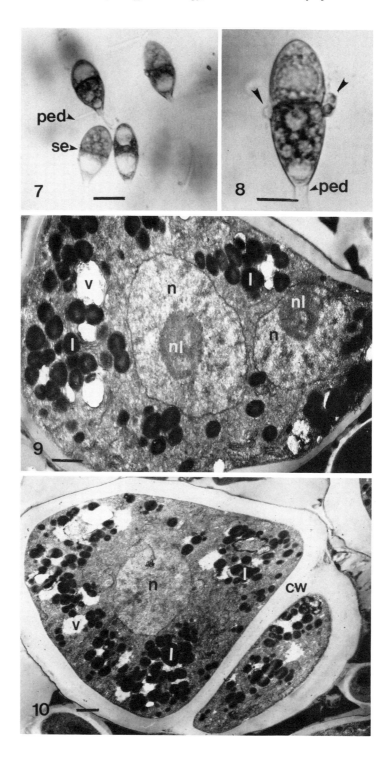

FIGURES 7 to 10. **Figure 7.** Bicellular teliospores with a long, hyaline pedicel. Bar = 20 μm. **Figure 8.** A teliospore with two globose germ tube initials (arrowheads) near the septum. Bar = 10 μm. **Figure 9.** A binucleate cell of a teliospore. A number of lipid bodies are present in the cytoplasm. Bar = 1 μm. **Figure 10.** A cell of a teliospore having a thick cell wall. A nucleus is visible in the cell. Bar = 1 μm.

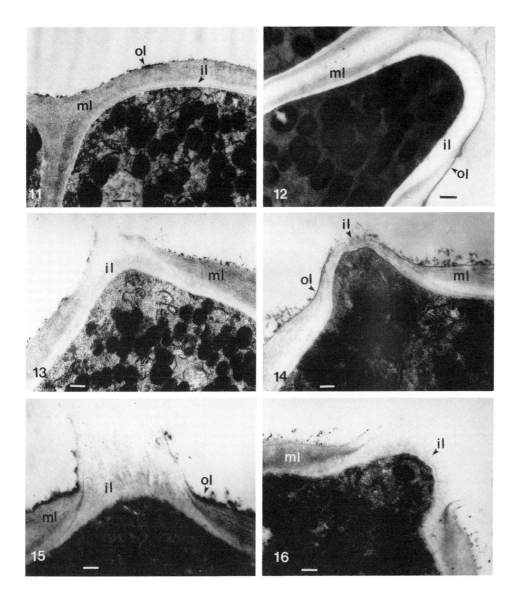

FIGURES 11 to 16. **Figure 11.** A part of a ungerminated teliospore. The cell wall consists of a thin electron-dense outer layer and thick middle and inner layers in which at least two layers are distinguished. Bar = 0.5 μm. **Figure 12.** The innermost layer of the teliospore cell has increased in thickness at the point of germ tube emergence of the teliospore. The outer layer still surrounds two cells in the spore. Bar = 0.5 μm. **Figures 13 and 14.** The outer layer of the teliospore cell has split and the components of the middle layer(s) have been released through the split. Bar = 0.5 μm. **Figure 15.** All the wall layers of the teliospore cell have been forced to break up by the appearance of a protrusion of cytoplasm. Bar = 0.5 μm. **Figure 16.** A fibrous layer which has apparently originated from the innermost layer of the spore covers the germ tube. The outer and middle layers apparently split at the site of germ tube emergence. Bar = 0.5 μm.

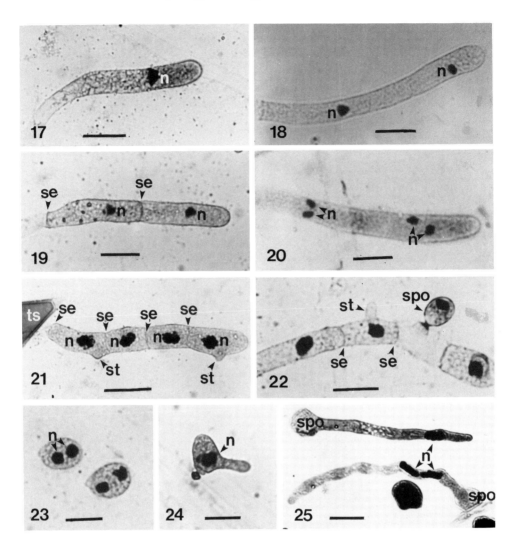

FIGURES 17 to 25.   Figure 17. A nucleus in a promycelium arising from a teliospore, stained by the Giemsa method. Bar = 10 μm. **Figure 18.** Two nuclei resulting from the first mitotic division in a promycelium. Bar = 10 μm. **Figure 19.** A septum that was laid down between two nuclei, separating the nuclei from each other into two cells. Bar = 10 μm. **Figure 20.** Two nuclei in each cell that result from the second mitotic division. Bar = 10 μm. **Figure 21.** Septa were produced after the second nuclear division, giving rise to a four-celled promycelium. Bar = 10 μm. **Figure 22.** A sporidium into which the cytoplasm of the mother cell migrated was seen on the second cell, and another sterigma was produced from the third cell. Bar = 10 μm. **Figure 23.** Binucleate sporidia soon after being discharged from the sterigmata. Bar = 10 μm. **Figure 24.** A germinating, uninucleate sporidium. Bar = 10 μm. **Figure 25.** Germ tubes of the sporidium in which two elongated nuclei are seen. Bar = 10 μm.

when the basidiospores initiate germination (Figure 24), or two nuclei independently migrate into a germ tube (Figure 25).

2. Fine Structure of Basidiospores

Observations by TEM revealed that basidiospores were either uni- or binucleate, and included organelles such as mitochondria, endoplasmic reticula (ER), ribosomes, and a large number of lipid bodies (Figures 26, 28, and 29).[15] In specimens fixed with $KMnO_4$, a thin, amorphous cell wall surrounds the cytoplasm of basidiospores (Figure 28), while in specimens fixed with glutaraldehyde and OsO_4, the cell wall is extremely electron translucent and difficult to define (Figure 26). When basidiospores germinate, the tip of the germ tubes is covered only with scanty fibrous wall materials (Figure 29). A large nucleus is always observed in the basidiospore (Figure 27) and it migrates into a germ tube, soon after the basidiospores initiate germination (Figure 30).

3. Appressorium Formation by Basidiospores and Penetration of the Host Epidermis

Upon the germination of a basidiospore, a well-defined appressorium is produced at the apex of an either short of elongated germ tube on the upper surface of the host, in this case the Japanese pear (Figures 31 and 32).[15] The subsequent penetration of the host was revealed by micromanipulation at the scanning electron microscope (SEM) level. When the appressorium was moved aside with a microneedle, an indentation of the host surface, caused by a penetration peg at the center of the appressorium, appeared at the site on which the appressorium had been attached (Figure 33). Wart-like structures which originally had covered the entire surface of the host epidermis disappeared from the indentation, suggesting that the appressorium either secreted degenerating enzyme(s) or exerted mechanical pressure which caused their collapse. A similar sign of wall degradation by the fungal enzymes was also revealed in an indentation of the inner side of the host cell wall, when the cell wall was reversed by micromanipulation (Figure 34).

C. PYCNIAL STAGE

Pycnia (= spermagonia) are structures that produce pycniospores (= spermatia), the monokaryotic gametes of rust fungi.[4,10] Within 2 weeks after infection by basidiospores, initials of pycnia appear on the upper surface of pear leaves. Hiratsuka and Cummins[16] distinguished 11 types of pycnial morphology using light microscopy. Following their description, *G. asiaticum* produces pycnia that are subepidermal and determinate and have strongly convex hymenia with well-developed paraphyses.

1. Formation of Pycnia and Pycniospores

A penetration peg emerging from an appressorium of the basidiospore develops into a binucleate haustorium in the upper epidermis of the pear leaf (Figures 35 and 36). Whether binucleate basidiospores produce the binucleate haustorium or uninucleate basidiospores produce it after mitosis at the certain stage is still unknown. At high magnification, electron microscopy reveals a very thin, electron-lucent zone corresponding in an encapsulation, approximately 0.05 μm thick, which surrounds the haustorium (Figure 37). From this haustorial apparatus, an intracellular hypha emerges, further elongates downward, breaches the lower cell wall of the epidermis, and then extends into intercellular spaces within the palisade parenchyma tissue (Figure 38). Many haustoria and intracellular hyphae which are elongated from the haustoria are produced in palisade parenchyma cells (Figure 39). Again, the haustoria are surrounded by a very thin encapsulation, as illustrated in Figurre 37. Such an encepsulation is very poorly developed compared to that of the uredial haustoria of other rusts.[4] Furthermore, unlike the haustoria, which are usually produced by other obligate fungi, those of *G. asiaticum* subsequently produce intercellular hyphae that grow into the underlying mesophyll tissues. Thus, Kohno et al.[17] defined this structure as an intracellular hypha rather than an haustorium. This type of parasitism may resemble that of some anthracnose fungi.[18]

By 5 to 7 d after basidiospore infection, many hyphae grow inter- and intracellularly within palisade parenchyma tissue of the pear leaf (Figure 40). Pycnial initials form at the subepidermal spaces, resulting in a pushing up of the epidermis (Figure 41). As the projections extend, the epidermis just above the ostiole of the flask-shaped pycnia are raised (Figure 42) and finally rupture to expose abundant trichomes, called paraphyses or

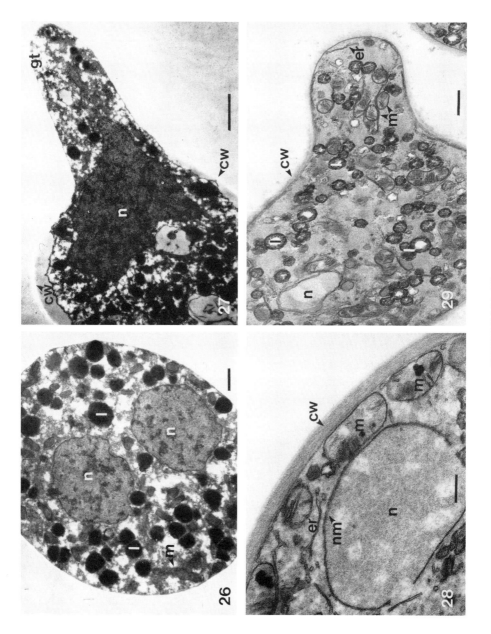

FIGURES 26 to 29

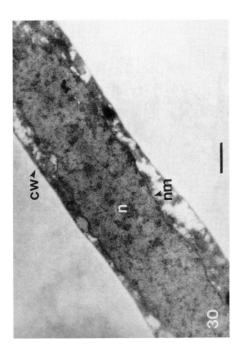

FIGURE 30.

FIGURES 26 to 30. **Figure 26.** An electron micrograph showing a binucleate sporidum. A large number of lipid bodies are present in the cytoplasm; fixed with glutaraldehyde and OsO$_4$. Bar = 1 μm. **Figure 27.** A large nucleus has migrated into a germ tube of a sporidium. Bar = 1 μm. **Figure 28.** A sporidium that is surrounded by a thin cell wall, fixed with KMnO$_4$. Bar = 0.5 μm. **Figure 29.** A germinating sporidium having a number of mitochondria and lipid bodies. Bar = 1 μm. **Figure 30.** An elongated nucleus in a germ tube of a sporidium. Bar = 1 μm.

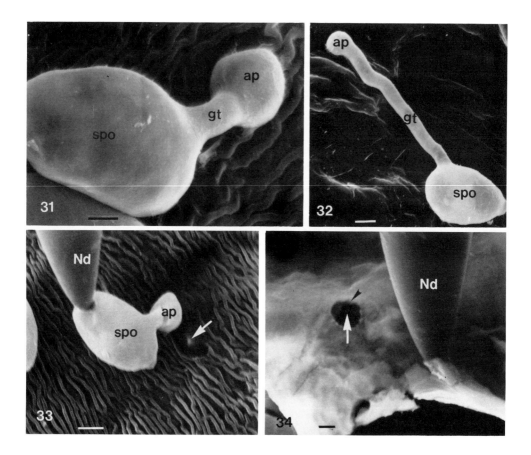

FIGURES 31 to 34. **Figure 31.** A SEM showing an appressorium produced at the tip of a germ tube of a sporidium. Bar = 1 μm. **Figure 32.** An appressorium formed at the tip of an elongating germ tube of a sporidium. Bar = 2 μm. **Figure 33.** A track of the appressorium was exposed on a surface of host plant by micromanipulation. White dot (arrow) in the track indicates a pentration site. Bar = 2 μm. **Figure 34.** The inner surface of the epidermal cell of the host cell just below the track in Figure 33, reversed by micromanipulation. An arrowhead indicates a pentration pore; a dark area surrounding the pore probably resulted from enzymic degradation by the penetration peg. Bar = 1 μm.

periphyses, which extend from the ostiole (Figure 43). The primary function of paraphyses is to rupture the host epidermis during ontogeny of the pycnium and flexuous hyphae. The flexuous hyphae are the female structures onto which the male pycniospores attach and transfer their nuclei during plasmogamy.[4] In the case of *G. asiaticum*, it is hard to distinguish paraphyses and flexuous hyphae based on their morphology. Similar to most other rust fungi, flexuous hyphae seem to occur intermixed among the paraphyses.[4]

The hymenium at the basal part of a mature pycnia is composed of small cells that tightly attach to each other (Figure 44). Each cell contains a large nucleus. Pycniosporophores and ostiolar paraphyses consisting of long cells with an elongated nucleus differentiate in the hymenial cells (Figure 45). Pycniospores are produced in basipetal succession from the tips of the sporophores; each one is uninucleate following the mitotic division of the nucleus in the sporophore prior to spore cleavage (Figure 46). Immediately prior to formation of the first pycniospore the nucleus in the sporophore undergoes a single mitotic division and the apex of the sporophore swells to form the pycniospore initial. One daughter nucleus migrates into the spore initial and the other remains in the sporophore. A septum is then formed at

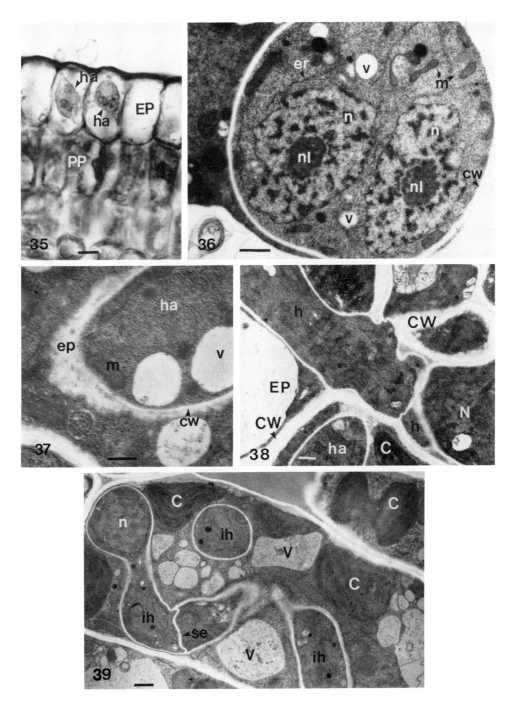

FIGURES 35 to 39. **Figure 35.** The binucleate haustoria formed in the upper epidermis of a pear leaf. Bar = 10 μm. **Figure 36.** An electron micrograph showing a binucleate haustorium surrounded by a thin encapsulation. Bar = 1 μm. **Figure 37.** An haustorium bordered by a thin encapsulation zone in an epidermal cell. Bar = 1 μm. **Figure 38.** An intracellular hypha formed in an epidermal cell which elongated into the palisade parenchyma tissue of a pear leaf as an intercellular hypha. Bar = 1 μm. **Figure 39.** Intracellular hyphae in palisade parenchyma tissue. Bar = 1 μm.

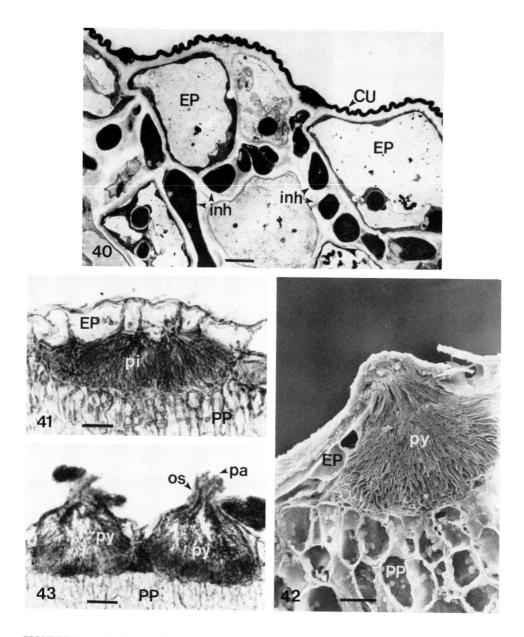

FIGURES 40 to 43. **Figure 40.** A number of hyphae gather in the intercellular space between the epidermis and the palisade parenchyma tissue. Bar = 1 μm. **Figure 41.** The pycnium initial formed between the epidermis and the palisade parenchyma tissue. Bar = 20 μm. **Figure 42.** An SEM of a cross section of an immature pycnium. Bar = 20 μm. **Figure 43.** Two pycnia formed on the upper surface of a pear leaf which have paraphyses emerging from an ostiole. Bar = 20 μm.

the base of the spore initial which separates the immature pycniospore from the pycnio-sporophore (Figure 47).[4,19] Upon cleavage of pycniospores from pycniosporophores, the former are released into the pyncial cavity in a matrix of honeydew which fills the cavity. The honeydew and the pycniospores exude through the pycnial ostiole onto the surface of the pear leaf (Figure 46).[4,19] Pycniospores surrounded with a very thin cell wall have a single, large nucleus (Figure 47).

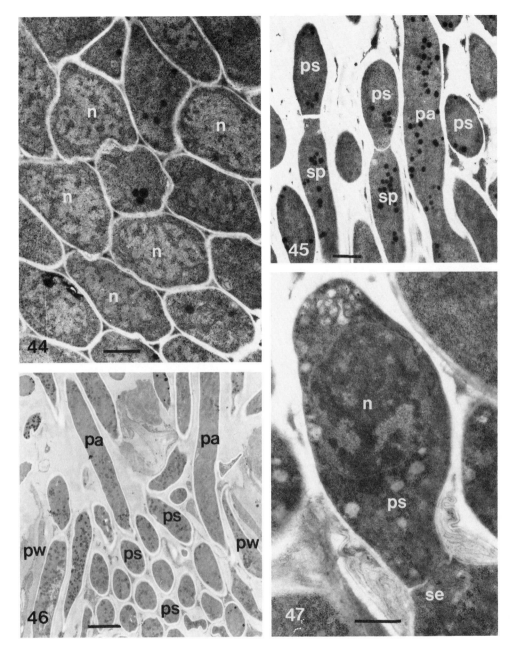

FIGURES 44 to 47. **Figure 44.** The hymenial cells with large nuclei which compose the basal part of a pycnium. Bar = 2 μm. **Figure 45.** Pycniospores produced in basipetal succession from the tip of the sporophores above the hymenial cells. Bar = 2 μm. **Figure 46.** Long paraphyses and pycniospores at the ostiolar region of a pycnium. Bar = 5 μm. **Figure 47.** An immature pycniospore with a large nucleus that has been separated from the sporophore by a thin septum. Bar = 1 μm.

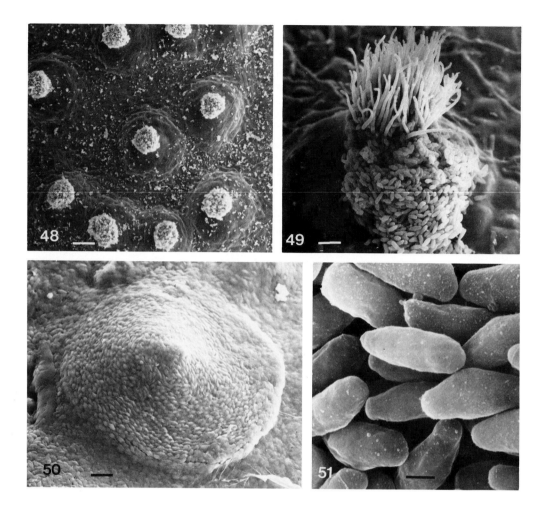

FIGURES 48 to 51.   Figure 48. The initial stage of pycnium emergence from the surface of a pear leaf. Several raised areas whose centers are covered with abundant pycniospores are seen on the leaf surface. Bar = 50 μm. **Figure 49.** Paraphyses emerging from an ostiolar opening which is covered with a large number of pycniospores. Bar = 10 μm. **Figure 50.** The entire surface of a pycnium, including paraphyses, is covered with pycniospores when the honeydew dries. Bar = 10 μm. **Figure 51.** Pycniospores with a smooth surface that have a slightly tapered end. Bar = 1 μm.

The pycniosporogenesis of *G. asiaticum* is basically similar to that of *G. juniperi-virginianae*.[20]

2. Surface View of Pycnia

On infected pear leaves collected at the stage shown in Figure 1, several raised areas, the center of which is covered with abundant pycniospores, can be observed by SEM (Figure 48). At a later stage, a great number of the ostiolar projections emerge from the raised center (Figure 49). When the honeydew is dried, the entire raised areas, including the ostiolar projections, are covered with aggregated pycniospores (Figure 50). Each area corresponds to the dark dots illustrated in Figure 2, which are distinguishable via the naked eye. Pycniospores with a smooth surface are ellipsoid with a slightly tapered end (Figure 51).

D. AECIAL STAGE

Aecia are sori that produce nonrepeating, vegetative aeciospores as a result of dikaryotization, and thus are typically associated with pycnia.[4,10]

1. Formation of Aecial Horns and Aeciospores

Within 2 to 3 weeks after the honeydew around the ostioles of pycnia on pear leaves is dried, the aecial primordium forms in the spongy parenchyma tissue just below the palisade parenchyma tissue (Figure 52). This primordium repeats cell division to expand the occupying area toward the low surface of the pear leaf (Figure 53). Within the spongy parenchyma tissues, aeciosporophores and aeciospores differentiate (Figure 54). A tubular, one-cell-thick peridium develops surrounding the entire aecial space in the spongy parenchyma tissue.

The base of the aecial stroma consists of a mass of interwoven cells of two main types: (1) uninucleate, often vacuolate cells (Figure 55) and (2) binucleate, elongated aeciosporophore cells, including either large or many small vacuoles (Figure 56).[21] With each successive separation of an aeciospore initial from the aeciosporophore, each aeciospore initial cleaves into alternating immature aeciospores and intercalary cells that reside between each of the aeciospores in the chain (Figure 57). Both structures are binucleate and separated by a thin septum. A cell wall of the aeciospore is thickened and at maturity a characteristic ornamentation develops on the spore surface. The intercalary cells remain thin walled, lack the surface ornamentation, and disintegrate prior to aeciospore discharge (Figure 58).

2. Surface View of Aecial Horns and Aeciospores

Aecia vary in their overall morphology from species to species. Among the whole rust fungi, aecia can be grouped into five major morphological types, depending primarily on the absence or presence of a peridium and the nature of the peridium, if present.[4] The horn-like aecium that is characteristic of *Gymnosporangium* spp. is called the roestelioid aecium. The aecial horn of *G. asiaticum* is 1 to 2 cm long and 0.1 to 0.2 cm in diameter, and thus readily observed (Figures 3 and 4).

When aecial horns are mature in the spongy parenchyma tissue of pear leaves, they forcibly break the epidermal tissue and emerge from the lower surface of the leaf (Figure 59). They continually elongate to form horn-like aecia with a tapered top (Figure 60). The peridial cells of the aecial side are elongated, 100 to 250 μm long, and arranged end to end, forming long files of cells that separate longitudinally to produce a characteristic fimbriate appearance (Figures 61 and 62). The peridial cells covering the apical portion of the aecium have a flat, disk shape similar to that of a hemocyte (Figure 59, arrowheads). Numerous aeciospores are packed in an aecial horn (Figure 61). The peridium functions as the protective coat for aeciospores and also exhibits hygroscopic properties which aid in spore release.[22]

The surfaces of aeciospores bear a characteristic ornamentation (Figures 63 and 64). As illustrated in Figure 58, an immature spore still attached to an intercalary cell has a smooth surface, although varrucae (ornamentation initials) have already formed in the cell wall. Thus, the surface ornamentation seems to develop with the maturation of spores. The ornamentation of the spore surface consists of unique, flower-like knobs (Figures 63 and 64). Each knob is comprised of a structure like coalesced petals with 3 to 10 small, bud-like projections (Figure 64). Dark, round areas appearing on the surface of an aeciospore, indicated by arrowheads in Fig. 63, represent the germ pores. Each spore has at least five to six germ pores.

E. INFECTION OF AN ALTERNATE HOST BY AECIOSPORES

Aeciospores deposited on the surface of juniper leaves produce a long germ tube under a favorable humid condition. When a germ tube encounters a stoma, a rectangular appres-

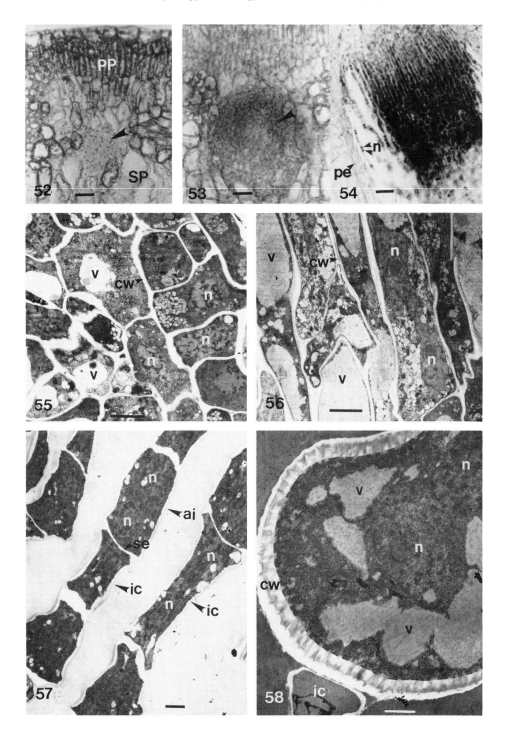

FIGURES 52 to 58

sorium is formed at the apex of the germ tube on guard cells of the stoma (Figure 65). After entering through the stomatal opening, binucleate hyphae emerge from the substomatal vesicle and grow in intercellular spaces inside the leaf (Figure 66). The hyphae are surrounded by a thick cell wall which appears as a double-layered wall (Figure 68). The cytoplasm within the hyphae is very dense, making their organelles indistinguishable, a situation that also occurs in *G. juniperi-virginianae*.[23] In parenchymatous cells, haustoria with a compact cytoplasm are formed (Figure 67), which are surrounded by a thick encapsulation, unlike those formed in pear leaves. Teliospore sori are produced from the dark brown scab-like structures that appear on the surface of infected leaves or twigs of junipers in early spring (Figure 69). A sorous is formed above the epidermal layer which has not been broken, but is a little swollen, as is the case in *G. clavipes*.[24]

IV. CONCLUDING REMARKS

As described in this chapter, the life cycle of *G. asiaticum* is fundamentally similar to that of *G. fuscum*,[2,8] *G. juniperi-virginianae*,[6,11,14,20,22,23] and *G. clavipes*.[13,24] Infection of pear and apple trees by these rust fungi causes more or less reduction of the photosynthetic activity of leaves and deformation of fruits, affecting fruit yield and commercial value.

Direct control of the rust of Japanese pear has been achieved by spraying juniperous trees with calcium polysulfide or mepronil in early spring before telia swell and/or by repeatedly spraying pear trees with mepronil, triadimefon, zineb, maneb, and other fungicides until 1.5 to 2 months preharvest. Indirect control has been achieved by eradicating alternate host plants, on which these fungi overwinter as a form of telia, within a radius of 1.5 km from the pear and apple orchards. However, the desirability of junipers as garden trees makes it difficult to remove them from private gardens.

FIGURES 52 to 58. Figure 52. Hyphae (arrowhead) gathering in the spongy parenchyma tissue below the palisade parenchyma tissue of a pear leaf at the initial stage of aecial horn formation. Bar = 20 μm. **Figure 53.** The aecial primordium (arrowhead) composed of numerous hyphae formed in the spongy parenchyma tissue. Bar = 20 μm. **Figure 54.** A cross section of an aecial horn consisting of a chain of binucleate aeciospores and intercalary cells. Bar = 20 μm. **Figure 55.** The base of the aecial stroma consisting of a mass of interwoven, uninucleate cells. Bar = 5 μm. **Figure 56.** Binucleate, elongated aeciosporophore cells including either large or many small vacuoles. Bar = 5 μm. **Figure 57.** Alternating immature aeciospores and intercalary cells which reside between each of the aeciospores in the chain. Both structures are binucleate and are separated by a thin septum. Bar = 2 μm. **Figure 58.** A binucleate aeciospore with a thick cell wall connected to a disintegrated intercalary cell. The initials of wall ornamentation are buried in the cell wall. Bar = 2 μm.

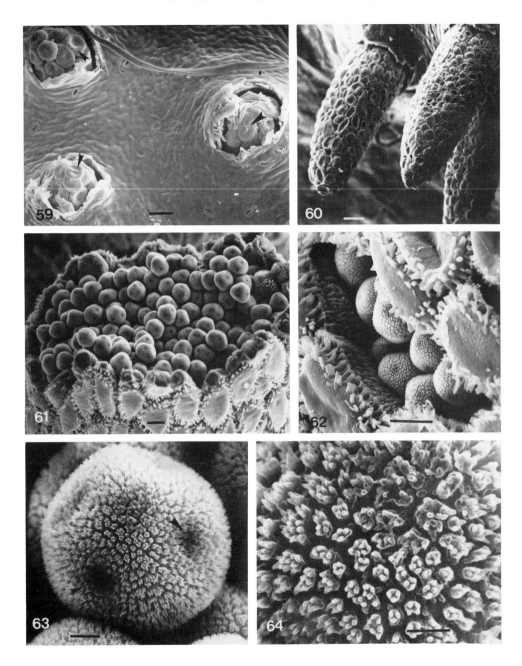

FIGURES 59 to 64. **Figure 59.** The initial stage of emergence of the aecial horns from the lower surface of a
pear leaf. Peridial cells (arrowheads) covering the tip of the aecial horns have a hemocyte-like flat shape. Bar =
30 μm. **Figure 60.** The aecial horns emerging from the lower surface of a pear leaf. Bar = 50 μm. **Figure 61.**
Numerous aeciospores packed in an aecial horn. Bar = 10 μm. **Figure 62.** The fimbriated peridial cells, one of
which has been removed to expose the packed aeciospores. Bar = 10 μm. **Figure 63.** Aeciospores covered with
the characteristic surface ornamentation. Dark areas (arrowheads) show the germination pores. Bar = 2 μm.
Figure 64. The enlarged surface ornamentation with flower-like appearance. Bar = 1 μm.

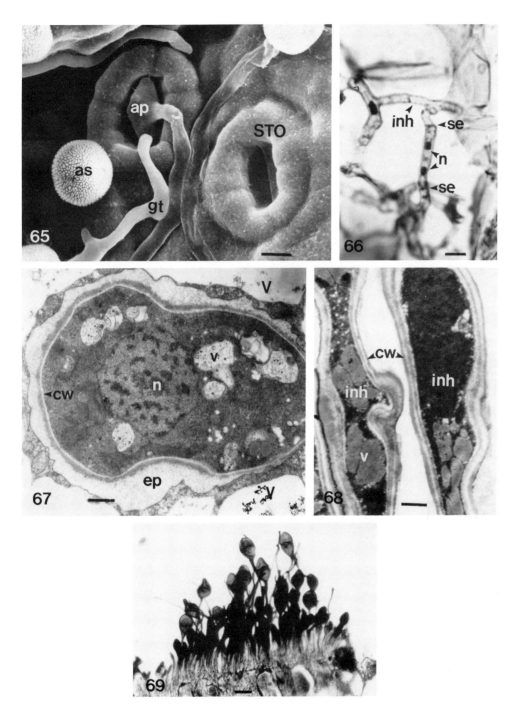

FIGURES 65 to 69. **Figure 65.** A rectangular appressorium formed at the apex of a germ tube of an aeciospore on guard cells of the stoma of a juniper leaf. Bar = 5 μm. **Figure 66.** Binucleate hyphae growing in the intercellular space of a juniper leaf. Bar = 5 μm. **Figure 67.** A haustorium enveloped by a thick encapsulation which was formed in a juniper leaf. Bar = 1 μm. **Figure 68.** Intercellular hyphae with a thick, layered cell wall in a juniper leaf. Bar = 1 μm. **Figure 69.** A teliospore sorus produced from the dark brown scab-like structure that appeared on the surface of an infected leaf of juniper. Bar = 20 μm.

REFERENCES

1. **Parmelee, J. A.,** The rusts, *Greenhouse-Garden-Grass,* 5, 11, 1965.
2. **Ziller, W. G.,** Pear rust (*Gymnosporangium fuscum*) in North America, *Plant Dis. Rep.,* 45, 90, 1961.
3. **Kitajima, H.,** Pear rust, *Agric. Hort.,* 51, 1541, 1976.
4. **Littlefield, L. J. and Heath, M. C.,** *Ultrastructure of Rust Fungi,* Academic Press, New York, 1979, 227.
5. **Petersen, R. H.,** The rust fungus life cycle, *Bot. Rev.,* 40, 453, 1974.
6. **Stevens, E.,** Cytological features of the life history of *Gymnosporangium juniperi-virginianae, Bot. Gaz.,* 89, 394, 1930.
7. **Kohno, M.,** unpublished data, 1981.
8. **Borno, C. and van der Kamp, B. J.,** Timing of infection and development of *Gymnosporangium fuscum* on junipers, *Can. J. Bot.,* 53, 1266, 1975.
9. **Kitajima, H.,** Pear rust, *Agric. Hortic.,* 52, 89, 1977.
10. **Hiratsuka, Y.,** The nuclear cycle and the terminiology of spore states in Uredinales, *Mycologia,* 65, 432, 1973.
11. **Mims, C. W., Seabury, F., and Thurston, E. L.,** Fine structure of teliospores of the cedar-apple rust *Gymnosporangium juniperi-virginianae, Can. J. Bot.,* 53, 544, 1975.
12. **Kohno, M., Nishimura, T., Ishizaki, H., and Kunoh, H.,** Ultrastructural changes of cell wall in germinating teliospore of *Gymnosporangium haraeanum* Sydow, *Trans. Mycol. Soc. Jpn.,* 16, 106, 1975.
13. **Mims, C. W.,** Ultrastructure of teliospore germination and basidiospore formation in the rust fungus *Gymnosporangium clavipes, Can. J. Bot.,* 59, 1041, 1981.
14. **Mims, C. W.,** Fine structure of basidiospores of the cedar-apple rust fungus *Gymnosporangium juniperi-virginianae, Can. J. Bot.,* 55, 1057, 1977.
15. **Kohno, M., Nishimura, T., Noda, M., Ishizaki, H., and Kunoh, H.,** Cytological studies on rust fungi. VII. The nuclear behavior of *Gymnosporangium asiaticum* Miyabe ex Yamada during the stages from teliospore germination through sporidium germination, *Trans. Mycol. Soc. Jpn.,* 18, 211, 1977.
16. **Hiratsuka, Y. and Cummins, G. B.,** Morphology of the spermagonia of the rust fungi, *Mycologia,* 55, 487, 1963.
17. **Kohno, M., Ishizaki, H., and Kunoh, H.,** Cytological studies on rust fungi. V. Intracellular hyphae of *Gymnosporangium haraeanum* Sydow in cells of Japanese pear leaves, *Ann. Phytopathol. Soc. Jpn.,* 42, 417, 1976.
18. **O'Connell, R. J., Bailey, J. A., and Richmond, D. V.,** Cytology and physiology of infection of *Phaseolus vulgaris* by *Colletotrichum lindemuthianum, Physiol. Plant Pathol.,* 27, 75, 1985.
19. **Kohno, M., Ishizaki, H., and Kunoh, H.,** Cytological studies on rust fungi. IX. Ultrastructure of pycnia of *Gymnosporangium asiaticum* Miyabe ex Yamada, *Bull. Fac. Agric. Mie Univ.,* 62, 1, 1981.
20. **Mims, C. W., Seabury, F., and Thurston, E. L.,** An ultrastructural study of spermatium formation in the rust fungus *Gymnosporangium juniperi-virginianae, Am. J. Bot.,* 63, 997, 1976.
21. **Kohno, M., Ishizaki, H., and Kunoh, H.,** Cytological studies on rust fungi. X. Ultrastructure of roestelium of *Gymnosporangium asiaticum* Miyabe ex Yamada, *Proc. Kansai Protect. Soc.,* 23, 20, 1981.
22. **Pady, S. M., Kramer, C. L., and Clary, R.,** Periodicity in aeciospore release in *Gymnosporangium juniperi-virginianea, Phytopathology,* 58, 329, 1968.
23. **Mims, C. W. and Glidewell, D. C.,** Some ultrastructural observations on the host-pathogen relationship within the telial gall of the rust fungus *Gymnosporangium juniperi-virginianae, Bot. Gaz.,* 139, 11, 1978.
24. **Dodge, B. O.,** Studies in the genus *Gymnosporangium.* IV. Distribution of the mycelium and the subcuticular origin of the telium in *G. clavipes, Am. J. Bot.,* 9, 354, 1922.

Chapter 5

CYTOLOGY, HISTOLOGY, AND HISTOCHEMISTRY OF FRUIT INFECTION BY *MONILINIA* SPECIES

Haydn J. Willetts and Suzanne Bullock

TABLE OF CONTENTS

I. INTRODUCTION

The fungal genus *Monilinia* is economically important since many of its species are troublesome pathogens of fruit trees of the plant family Rosaceae, particularly apples, pears, and stone fruits. Some *Monilinia* spp. also attack ornamental flowering fruit trees, mainly *Prunus* spp. and several related genera. Other species are pathogens of the Ericaceae, a family of low, creeping, almost herbaceous heath plants, but these are not considered in this chapter because they are not pathogens of trees.

Monilinia spp. produce dark brown, rotting areas on infected rosaceous tissues and consequently are often referred to as the brown rot fungi of fruits. The dark brown discoloration is due to the oxidation of phenolics of the host by polyphenoloxidases, probably produced by both the host and pathogen. The greatest damage is to fruits of pomaceous and drupaceous trees, which are long-lived perennials. Many of these are extensively cultivated or grow wild in large stands in temperate regions. Their fruits, especially those of commercial cultivars, are large, fleshy, and very rich in nutrients and thus provide ideal substrates for growth of the brown rot fungi.

Hyphae of *Monilinia* spp. permeate fleshy parts of fruits to produce large, confluent spheres of fungal tissue with viable fruit seeds remaining in their centers. Fruit tissue is only partially degraded by *Monilinia* spp. so that some host cells remain among the interwoven hyphae. Water is lost from infected fruits, which shrink and assume a mummified appearance. Hence, the dried out, partly degraded, colonized fruit is commonly referred to as a "fruit mummy."

In the temperate regions in which rosaceous trees grow, winters are often severe, with subzero temperatures, dry winds, and, in many areas, snow cover prevailing for long periods. Such adverse environmental conditions have resulted in the adaptation of mummified fruits as highly effective resting structures. Mummies can remain dormant or quiescent, sometimes for periods of several years. Then, when conditions become favorable for renewed growth, they use their stored reserves for the production of propagules which initiate a new cycle of infection. Thus, mummified fruits of rosaceous trees serve an important survival function and play a vital role in the life cycle of brown rot fungi.

Records of the brown rot diseases of fruits date back to the early European mycological literature, with the first description given by Persoon.[1] Many European mycologists studied these fungi, with the work of Woronin,[2] Aderhold and Ruhland,[3] and Wormald[4,5] providing the main descriptions of the causal organisms and an understanding of their taxonomy. Roberts and Dunegan[6] described how in North America, brown rot was recorded as troublesome on fruits at the beginning of the 19th century, but the disease was not recognized as being caused by a fungus until much later.[7]

An extensive literature has now accumulated on the brown rot *Monilinia* diseases. Early work in America was outlined by Norton et al.[8] and Roberts and Dunegan.[6] Harrison[9] published an account of brown rot and associated diseases in Australia and then extended his work to Europe and North America, where he studied the history, taxonomy, and nomenclature of the common *Monilinia* species.[10] Wormald[11] presented the first general comprehensive account of the common brown rot diseases, and the biology and control of the group were updated by Byrde and Willetts.[12] Recently, Batra[13] published a monograph on the ecology, biosystematics, and control of the genus *Monilinia*. A report on the *Monilinia* spp. of Japan, with a useful summary in English, was published by Harada.[14]

The purpose of this chapter is to outline some of the general characteristics of the brown rot *Monilinia* spp., to discuss how fruits are colonized, and then to describe the cytology, histology, and histochemistry of mummified fruits. Finally, the authors have highlighted some of the characteristics of both the host and fungus that contribute to the form and chemical composition of mummies of the common rot fungi.

II. GENERAL INFORMATION ON THE BROWN ROT DISEASES

A. TERMINOLOGY

In mycological terms a stroma (pl. stromata) is a compact structure produced by the interweaving of vegetative hyphae, sometimes including host tissue or portions of the substratum, and usually producing fructifications. Stromata are often sclerotium-like in form, and function as resting structures. According to this description, the mummy of brown rot fungi is a stroma and therefore these terms are synonymous.

In detailed studies on *Monilinia* spp. Honey[15] and Batra[13] distinguished between an ectostroma and an entostroma. The ectostroma is a small, discrete, nondifferentiated aggregation of hyphae that bears conidia. The ectostromatal tissue erupts through the surface of host tissue to facilitate spore production and dispersal. The entostroma is a larger, well-differentiated tissue that develops as a result of extensive vegetative growth. It consists of interwoven hyphae and host elements that form the medulla. This is surrounded dorsally and often ventrally by a continuous, heavily pigmented layer of fungal cells known as the rind (see Figures 5 to 7). The entostroma can germinate to produce apothecia, conidia, and/or mycelia. The latter often form extensive, external, mycelial wefts over the surface of mummies.[16] In this chapter, and in almost all other *Monilinia* studies, the terms "stroma" and "mummy" refer to the entostroma.

Other terms used for the stroma or mummy include hollow-spheroid sclerotium,[17] pseudosclerotium,[13,15] stromatal anamorph,[18,19] and stromatized or mummified fruit.[11,12] Based on criteria such as the tissue from which mummies have been derived, their ontogeny, and their position of development in relation to host substrata, the authors do not regard mummies as true sclerotia of the type produced by species of *Sclerotinia* and *Botrytis*. The authors agree with the conclusions of Honey[15] and Batra[13] that the mummy is a pseudosclerotium. The common, nondisjunctor *Monilinia* spp., referred to in the next section, normally (although not always) produce hollow and spheroid mummies, while Batra[13] found that some *Monilinia* spp. produce solid stromata with only a dorsal rind. Therefore, the term "hollow-spheroid sclerotium", first introduced by Whetzel,[17] is a misleading and incorrect name for mummies produced by the genus *Monilina*.

B. CAUSAL ORGANISMS

The genus *Monilinia* belongs to the family Sclerotiniaceae of the Discomycetes, a subdivision of the Ascomycotina. The genus is characterized by the production of conidial anamorphs consisting of bead-like (monilioid) chains of conidia (Figure 1); stromatal anamorphs in the form of compact stromatic tissue in fruits; apothecial ascomata (Figure 2), bearing inoperculate asci containing ellipsoidal ascospores, and globose microconidia, which have been shown to function as spermatia in some sclerotiniaceous fungi, but not in *Monilinia* spp.[12]

Honey[20] separated the genus into two sections, the Junctoriae and Disjunctoriae, according to the absence or presence, respectively, of small structures (disjunctors) between adjacent conidia in spore chains. Disjunctors are adaptations that bring about fragmentation of conidial chains and facilitate dispersal of spores by wind. The disjunctor species are morphologically and physiologically more specialized than those without disjunctors.[12,13]

In this chapter, the term "brown rot fungi" refers to the three nondisjunctor *Monilinia* (form genus *Monilia*) species: *Monilinia fructigena* (Aderh. & Ruhl.) Honey (≡ *Sclerotinia fructigena* Aderh. & Ruhl.); *Monilinia laxa* (Aderh. & Ruhl.) Honey (≡ *Sclerotinia laxa* Aderh. & Ruhl.); and *Monilinia fructicola* (Wint.) Honey (≡ *Sclerotinia fructicola* (Wint.) Rehm.). In some earlier reviews the fungus *Monilia cinerea* forma *mali* Wormald [≡ *Monilinia (Sclerotinia) laxa* (Aderh. & Ruhl.) Honey forma *mali* Wormald sensu Harrison], which causes blossom wilt and canker of apple trees in Europe, was also included as one

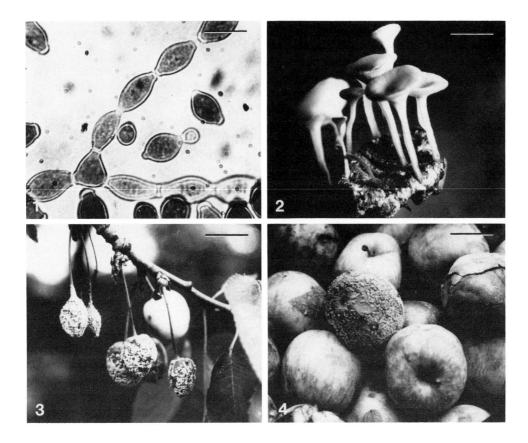

FIGURES 1 to 4. General morphological features of *Monilinia* spp. **Figure 1.** Chains of conidia of the non-disjunctor species *M. fructicola*. Stained with lactophenol cotton blue. Scale = 17 μm. **Figure 2.** Apothecia of *M. fructicola* on a portion of mummified stone fruit. Scale = 9 mm. **Figure 3.** Cherries still on the tree showing various stages of infection and mummification by *M. fructicola*. Scale = 15 mm. **Figure 4.** Apples in storage infected by *M. fructigena*. Scale = 15 mm. (Figure 1 is from Willetts, H. J. and Calonge, F. D., *New Phytologist,* 68, 123, 1969. With permission.)

of the common brown rot fungi. However, it is rarely seen in the field because of the effectiveness of current control methods and therefore is not considered further in this chapter.

Identification of the brown rot species is based on morphological and physiological characteristics and host range.[12,13] In some instances confirmation of identification depends on enzyme patterns obtained from isoelectric focusing[21] and electrophoretic[22] techniques.

The main hosts, tissues infected, and general distribution of the common brown rot fungi are given in Table 1. Note that the fungi do not produce disjunctors, are pathogenic on several hosts, normally colonize ripe fruits, and have relatively wide regional distributions. They are comparatively unspecialized pathogens.

The disjunctor *Monilinia* spp. differ from the nondisjunctor forms (the common brown rot fungi) in that they have narrow host ranges, are of restricted regional distribution, and attack immature fruits and leaves.[13] Two important disjunctor species that have been reported from eastern Asia only are *Monilinia mali* (Tak.) Whetzel, which is a severe pathogen of apple and Japanese crab apple [*Malus sieboldii* (Regel) Rehder],[23] and *Monilinia kusanoi* (Tak.) Yamamoto, which attacks several *Prunus* spp.[14] Equivalent diseases reported from Europe only are *Monilinia linhartiana* (Prill. & Delacr.) Dennis [≡ *Monilinia cydoniae* (Schell.) Whetzel] on quince; *Monilinia johnsonii* (Ellis & Everh.) Honey on hawthorn; and *Monilinai mespili* (Schell.) Whetzel on medlar.[13]

TABLE 1
Hosts, Tissues Attacked, and Regional Distributions of the Common Brown Rot Fungi

	Species		
	M. fructicola	*M. fructigena*	*M. laxa*
Hosts	*Prunus* spp., quince, almond; sometimes apple and pear, especially injured fruits, but losses on the latter normally light	Common and severe on apples and pears; occasionally on stone fruits; various trees infected including quince, loquat, medlar, hazel, and *Sorbus commixta*	*Prunus* spp., quince, and occasionally pear and almond
Plant tissues attacked	Blight of blossoms and twigs; cankers on branches of trees sometimes severe; severe rot of ripe stone fruits; latency in young fruits	Cankers of branches; destructive fruit rot of apples and pears; blossoms not normally attacked because conidia are usually produced later in season	Blossom wilt severe; generally regarded as a pathogen of twigs rather than fruit, but rot of ripe fruits sometimes serious
Regional distributions	Originally America, but now also Africa (probably pre-1931); Australasia (turn of the century), and eastern Asia (probably about 1960); not in Europe	Essentially an Old World species with wide distribution; reported in North America, late 1970s; not yet recorded in Australasia	Essentially an Old World species with wide distribution; established in North America, 1920s and Australasia, late 1960s

C. LIFE CYCLES

The common brown rot species readily produce conidia, stromata, and microconidia in the field and culture, but conidial production by some of the disjunctor species requires exacting light and nutrient conditions.[13,14] Apothecial production by *M. fructicola* has been frequently reported, although production is erratic. Apothecia of *M. fructigena* have been observed only occasionally, and there are very few records of apothecial production by *M. laxa.*[24]

Conidia do not normally survive through winter; however, mycelium can overwinter in fruit mummies still hanging on the tree (Figure 3), or in those which have fallen to the ground, and in infected twigs, peduncles, and cankers on branches. In the spring, mycelia in host tissue produce conidiophores and large numbers of conidia. If apothecia develop, ascospores also provide primary spring inoculum. Both spore types infect flowers and the resulting mycelium spreads via blossoms to fruit peduncles and young fruits, and cankers may develop on stems. Mycelium of *M. fructicola* remains dormant or quiescent in unripe fruits until ripening begins. Although the common brown rot fungi rarely infect leaves directly, leaves may wither due to blighting of twigs. During the summer when environmental conditions are favorable for sporulation several crops of conidia are produced on infected tissues. Large numbers of microconidia are also produced on the surface or in cavities of well-developed fruit mummies. Fruit infection normally takes place just before or at the time of harvesting and rotting begins while fruits are still hanging on the tree. Apparently healthy fruits, however, often develop brown rot lesions after harvesting. Under prevailing storage conditions, infection spreads quickly through individual fruits and by mycelial contact to other fruits in the consignment (Figure 4). In the field, mummified fruits pass through a resting period and then germinate to produce spores and/or mycelia, which initiate a new cycle of infection.

III. CYTOLOGY OF VEGETATIVE HYPHAE

Regardless of whether conidia, ascospores, or mycelia serve as inoculum, hyphae are ultimately responsible for the colonization of host tissue. Thus, it is appropriate to briefly outline the cytology of vegetative hyphae of the brown rot fungi.

Willetts and Calonge[25] described the walls of vegetative hyphae of *M. fructigena* and *M. laxa* as being about 0.15 μm thick, with an electron-transparent inner layer and an electron-dense outer layer. It is now known that this outer layer contains glucans (see Section V). Centripetally produced septa develop at regular intervals along hyphae. Each septum has a simple pore in the center that permits free movement of cytoplasm, including large organelles, along hyphae. Anastomoses are common between hyphae, further increasing cytoplasmic interchange throughout the mycelia, except when septa are plugged.

In an ultrastructural study of vegetative hyphae of *M. fructigena* Najim and Turian[26] observed a well-delimited "Spitzenkörper" at hyphal tips. Each Spitzenkörper was made up of a microvesicular aggregate surrounded by vesicles that fused with the plasmalemma. Behind the tip, mitochondria with their long axes oriented along the hyphae, were associated with numerous and often large bodies which were assumed to contain lipid material. Najim and Turian[26] noted that the endoplasmic reticulum (ER) extended into pseudo-Golgi cisternae, which produced the apical vesicles. They also described an elaborate system of microtubules and microbodies with pseudocrystalline inclusions.

Hoffmann[27] observed as many as 100 nuclei in tip cells of vegetative hyphae of *M. laxa* and *M. fructigena*. Nucleoli are centrally placed within the nuclei.[28] Hall[29] reported that the nuclei of *M. fructicola* have four chromosomes; chromosome numbers of other *Monilinia* spp. have not been determined.

In summary, the cytological features of hyphae of the brown rot fungi are very similar to those of other ascomycetous fungi.

IV. INFECTION OF HOST TISSUES

A. PENETRATION

Conidia of *M. fructigena* are not normally produced until after blossoming and therefore do not infect flowers. However, germ tubes of conidia of both *M. fructicola* and *M. laxa* can penetrate through any part of a flower. Entry is usually through stigmas and stamens so that open flowers are more readily infected than those that are unopened. Floral parts are moist with an abundance of exudates that provide exogenous nutrients for growth of conidial germ tubes. This external supply of nutrients is necessary as conidia of the brown rot fungi contain insufficient reserves for germ tube differentiation and penetration of host surfaces. After penetration of flowers via stigmas or stamens, the mycelium grows through the ovary chamber and pedicel to the woody tissue beyond. After the mycelium of *M. fructicola* has become established in floral tissue, formation of an abscission layer is inhibited and infected blossoms and young fruits remain on the tree. Ascospore infection of flowers by the brown rot fungi probably takes place in a similar manner, but relevant information is scarce due to the less frequent production of apothecial ascomata.

Infection of fruits by conidial germ tubes and mycelia is normally through wounds, with the water and abundant nutrients that accumulate in damaged areas on fruits providing favorable conditions for spore germination and mycelial growth. However, entry of fruits has also been reported through undamaged surfaces, including direct penetration of the cuticle and structures such as hair sockets and lenticels.[30,31] Insects may serve as vectors of brown rot inocula, with special relationships developing between certain insects and fruit infection.[13]

B. COLONIZATION OF FRUIT TISSUE

After the brown rot fungi have gained entry into susceptible plant tissue, their mycelia spread around the site of infection. A general understanding of events associated with colonization of hosts has been obtained from studies using light and electron microscopy and many biochemical techniques, mainly on *M. fructigena.*

1. Morphological Observations

Early light microscope work indicated that in fruit tissue the common brown rot fungi grew intercellularly in the middle lamellae between cell walls.[12] Further investigations showed that intracellular hyphae were also present, usually in the space between host wall and the collapsed host protoplast, but never passing directly through the protoplast itself. Some workers challenge whether this is true intracellular growth.

Calonge et al.[32] used the transmission electron microscope (TEM) to study infection of apple and pear tissue by *M. fructigena.* They reported both inter- and intracellular hyphal growth and very convoluted plasmalemmae and numerous multivesicular bodies in some infecting hyphae. The host cell walls around intercellular hyphae appeared to be much less dense than in uninfected tissue, indicating likely degradation of host wall material by the fungus.

Using histochemical and TEM techniques, Hislop et al.[33] investigated the secretion of the pectolytic wall-degrading enzyme α-L-arabinofuranosidase (AF) by *M. fructigena.* They observed membranous spherosome-like structures, similar to those seen by Calonge et al.[32] Several of their micrographs showed what appeared to be a sequence in movement of the enzyme from the spherosome-like bodies, through the plasmalemma of the fungal cell, presumably en route to the site of degradation of the middle lamellae between host cell walls. Byrde[34] discussed the movement of fungal "pectinases" from ribosome to plant cell wall and referred to many studies on enzymes of the brown rot fungi and their role in plant pathogenesis.

Pring et al.[35] studied in detail the colonization of pear tissue by *M. fructigena.* Material was examined with Nomarski interference contrast optics, TEM, and scanning electron microscopy (SEM), coupled with etching with sodium ethoxide solution.

Initially, hyphae grew intercellularly with host cell walls separating ahead of invading hyphae in the plane of middle lamellae. SEM observations on etched infected tissue revealed tunnels between adjacent cells. Some of these tunnels were empty and others had hyphae still in place. Protoplasts of host cells adjacent to intercellular hyphae collapsed and withdrew from the cell walls. Some hyphae penetrated cell walls, growing through the cell lumen, and pushing aside collapsed protoplasts. At the sites of penetration, cell walls had jagged margins around the invading hyphae.

Three different types of hyphae were observed in the colonized pear tissue. One type of about 7 μm diameter was similar in size and ultrastructure to the vegetative hyphae of *M. fructigena.*[26] They grew between cells and in intercellular spaces. They had very convoluted plasmalemmae, prominent, elongated mitochondria and multivesicular, lomasome-like bodies with associated vesicles. The lomasome-like bodies were similar in appearance to those reported by Calonge et al.[32] and Hislop et al.[33] Presumably, these hyphae are the first to colonize host tissue, with the lomasomes producing enzymes that diffuse ahead of the hyphae, bringing about dissolution of the middle lamella. Staining in plant cell walls adjacent to these hyphae showed fibrils oriented at right angles to the convoluted cell membranes. Pring et al.[35] suggested that localized enzyme activity was probably responsible for the increased staining of the walls.

The second type of hypha had the largest diameter (16 to 20 μm) and grew intercellularly in regions that were well colonized. They formed the main groundwork of the hyphal tissue within fruits. Their cytoplasm contained irregular-shaped mitochondria, nuclei, ER, and numerous large vacuoles that contained reserve materials (see Section V), indicating a storage function for the hyphae. The large hyphae were often "rounded-off" at septa forming separate segments.

The final type of hypha was the smallest and found in highly colonized regions of the host. They had a diameter of 3 to 5 μm and arose as branches from the largest hyphae. Numerous organelles were observed in them, including irregular-shaped mitochondria, evenly dispersed ribosomes, convoluted plasma membranes, and some large vacuoles, but no lomasomes. Their ultrastructure indicates high metabolic activity and they may serve to absorb nutrients that accumulate in colonized tissue as a result of dissolution of cell walls and leakage from and breakdown of cell protoplasts. Pring et al.[35] referred to them as "ramifying" hyphae, similar to those produced by *Sclerotinia* spp. in host tissue.[36,37]

Thus, colonization of fruits by *M. fructigena* involves three hyphal types that differ in size, ultrastructure, and function. Hyphae with characteristics similar to those of vegetative hyphae are the initial colonizers of fruit tissue and appear to serve an enzyme-secreting function. The activities of these enzymes result in degradation of host tissue and liberation of nutrients into the infection court from which they are absorbed by the fungus and utilized for mycelial growth. Hyphae permeate through host tissue where they hold together the remnants of host cells isolated by the dissolution of their middle lamellae. Interweaving of these hyphae produces a compact stroma that takes the general shape of the fruit. Slow air drying causes shrinkage of the stromatized fruit and a fruit mummy develops. The main groundwork of stromata consists of large hyphae and their small branches. The numerous organelles of the latter are an indication of high metabolic activity, while their convoluted cell membranes could be associated with nutrient uptake from the infection court. During the early stages of stroma differentiation, most nutrients are utilized for mycelial growth. However, as structural development slows, surplus nutrients are deposited as intra- and extracellular reserves, particularly in large hyphae which serve both structural and storage functions in mature stromata. Therefore, during colonization of fruits and in mature fruit mummies, hyphae of *M. fructigena* show considerable morphological and physiological specialization.

2. Enzymes in Pathogenesis

Pectolytic enzymes degrade pectin of the middle lamellae and play an important role in colonization of fruits by brown rot fungi. Several pectolytic enzymes accumulate in cultures of *Monilinia* spp. and are present and active in host tissues infected by these fungi.[12,38] Some exist in several molecular forms, and pathogens with the greatest number of multiple enzymic forms show the widest host and tissue specificities.[12] Specific roles of the different forms cannot yet be attributed with any degree of certainty, however.

Four wall-degrading enzymes of the brown rot fungi have been studied in detail, three pectolytic enzymes that act on the main backbone of the amorphous pectin matrix of the middle lamella and one, AF, responsible for degrading arabans, which are neutral sugars associated with fruit pectin. The pectolytic enzymes are polygalacturonase (PG), which hydrolytically attacks polygalacturonic acid; pectin lyase (PL) (pectin methyl-*trans*-eliminase), which attacks a polygalacturonic acid of a high degree of esterification; and pectin esterase (PE) (pectin methylesterase), which acts on esterified carboxyl groups of galacturonic acid-liberating methoxyl groups.

Bateman and Basham[39] concluded that endo-PG and endo-PL, which act randomly on sugar polymers, are mainly responsible for wall degradation by plant pathogens. Pectolytic enzymes of the fungus also cause physiological changes to plasma membranes of host cells, resulting in leakage of electrolytes from the protoplasts and cell death.[34]

Cellulases and hemicellulases are either not produced by the common brown rot fungi, or their activities are low.[12] This accounts, at least partially, for the incomplete degradation of host cell walls. Partial dissolution of host tissue is also associated with the accumulation in infected fruits of toxic oxidized phenolics, which have an inhibitory effect on many hydrolytic enzymes. Both lack cellulolytic enzymes and the inhibitory effects of oxidized phenolics are important factors in the dry type of rot that is characteristic of infection of fruits by brown rot fungi. By contrast, *Sclerotinia* spp. that secrete cellulolytic enzymes normally produce soft, watery rots in susceptible plants.

Only limited information exists on other enzymes participating in pathogenesis by brown rot fungi,[12] but obviously many are involved both in the infection court and within hyphae. Proteases, lipases, phosphatases, and enzymes metabolizing carbohydrates are presumably highly active in the accumulation and depletion of reserves (see Section V).

3. Factors Affecting Colonization

Many factors have been reported to delay or prevent infection of fruits by the brown rot fungi.[12] Host resistance can be affected by physical, chemical, or physiological characteristics such as cork production, thickness of fruit skins, texture of ripe fruits, date of flowering, and gumming up of wounds. Sugar content, acidity of juice and phenolic content of fruits, and the production of phytoalexins and toxins by the host also affect colonization. There is also a considerable literature[12] on quiescence and latency of *M. fructicola* during which mycelial growth is suspended until fruits begin to ripen; however, the reasons for this temporary cessation of fungal activity have not been fully explained. Unlike the common brown rot fungi, the disjunctor species colonize unripe fruits and leaves. Unpublished data (H.J.W.)[65] indicate that there are consistent differences between the enzyme patterns of some pectolytic enzymes of nondisjunctor and disjunctor species that could be associated with the different hosts of the two groups.

V. ONTOGENY, HISTOLOGY, AND HISTOCHEMISTRY OF INFECTED FRUITS

Two regions or zones can usually be distinguished in stromata of the brown rot fungi, according to hyphal density and the amount of pigmentation present. A black or dark-colored zone, formed by close-fitting hyphal tips, covers the dorsal and ventral surfaces of the stroma, although sometimes a ventral zone is not produced by *M. laxa* and *M. fructigena*.[16,40] Conventionally, the surface zones are referred to as rinds (Figures 5 to 7). The main part of the stroma consists of a medulla of loosely interwoven, unpigmented hyphae (Figure 5). Sometimes a cortical region, intermediate in hyphal density and degree of pigmentation, can be distinguished.

An understanding of the morphogenesis, morphology, and composition of stromata of *Monilinia* spp. comes from observations using light microscopy,[13,16,40] SEM,[41] TEM,[18,19,25] and histochemical techniques.[18,19] In addition, the authors possess considerable unpublished ultrastructural and histochemical data on stromata of *M. fructicola* and *M. laxa* produced in agar cultures and in a variety of stone fruits. The authors' findings supplement published electron microscope and histochemical information. Also, data on sclerotia of other sclerotiniaceous genera can be referred to for comparison to stromata of *Monilinia* fungi. Individual sclerotium-producing species that have been studied include *Sclerotinia sclerotiorum*,[42,43] *S. minor*,[44,45] *Botrytis* spp.,[46-48] and *Sclerotium cepivorum*. While Kohn and Grenville[18] studied many sclerotiniaceous species,[49] the authors have used information from all these sources to describe ontogenetic, histological, and histochemical characteristics of stromata of the brown rot fungi.

A. ONTOGENY OF STROMATA

Willetts[16,40] studied the development of stromata of the three brown rot species in culture and in fruits. The stroma of *M. fructigena* is formed by the interweaving and cross-linking of strands consisting of several hyphae. This results in an open basketwork effect, except at stromatal surfaces, where greater branching gives a more compact tissue. The stromata of *M. fructicola* and *M. laxa* develop by hyphal branching to form small initials that coalesce. According to the number of initials involved, either a large continuous stroma or several small discrete stromata are produced.[40] Similar types of development have been reported for the stromata of disjunctor species.[13] The interweaving of hyphae within host tissue inevitably results in the inclusion of host cells within stromatal tissue (Figure 5), but nutrients remaining in the cells are unlikely to be used by the fungus as reserve materials for subsequent growth.

Morphogenetic factors do not appear to be involved in the initiation and differentiation of stromata, as mycelial growth takes place in a haphazard manner. Most hyphae are in contact with the nutrient-rich substrate and nutrients are translocated along a decreasing concentration gradient to metabolically active sites.[50] Even during early stages of stroma differentiation, not all translocated nutrients are utilized for hyphal growth. These surplus materials must be prevented from accumulating in growth centers if movement of nutrients into these centers is to be maintained by concentration gradients. During stromatal development liquid droplets form on exposed surfaces. Water in these droplets is probably forced out by high internal hydrostatic pressure and during the process it is ultrafiltered. The exudates from sclerotia of *Sclerotinia* and *Sclerotium* spp. contain a variety of inorganic ions and organic compounds, usually in low concentrations, and it has been suggested that exudation of droplets is involved in the maintenance of internal physiological balance within growing sclerotia.[51] However, Cooke and Al-Hamdani[52] proposed that exudation is essentially a means of expelling water from sclerotia and, while it is inevitable that nutrients will accumulate in droplets, the process is not associated with the maintenance of internal physiological balance. Jennings[53] tentatively concluded that droplets provide a reservoir of water that sustains hyphal growth over substrates of low water potential when supplies via the normal translocation route are not available. Obviously the role of exudation is still subject to controversy.

When nutrient requirements for growth are reduced, the conversion of surplus soluble materials to insoluble reserves can assist in the maintenance of physiological balance in stromatal tissues. The insoluble deposits would not directly affect cellular processes, yet are a readily available source of endogenous nutrients and energy for subsequent periods of rest, tissue repair, mycelial growth, or propagule production. When required, the insoluble reserves are reconverted to soluble forms that generate high osmotic concentrations, providing the energy for translocation of materials to metabolically active regions created by the initiation of apothecial and conidial primordia or mycelial growth.

B. HISTOLOGY OF STROMATA

The zones discernible in stromata are due to different hyphal densities with greatest density usually at outer surfaces of the structure (Figure 5). In stromata of the brown rot fungi the outermost zone or rind is usually several cells wide, forming a continuous, closely packed layer of short-celled hyphae having swollen hyphal tips that protrude through the stromatal surface.[18,41] Mature, fully differentiated rinds are usually heavily pigmented due to the accumulation of oxidized phenolic substances or melanin in walls, between the cells and over the outer surface (Figures 6 and 7). In the cytoplasm of mature rind cells of stromata of *M. fructicola* and *M. laxa,* nuclei, mitochondria, some irregular ER, and sometimes large storage vacuoles have been reported[19] (see Figure 8). During resting periods, however, there

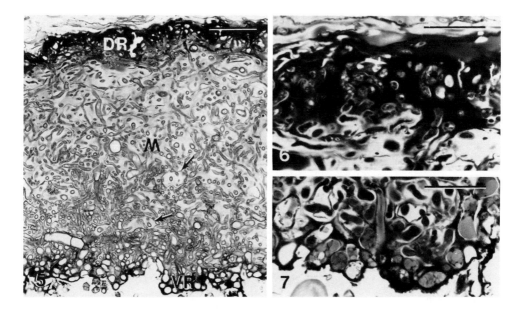

FIGURES 5 to 7. Liight micrographs of *M. fructicola* in peach mummies. **Figure 5.** Section showing a well-developed, pigmented dorsal rind (DR), a lesser developed ventral rind (VR), and a broad medulla (M) containing host cells (arrows). Section stained with PAS reagents. Scale = 50 μm. **Figure 6.** Section showing a dorsal rind several cells wide with heavy pigmentation between cells and over the outer surface. Section stained with amido black 10B. Scale = 25 μm. **Figure 7.** Section showing a ventral rind consisting of a narrow layer of cells with considerably less pigment than the dorsal rind. Section stained with amido black 10B. Scale = 25 μm.

is often loss of contents of rind cells (Figure 9). Ventral rind cells are normally less densely packed together, but similar in most other respects to those of the dorsal rind (Figures 6 and 7).

When present in stromata of brown rot fungi the cortex is irregular, poorly differentiated, and identified mainly by hyphae running perpendicular to the rind surface.[18] The outer and inner boundaries of the cortex are indistinct (Figure 5). Black pigmentation often extends from the rind layer into the cortex, accumulating in an extracellular matrix which extends around both cortical and medullary hyphae (Figure 5). The cytoplasmic contents of cortical cells are essentially the same as those of rind cells.

The medulla forms the main part of the stromata and consists of loosely interwoven hyphae embedded in a continuous extracellular matrix (Figures 5, 10, and 11). Remnants of host cells are frequently observed within (Figures 5 and 10). With the light microscope, Willetts and Calonge[25] observed small (diameter 1 to 2 μm), thin-walled hyphae forming a network among the main groundwork of stromata of *M. fructigena* and *M. laxa*. With the TEM they observed clearly defined organelles in the small hyphae, while the other hyphae contained large vacuoles and what were thought to be lipid bodies. Hyphae with these ultrastructural characteristics were reported by Pring et al.[35] in the stromata of *M. fructigena* and by Kohn and Grenville[19] in the stromata of *M. fructicola*.

Byrde and Willetts[12] described how the continuous rind of close-fitting, darkly pigmented cells which differentiates over exposed stromatal surfaces serves as a barrier protecting medullary cells against dehydration. Also, the melanin pigments in the rind provide protection against harmful solar radiations and the lytic action of enzymes produced by antagonistic microorganisms. Recent work by Young and Ashford,[54] who used the apoplastic tracer sulforhodamine G to study the permeability of sclerotia of *Sclerotinia minor,* found that

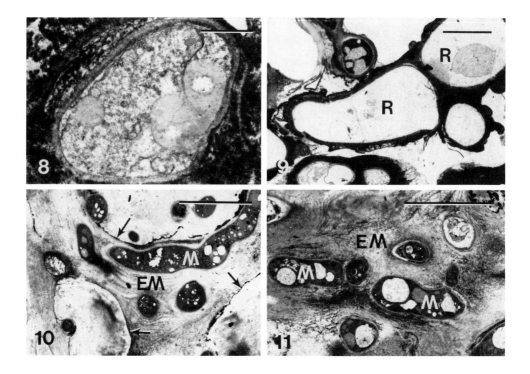

FIGURES 8 to 11.   TEMs of rind and medullary cells of *M. laxa* and *M. fructicola* in stone fruits. **Figure 8.** Rind cell of *M. laxa* in nectarine, showing cytoplasmic contents. Scale = 2.5 μm. **Figure 9.** Mature rind cells (R) of *M. fructicola* in peach, showing loss of contents. Scale = 5 μm. **Figure 10.** Medullary hyphae (M) of *M. fructicola* in peach, showing remnants of host cell walls (arrows) in the extracellular matrix (EM). Scale = 10 μm. **Figure 11.** Medullary hyphae (M) of *M. fructicola* in nectarine surrounded by extracellular matrix (EM). Scale = 10 μm.

reduction in permeability corresponded with the deposition of pigment in the rind rather than differentiation of the rind as a continuous layer. Their work demonstrated the effectiveness of a fully differentiated rind as a barrier capable of reducing permeability. The formation of stromata in host tissue and their compact nature give some protection against adverse external conditions, while the structure of individual hyphae is also a factor in resistance.[55] Morphological features are probably of lesser importance than physiological and biochemical adaptations, however.

C. HISTOCHEMISTRY AND ULTRASTRUCTURE OF STROMATAL RESERVES

The recent work of Kohn and Grenville[18,19] and the authors' unpublished data indicate that considerable amounts of nutrients accumulate in stromata of the brown rot fungi. Some differences are apparent in the abundance of reserves reported by Kohn and Grenville and in our studies, but the inconsistencies can be accounted for by variation between isolates, conditions of incubation, and the stages of stroma maturity examined. The authors studied fully differentiated stromata of *M. fructicola* produced in stone fruits. Stromatal hyphae produced in all these hosts accumulate nutrients intracellularly as cytoplasmic reserves. Unfortunately, no information can be found on the deposition of reserves during stroma differentiation; however, the processes of reserve deposition in *Monilinia* spp. are probably similar to those in sclerotia of *Sclerotinia minor*[45] and *Botrytis* spp.[47]

The main intracellular reserves that have been detected in mature stromata of *Monilinia* spp. are glycogen, protein, polyphosphate, and lipid. Considerable amounts of glucans are

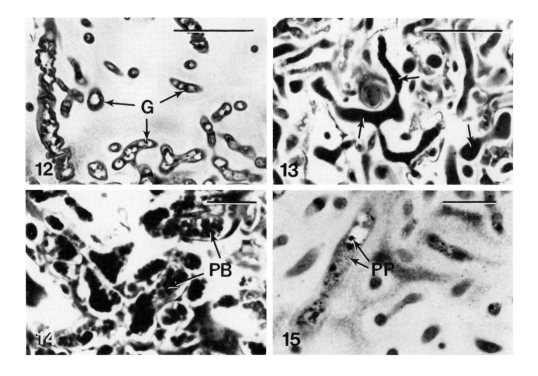

FIGURES 12 to 15. Light micrographs showing histochemical reactions of *M. fructicola* on agar and in stone fruits. **Figure 12.** Medullary hyphae from a stroma cultured on agar showing glycogen (G) present in the cytoplasm. Stained with PAS reagents. Scale = 20 μm. **Figure 13.** Medullary hyphae in peach, showing the presence of cytoplasmic proteins (arrows). Stained with amido black 10B. Scale = 20 μm. **Figure 14.** Medullary hyphae in plum, showing protein bodies (PB) within the cytoplasm. Stained with amido black 10B. Scale = 10 μm. **Figure 15.** Medullary hyphae from a stroma cultured on agar, showing polyphosphate granules (PP) of varying sizes in vacuoles and the cytoplasm. Stained with toluidine blue O. Scale = 10 μm.

found in hyphal walls and in the extensive extracellular matrix around cortical and medullary cells. The chitin of walls also contains large amounts of carbohydrate, while soluble protein has been reported in walls and in the extracellular matrix. Some of the main histochemical techniques used to identify these reserves are outlined in the Appendix. These techniques have been compiled from methods used in the authors' laboratory in histochemical investigations on several sclerotiniaceous fungi.[45,47,48]

1. Cytoplasmic Reserves
a. Glycogen

Kohn and Grenville[18] reported that with periodic acid-Schiff's reagent (PAS) alone, the cytoplasm of almost all stromatal anamorphs of the Sclerotiniaceae they studied stained in a reticulate pattern. The reticulation was removed by treatment with amylase, indicating that glycogen was present in the cytoplasm. They did find, however, that *M. fructicola* was one of two exceptions, with staining for intracellular carbohydrates occurring in large numbers of cytoplasmic vacuoles. In unpublished observations on stromata of *M. fructicola* grown in agar cultures the authors did not see glycogen in vacuoles, but it was present in the cytoplasm of some hyphae (Figure 12). Little or no glycogen was produced in the older stromata in stone fruits. Earlier findings on sclerotia of *S. minor*[45] indicate that during sclerotial differentiation glycogen is normally present in cytoplasm, but decreases in abundance during maturation when it is utilized in the synthesis of other reserves, such as protein bodies.

Glycogen, when viewed with the TEM, normally appears as electron-dense, rosette-shaped bodies;[56-58] however, the authors have not observed such bodies in the stromatal hyphae of any sclerotiniaceous fungi, including *Monilinia* spp. The bodies have not been reported by Kohn and Grenville,[18] despite histochemical techniques indicating the presence of glycogen. Granular deposits that are too large for ribosomes are dispersed in the cytoplasm of some stromatal cells. These granules are probably glycogen.

b. Proteins

With both amido black 10B and acid fuchsin, Kohn and Grenville[18] obtained intense staining for cytoplasmic protein in rind, cortical, and medullary cells of stromata of *M. fructicola*. They also reported protein bodies in all stromatal tissues with the largest number present in rind. The authors also detected cytoplasmic protein (Figure 13) and protein bodies throughout the stromata (Figure 14), but numbers of the latter were variable. The numbers of protein bodies were less than in sclerotia of the other sclerotiniaceous fungi the authors have studied.[45,47,48]

c. Polyphosphate

Kohn and Grenville[18] reported many strongly staining metachromatic granules in most rind cells and few weakly staining granules in less than half the cortical and medullary cells of the stromata of *M. fructicola*. The granules were probably polyphosphate, but not all the histochemical tests needed to confirm this were carried out. In stromata of *M. fructicola* produced in agar cultures the authors observed and confirmed the presence of polyphosphate granules in some rind cells and most medullary hyphae. Some granules were large and stained red, while others were small and stained purple (Figure 15). The amount and staining characteristics of polyphosphate observed in chemically fixed material, however, must be treated with caution. Recent unpublished work of Young et al.[66] indicates that sclerotial material prepared for histochemical study by freeze substitution contains much more polyphosphate than chemically fixed material.

Ultrastructurally, polyphosphate granules are of two types, single homogenous, very electron-dense spherical structures within (but not filling) a vacuole, or many smaller structures embedded in a groundwork of heterogenous, metachromatic material within a single vacuole.[18]

d. Lipids

Stromata of *M. fructicola* are different from most stromata and sclerotia of the Sclerotiniaceae in that they contain considerable amounts of lipid. Kohn and Grenville[18] observed lipid bodies in rind, cortical, and medullary cells of this species. The authors conducted histochemical studies that confirmed these results.

With TEM, lipid bodies are of medium electron density, round, and variable in size.[18]

2. Cell Walls

Jones et al.[59] used infrared (IR) and biochemical analyses to study the composition of hyphal walls of stromata of *M. fructigena* and sclerotia of several other sclerotiniaceous fungi. They determined that the major constituents of the wall are chitin and β-glucans with β-1,3- and β-1,6-linkages.

Kohn and Grenville[18] used PAS reagents and acriflavin-HCl (fluorescent Schiff) in histochemical studies on stromatal walls of *M. fructicola*. In addition to these stains, the authors investigated the walls with aniline blue and calcofluor white M2R (Figure 16). The results of these tests indicate that the walls contain a β-1,3-linked glucan and probably a PAS-positive glucan with a β-1,6-linkage. These findings agree with those of Jones et al.[59]

No histochemical tests for chitin have been reported for the brown rot fungi; however, from the biochemical data on *M. fructigena*[59] and the histochemical results from *Sclerotinia minor*[45] and *Botrytis* spp.,[47] it seems reasonable to assume that chitin is also a major component of walls of the brown rot species. Protein was detected with amido black 10B in walls of *M. fructicola* by the authors and by Kohn and Grenville.[18]

3. Extracellular Matrix

A copious matrix of highly hydrated material fills all interhyphal spaces between stromatal hyphae of the brown rot fungi. Saito[42,43] interpreted the matrix in the sclerotia of *Sclerotinia sclerotiorum* as being a fibrous outer wall layer. However, Bullock et al.[45] concluded that in sclerotia of *S. minor* the matrix is an extracellular material secreted by sclerotial hyphae. Kohn and Grenville[18] accepted this interpretation for the matrix of *M. fructicola* and it now seems that most workers regard the matrix as extracellular and not part of the wall.

Biochemical analyses have resulted in conflicting reports on the composition of the extracellular matrix of the brown rot fungi. Feather and Malek[60] analyzed the extracellular mucilage of *M. fructicola* and suggested that it is a highly branched glucose polymer with mainly 1,3- and some 1,4-linkages. They failed to detect any other sugars. Archer et al.[61] found that the mucilage of *M. fructigena* was mostly a glucan with traces of galactose and mannose. The polysaccharide was polydisperse and highly viscous. Periodate oxidation studies indicated a branched structure, but contrary to the previous report, mainly of 1,4-linkages.

Histochemical studies on stromata of both *M. fructicola* and *M. laxa* by Kohn and Grenville[18] and the authors resulted in similar but weaker staining reactions for the matrix compared with walls. These results indicate that the matrix contains β-1,3 glucans and small amounts of one or more PAS-positive polysaccharides which could have β-1,6- or β-1,4-linkages or possibly a mixture of both. As with walls, some protein was detected in the matrix.

The authors observed host cells in the extracellular matrix of the medulla of stromata of both *M. fructicola* and *M. laxa* produced in stone fruits. The walls of the host cells usually retained their integrity, reacted positively with PAS reagents (Figure 17), and fluoresced strongly (Figure 18), indicating considerable carbohydrate materials were still present. The host cells, however, were either devoid of contents or contained shriveled protoplasts.

General survival and morphogenetic roles[12,55] have been attributed to the extracellular matrix which is secreted by stromatal hyphae. The mucilaginous matrix readily absorbs and retains water which serves as a reservoir during prolonged resting periods and germination. The hydrated material protects hyphal surfaces against dehydration, while mucilage in cells tends to stabilize intracellular proteins and prevent them from being denatured when subjected to temperature extremes and desiccating conditions. During the ontogeny of stromata, the matrix binds stromatal hyphae together[55] and possibly serves in the transmission of stimuli between individual hyphae.[62]

4. Phenolic Compounds

A strong blue-green staining reaction indicating the presence of phenolics was obtained with toluidine blue O by Kohn and Grenville[18] and the authors in rind cell walls, and in the extracellular matrix around the rind and the cortical hyphae of *M. fructicola* and *M. laxa*. A less intense reaction occurred in the walls of cortical and medullary hyphae. The authors also observed a strong reaction for phenolics in walls of host cells incorporated in stromatal tissue (Figure 19).

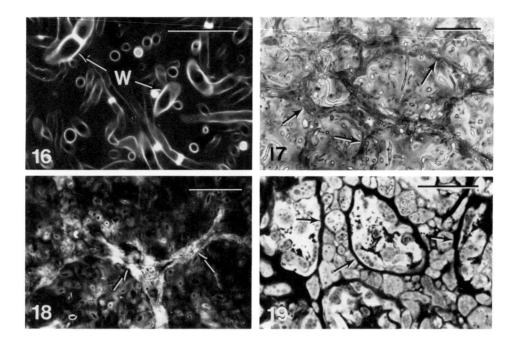

FIGURES 16 to 19.   Light micrographs of stromata of *M. fructicola* and *M. laxa* in stone fruits. **Figure 16.** Medullary hyphae of *M. laxa* in plum, showing strong fluorescence of walls (W). Stained with calcofluor white M2R. Scale = 20 μm. **Figure 17.** Medulla of *M. fructicola* in nectarine, showing remnants of host cell walls (arrows). Stained with PAS reagents. Scale = 50 μm. **Figure 18.** Medulla of *M. fructicola* in plum, showing strong fluorescence by remnants of host cell walls (arrows). Stained with PAS reagents and viewed by fluorescence microscopy. Scale = 50 μm. **Figure 19.** Medulla of *M. fructicola* in plum, showing strong phenolic reaction by remnants of host cell walls (arrows). Stained with toluidine blue O. Scale = 25 μm.

VI. UTILIZATION OF STROMATAL RESERVES

No published information exists on the utilization of reserves during germination of the stromata of *Monilinia* spp.; however, the authors' unpublished data on ultrastructural and histochemical changes in the stromata of *M. fructicola* and *M. laxa,* which produced extensive external mycelial wefts over stromatal surfaces or large numbers of microconidia, indicate considerable depletion of intracellular reserves. After production of these structures glycogen, cytoplasmic proteins, protein bodies, polyphosphate granules, and lipid bodies were no longer present. Also, the extracellualr matrix was depleted in most stromata, although the walls of the stromatal cells were not degraded to any great extent.

Sclerotia of *Sclerotinia minor* have similar reserves to those of the stromata of *M. fructicola*. In a study of carpogenic germination of *S. minor* sclerotia, Bullock et al.[63] found that β-glucans of both walls and extracellualr matrix and all intracellular reserves were utilized during the germination process. Walls were more resistant to hydrolysis than the matrix, but they too were gradually degraded. When germination was complete, the rind was still intact around a virtually empty sclerotium. The melanin pigment in the walls of rind cells probably prevented dissolution of the layer. Corresponding with the loss of reserves from sclerotia of *Sclerotinia* spp. during carpogenic germination, Jayachandran[64] observed the accumulation of similar reserves in the hymenia of apothecia. Parallel studies on conidiogenic germination of sclerotia of *Botrytis cinerea*[48] and the production of eruptive mycelial plugs by *Sclerotium cepivorum*[49] showed patterns of reserve mobilization similar to those of *Sclerotinia minor* undergoing carpogenic germination.[63]

It seems reasonable to assume from the similarities between stromata of the brown rot fungi and sclerotia of *B. cinerea* and *Sclerotinia* spp. in regard to structure, distribution, and manner of degradation of their reserves that the former also utilize their reserves in propagule production. Therefore, materials deposited in hyphae provide an abundant and assured endogenous supply of nutrients that are available for the production in the spring of numerous apothecia on a single, large fruit mummy or, alternatively, several crops of conidia during the fruit-growing season. Both types of germination are extremely demanding in nutrient and energy requirements and the enormous numbers of propagules produced are a reflection of the large amounts of nutrients available in fruits of rosaceous plants, the size of stromata, the quantities and variety of materials stored in them, and the nutritional self-sufficiency of stromata.

VII. CONCLUSIONS

During the course of evolution of the host-pathogen relationship between the brown rot fungi and pomaceous and drupaceous trees infected fruits have become adapted to serve as highly effective survival and reproductive structures. Because of the characteristics of fruit tissue and various features of the brown rot fungi, mummies have assumed a central role in the life cycles of these fungi and contribute greatly to their success as pathogens.

Stromata of the brown rot fungi provide excellent models for studies on the role of intra- and extracellular reserves in survival, but there have been very few attempts to use stromata for this purpose. Studies on higher plants and animals indicate that various substances protect cells from severe environmental conditions such as desiccation and extremes of temperature. Basic mechanisms of survival against these factors are probably alike for all organisms.

Studies indicate that high concentrations of solutes within a cell depress the freezing point of its contents, thereby increasing resistance against dehydration and also reducing the formation of intracellular ice crystals that cause physical damage to membranes and cell contents. Certain substances, particularly carbohydrates and lipids, protect both surfaces and the interior of cells, while cold-hardening is associated with modifications of proteins that reduce sensitivity to cold and increase the water-holding capacity of macromolecules.[12,55] Considerable useful data on such aspects of survival could be obtained from studies on stromata of the brown rot fungi.

Monilinia spp. form a discrete group of fungi that illustrate host-pathogen relationships ranging from unspecialized to highly specialized interactions.[12,13] Some species have wide host ranges, other are confined to a single host and certain tissues of that host, while a few species require two hosts to complete their life cycles. Furthermore, the extremely limited regional distribution of some species appears to be closely related to host specialization and/ or their inability to become established in many areas. Because of these characteristics the genus has great potential for studies on fungal evolution, biological control, and relationships of enzymes and isozymes to host and tissue specificities. The latter applies particularly to wall-degrading enzymes.

VIII. SUMMARY

Monilinia spp. cause considerable damage to fruits of rosaceous trees. *M. fructicola,* *M. fructigena,* and *M. laxa* are the most common and extensively studied species and are important pathogens of stone fruits, apples, and pears. Infection of fruits is normally through damaged areas. Infecting hyphae secrete pectolytic enzymes that degrade middle lamellae and consequently early colonization is mainly intercellular. Hyphae permeate through fruits to form large, compact fungal structures which contain partly degraded host tissue. The colonized fruits are termed mummies or stromata. Incomplete breakdown of fruit tissue is due to the lack of cellulase production by *Monilinia* spp. and inactivation of hydrolytic enzymes of the fungi by oxidized phenolics which accumulate in infection courts. Abundant insoluble nutrient reserves are deposited during stromatal development. The main intracellular reserves that have been detected in mature stromata include glycogen, protein, polyphosphate, and lipid. Considerable amounts of glucans are found in hyphal walls and in the extensive extracellular matrix around stromatal hyphae. The chitin in walls also provides a source of carbohydrate material Stromata of *Monilinia* spp. possess many adaptations that enable them to survive periods of adverse environmental conditions and when susceptible hosts are not available. After a period of inactivity the accumulated reserves are utilized in the production of conidia, apothecia, and/or mycelia. Thus, stromata or fruit mummies of *Monilinia* spp. have assumed a central role in the life cycles of these fungi.

ACKNOWLEDGMENTS

Some of this work was supported by a grant from the Australian Research Grants Scheme to H.J.W. We are grateful to Professor Anne E. Ashford and Drs. R. J. W. Byrde and J. M. Leis for comments on the manuscript.

APPENDIX

Unless otherwise stated, the histochemical procedures are carried out on material fixed in glutaraldehyde, dehydrated, and embedded in glycol methacrylate (GMA), as outlined by Bullock et al.[44,45] Sections are cut 0.5 to 1.5 μm thick and mounted on glass slides in immersion oil. For fluorescence microscopy, the oil should be fluorescence free. For determination of lipids, unfixed material sectioned on a cryomicrotome to a thickness of 10 μm must be used.

POLYSACCHARIDES
Periodic acid Schiff (PAS) Reaction[47]

Procedure:

1. Block aldehyde groups by immersing sections in 0.1% 2,4-dinitrophenylhydrazine in 15% acetic acid for 45 min.
2. Rinse in running water, 30 min.
3. Immerse sections in 1% periodic acid for 10 min.
4. Rinse in running water, 15 min.
5. Immerse in Schiff reagent for 20 min.
6. Rinse in running water, 5 min.

As a control, omit the periodic acid step.

Results: Carbohydrates with hydroxyl groups on vicinal carbon atoms stain red.

PAS Fluorescence[45]

Procedure:

1. Stain sections as above.
2. View with fluorescence microscopy, using filter combination BP546/7, FT580, and LP590.

Results: The sensitivity of the PAS reaction is enhanced.

Amylase Digestion[45]

Procedure:

1. Immerse sections in 1% *Bacillus subtilis* α-amylase in 0.1 *M* phosphate buffer, pH 6.5, for 3 h at room temperature. Alternatively immerse in human saliva for 3 h at room temperature.
2. Rinse in distilled water.
3. Stain with PAS reagents.

Results: Glycogen is digested, giving a negative staining reaction with PAS.

Acriflavin-HCl[18]

Procedure:

1. Immerse sections in 1% aqueous periodic acid for 10 min.
2. Rinse briefly in distilled water.
3. Stain in acriflavin-HCl for 20 min.
4. Wash twice in 1% HCl in 95% ethanol for 5 min each.
5. Rinse in 100% ethanol.
6. Examine by fluorescence microscopy using filter combinations BP450–490 and LP520.

Results: PAS positive substances fluoresce bright yellow, other substances fluoresce green.

Aniline Blue[45]

Procedure:

1. Stain sections in 0.01% aniline blue in 0.6 *M* potassium phosphate buffer, pH 8.5, for 10 min.
2. Blot dry before mounting in immersion oil.
3. Examine by fluorescence microscopy using filter combinations, BG12, FT510, and 50.

Results: β-1,3 glucans fluoresce yellow, but the reaction may not be specific for this material.

Calcofluor White M2R[45]

Procedure:

1.   Stain sections for 5 to 10 min in 0.1% aqueous calcofluor white M2R.
2.   Rinse briefly in distilled water.
3.   Examine by fluorescence microscopy using filter combinations, UG1, FT420, and LP418.

Results: Polysaccharides in cell walls and extracellular gels, including β-glucans, fluoresce blue.

Chitosan Reaction[47]

Procedure:

1.   Cut GMA sections 4 mm thick.
2.   Autoclave in saturated KOH for 2 h.
3.   Rinse in distilled water.
4.   Mount on slides in IKI in 1% H_2SO_4.

Results: Chitin is converted to chitosan by the KOH treatment. Chitosan stains violet in IKI.

PROTEINS
Amido Black 10B[45]

Procedure:

1.   Stain sections in freshly filtered 1% amido black 10B in 7% acetic acid for 10 min.
2.   Rinse briefly in distilled water.

Results: Basic substances, including proteins, stain blue.

Protease Digestion[47]

Procedure:

1.   Immerse sections in 1% pronase B in 0.1 M phosphate buffer, pH 7.5, for 2 d at 37°C.
2.   Rinse thoroughly in water.
3.   Stain with amido black 10B.

Results: Proteins are digested, resulting in a negative staining reaction with amido black 10B.

Acid Fuchsin[45]

Procedure:

1. Stain sections in 0.005% acid fuchsin in 1% acetic acid, or in 1% acid fuchsin in distilled water, for 30 min.
2. Rinse in distilled water for 2 min.

Results: Proteins stain red.

POLYPHOSPHATES
Toluidine Blue O[47]

Procedure:

1. Stain sections in 0.05 to 0.5% toluidine blue O in 0.1 *M* acetate buffer at pH 4.4 for up to 2 h.
2. Rinse in distilled water for 2 min.
3. Stain other sections in toluidine blue O, acidified to pH 1.0 with HCl, as above.

Results: At pH 4.4, nucleic acid, polyphosphates, polysulfates, and acid polysaccharides stain shades of blue or red. Polyphenols and lignin stain green or blue-green. At pH 1.0, only polyphosphates and polysulfates stain.

Trichloroacetic Acid (TCA) Extraction[45]

Procedure:

1. Immerse sections in 10% TCA at 4°C for 6 h.
2. Stain with toluidine blue O, pH 1.0.

Results: Polyphosphates are extracted by TCA, polysulfates are not. A negative reaction after staining with toluidine blue O indicates removal of polyphosphates.

LIPIDS
Sudan Black B[47]

Procedure:

1. Stain unfixed, frozen sections in filtered saturated Sudan Black B in 70% ethanol for 10 min.
2. Rinse very briefly in 70% ethanol.
3. Mount in glycerol-gelatin.

Results: Acid and neutral lipids stain blue.

REFERENCES

1. **Persoon, C. H.,** *Torula fructigena,* in *Observationes Mycologicae,* Lipsiae, 1796, 26.
2. **Woronin, M.,** Über *Sclerotinia cinerea* und *Sclerotinia fructigena, Mem. Acad. Sci. St.-Peter. VIIIe sér.,* 10, Phys. Math., 1, 1900.
3. **Aderhold, R. and Ruhland, W.,** Zur Kenntnis der Obstbaum-Sklerotinien, *Arb. Biol. Abt. Land Forstwirtsch.,* 4, 427, 1905.
4. **Wormald, H.,** The "brown rot" diseases of fruit trees, with special reference to two biologic forms of *Monilia cinerea* Bon. I., *Ann Bot. (London),* 33, 361, 1919.
5. **Wormald, H.,** The "brown rot" diseases of fruit trees, with special reference to two biologic forms of *Monilia cinerea* Bon. II., *Ann. Bot. (London),* 34, 143, 1920.
6. **Roberts, J. W. and Dunegan, J. C.,** Peach Brown Rot, *U.S. Dept. Agric. Tech. Bull.,* No. 328, Washington, D.C., 1932, 59.
7. **Kirtland, J. R.,** Premature decay of the plum, *Florist Hort. J. (Philadelphia),* 4, 34, 1855.
8. **Norton, J. B. S., Ezekiel, W. N., and Jehle, R. A.,** Fruit-rotting Sclerotinias. I. Apothecia of the brown-rot fungus, *Mo. Agric. Exp. Stn. Bull.,* 256, 1, 1923.
9. **Harrison, T. H.,** Brown rot fruits, and associated diseases, in Australia. I. History of the disease and determination of the causal organism, *J. R. Soc. N.S.W.,* 62, 99, 1928.
10. **Harrison, T. H.,** Brown rot of fruits and associated diseases of deciduous fruit trees. I. Historical review, and critical remarks concerning taxonomy and nomenclature of the causal organisms, *J. R. Soc. N.S.W.,* 67, 132, 1933.
11. **Wormald, H.,** The brown rot diseases of fruit trees, Tech. Bull. No. 3, Ministry of Agriculture Fisheries, and Food, Her Majesty's Stationery Office, London, 1954.
12. **Byrde, R. J. W. and Willetts, H. J.,** *The Brown Rot Fungi of Fruit: Their Biology and Control,* Pergamon Press, Oxford, 1977.
13. **Batra, L. R.,** *World Species of Monilinia (Fungi): Their Ecology, Biosystematics and Control,* Cramer, Berlin, 1991.
14. **Harada, Y.,** Studies on the Japanese species of *Monilinia* (Sclerotiniaceae), *Bull. Fac. Agric. Hirosaki Univ.,* 27, 1977.
15. **Honey, E. E.,** The monilioid species of *Sclerotinia, Mycologia,* 20, 127, 1928.
16. **Willetts, H. J.,** The development of stromata of *Sclerotinia fructicola* and related species. II. In fruits, *Trans. Br. Mycol. Soc.,* 51, 633, 1968.
17. **Whetzel, H. H.,** A synopsis of the genera and species of the Sclerotiniaceae, a family of stromatic inoperculate Discomycetes, *Mycologia,* 37, 648, 1945.
18. **Kohn, L. M. and Grenville, D. J.,** Anatomy and histochemistry of stromatal anamorphs in the Sclerotiniaceae, *Can. J. Bot.,* 67, 371, 1989.
19. **Kohn, L. M. and Grenville, D. J.,** Ultrastructure of stromatal anamorphs in the Sclerotiniaceae, *Can. J. Bot.,* 67, 394, 1989.
20. **Honey, E. E.,** North American species of *Monilinia.* I. Occurrence, grouping and life-histories, *Am. J. Bot.,* 20, 100, 1936.
21. **Willetts, H. J., Byrde, R. J. W., Fielding, A. H., and Wong, A. -L.,** The taxonomy of the brown rot fungi (*Monilinia* spp.) related to their extracellular cell wall-degrading enzymes, *J. Gen. Microbiol.,* 103, 77, 1977.
22. **Penrose, L. J., Tarran, J., and Wong, A. -L.,** First record of *Sclerotinia laxa* Aderh. & Ruhl. in New South Wales: differentiation from *S. fructicola* (Wint.) Rehm by cultural characteristics and electrophoresis, *Aust. J. Agric. Res.,* 27, 547, 1976.
23. **Shima, Y.,** Studies on the young fruit-rot of apple-tree, *J. Fac. Agric. Hokkaido Univ.,* 39, 143, 1936.
24. **Willetts, H. J. and Harada, Y.,** A review of apothecial production by *Monilinia* fungi in Japan, *Mycologia,* 76, 314, 1984.
25. **Willetts, H. J. and Calonge, F. D.,** The ultrastructure of the stroma of the brown rot fungi, *Arch. Mikrobiol.,* 64, 279, 1969.
26. **Najim, I. and Turian, G.,** Ultrastructure de l'hyphe vegetatif de *Sclerotinia fructigena, Can. J. Bot.,* 57, 1299, 1979.
27. **Hoffmann, G. M.,** Kernverhältnisse bei *Monilinia fructigena* und *M. laxa, Phytopathol. Z.,* 68, 143, 1970.
28. **Heuberger, J. W.,** Fruit-rotting Sclerotinias. IV. A cytological study of *Sclerotinia fructicola* (Wint.) Rehm, *Mo. Agric. Exp. Stn. Bull.,* 371, 167, 1934.
29. **Hall, R.,** Cytology of the asexual stages of the Australian brown rot fungus, *Monilinia fructicola, Cytologia,* 28, 181, 1963.
30. **Moore, M. H.,** Brown rot of apples: notes on infection associated with hail bruises and lenticels, *Rep. East Malling Res. Stn.,* No. 1950, 131, 1951.

31. **Hall, R.,** Pathogenicity of *Monilinia fructicola*. II. Penetration of peach leaf and fruit, *Phytopathol. Z.,* 72, 281, 1971.
32. **Calonge, F. D., Fielding, A. H., Byrde, R. J. W., and Akinrefon, O. A.,** Changes in ultrastructure following fungal invasion and the possible relevance of extracellular enzymes, *J. Exp. Bot.,* 20, 350, 1969.
33. **Hislop, E. C., Barnaby, V. M., Shellis, C., and Laborda, F.,** Localization of α-L-arabinofuranosidase and phosphatase in mycelium of *Sclerotinia fructigena, J. Gen. Microbiol.,* 81, 79, 1974.
34. **Byrde, R. J. W.,** Fungal "pectinases": from ribosome to plant cell wall, *Trans. Br. Mycol. Soc.,* 79, 1, 1982.
35. **Pring, R. J., Byrde, R. J. W., and Willetts, H. J.,** An ultrastructural study of the infection of pear fruit by *Monilinia fructigena, Physiol. Plant Pathol.,* 19, 1, 1981.
36. **Lumsden, R. D. and Dow, R. L.,** Histopathology of *Sclerotinia sclerotiorum* infection of bean, *Phytopathology,* 63, 708, 1973.
37. **Lumsden, R. D.,** Histology and physiology of pathogenesis in plant diseases caused by *Sclerotinia* species, *Phytopathology,* 69, 890, 1979.
38. **Byrde, R. J. W., Fielding, A. H., Archer, S. A., and Davies, E.,** The role of extracellular enzymes in the rotting of fruit tissue by *Sclerotinia fructigena,* in *Fungal Pathogenicity and the Plant's Response,* Byrde, R. J. W. and Cutting, C. V., Eds., Academic Press, London, 1973, 39.
39. **Bateman, D. F. and Basham, H. G.,** Degradation of plant cell walls and membranes by microbial enzymes, in *Physiological Plant Pathology,* Heitefuss, R. and Williams, P. H., Eds., Springer-Verlag, Berlin, 1976, 316.
40. **Willetts, H. J.,** The development of stromata of *Sclerotinia fructicola* and related species. I. In culture, *Trans. Br. Mycol. Soc.,* 51, 625, 1968.
41. **Willetts, H. J.,** Stromatal rind formation in the brown rot fungi, *J. Gen. Microbiol.,* 52, 271, 1968.
42. **Saito, I.,** Ultrastructural aspects of the maturation of sclerotia of *Sclerotinia sclerotiorum* (Lib.) de Bary, *Trans. Mycol. Soc. Jpn.,* 15, 384, 1974.
43. **Saito, I.,** Studies on the maturation and germination of sclerotia of *Sclerotinia sclerotiorum* (Lib.) de Bary, a causal fungus of bean stem rot, *Rep. Hokkaido Pref. Agric. Exp. Stn.,* 26, 1, 1977.
44. **Bullock, S., Willetts, H. J., and Ashford, A. E.,** The structure and histochemistry of sclerotia of *Sclerotinia minor* Jagger. I. Light and electron microscope studies on sclerotial development, *Protoplasma,* 104, 315, 1980.
45. **Bullock, S., Ashford, A. E., and Willetts, H. J.,** The structure and histochemistry of sclerotia of *Sclerotinia minor* Jagger. II. Histochemistry of extracellular substances and cytoplasmic reserves, *Protoplasma,* 104, 333, 1980.
46. **Willetts, H. J. and Bullock, S.,** Studies on the ontogeny and ultrastructure of the sclerotium of *Botrytis cinerea* Pers. ex Nocca & Balbis, *Can. J. Microbiol.,* 28, 1347, 1982.
47. **Backhouse, D. and Willetts, H. J.,** A histochemical study of sclerotia of *Botrytis cinerea* and *Botrytis fabae, Can. J. Microbiol.,* 30, 171, 1984.
48. **Backhouse, D. and Willetts, H. J.,** Histochemical changes during conidiogenic germination of sclerotia of *Botrytis cinerea, Can. J. Microbiol.,* 31, 282, 1985.
49. **Backhouse, D. and Stewart, A.,** Anatomy and histochemistry of resting and germinating sclerotia of *Sclerotium cepivorum, Trans. Br. Mycol. Soc.,* 89, 561, 1987.
50. **Willetts, H. J.,** Sclerotium formation, in *The Filamentous Fungi,* Vol. 3, *Developmental Mycology,* Smith, J. E. and Berry, D. R., Eds., Edward Arnold, London, 1978, chap. 10.
51. **Colotelo, N.,** Fungal exudates, *Can. J. Bot.,* 24, 1173, 1978.
52. **Cooke, R. C. and Al-Hamdani, A. M.,** Water relations of sclerotia and other infective structures in *Water, Fungi and Plants, British Mycological Symposium II,* Ayres, P. G. and Boddy, L., Eds., Cambridge University Press, Cambridge, U.K., 1986, 49.
53. **Jennings, D. H.,** The role of droplets in helping to maintain a constant growth rate of aerial hyphae, *Mycol. Res.,* 95, 883, 1991.
54. **Young, N. and Ashford, A. E.,** Changes during development in the permeability of sclerotia of *Sclerotinia minor* to an apoplastic tracer, *Protoplasma,* 167, 205, 1992.
55. **Willetts, H. J.,** The survival of fungal sclerotia under adverse environmental conditions, *Biol. Rev.,* 47, 515, 1971.
56. **Fawcett, D. W.,** The identification of particulate glycogen and ribonucleoprotein granules in electron micrographs, *J. Histochem. Cytochem.,* 6, 95, 1958.
57. **Revel, J. P., Napolitano, L., and Fawcett, D. W.,** Identification of glycogen in electron micrographs of thin tissue sections, *J. Biophys. Biochem. Cytol.,* 8, 578, 1960.
58. **Bracker, C. E.,** Ultrastructure of fungi, *Annu. Rev. Phytopathol.,* 5, 343, 1967.
59. **Jones, D., Farmer, V. C., Bacon, J. S. D., and Wilson, M. J.,** Comparison of ultrastructure and chemical components of cell walls of certain plant pathogenic fungi, *Trans. Br. Mycol. Soc.,* 59, 11, 1972.

60. **Feather, M. S. and Malek, A.,** A highly branched extracellular D-glucan from *Monilinia fructicola, Biochim. Biophys. Acta,* 264, 103, 1972.
61. **Archer, S. A., Clamp, J. R., and Migliore, D.,** Isolation and partial characterization of an extracellular branched D-glucan from *Monilinia fructigena, J. Gen. Microbiol.,* 102, 157, 1977.
62. **Reijnders, A. F. and Moore, D.,** Developmental biology of agarics — an overview, in *Developmental Biology of Higher Fungi,* Moore, D., Casselton, L. A., Wood, D. A., and Frankland, J. C., Eds., Cambridge University Press, Cambridge, U.K., 1985, chap. 27.
63. **Bullock, S., Willetts, H. J., and Ashford, A. E.,** The structure and histochemistry of sclerotia of *Sclerotinia minor* Jagger. III. Changes in ultrastructure and loss of reserve materials during carpogenic germination, *Protoplasma,* 117, 214, 1983.
64. **Jayachandran, M.,** Studies on the Sclerotia and Apothecia of *Sclerotinia* Species, M. Sc. thesis, University of New South Wales, Kensington, Australia, 1983.
65. **Willetts, H. J.,** unpublished data.
66. **Young, N., Ashford, A. E., and Bullock, S.,** unpublished data.

Chapter 6

THE PATHOLOGICAL ANATOMY OF *ZYGOPHIALA JAMAICENSIS* ON FRUIT SURFACES

Hideo Nasu and Hitoshi Kunoh

TABLE OF CONTENTS

0-8493-2939-6/93/$0.00 + $.50

I. INTRODUCTION

A new speckle disease of banana leaves in Jamaica was first reported by Martyn[1] in 1945. Seven years later, a new disease called greasy blotch of carnation was found in California.[2] Almost simultaneously, flyspeck of apple, whose symptoms are different from that of carnation greasy blotch, was studied by Durbin and Snyder.[3] In those days, investigators thought that all these diseases were caused by different pathogens, although at present it is known that *Zygophiala jamaicensis* Mason is a common pathogen of all of these diseases. This confusion might have resulted from an inevitable consequence of the circumstances as described by Baker et al.[4]: "A fungus inadvertently may be listed in the literature under different binomials on its several hosts, particularly if its spore stages and the symptoms produced vary with the host. Recognition that the pathogens are identical then may be delayed because an investigator seldom studies a sufficiently wide range of hosts. Such was the case for *Z. jamaicensis*."

In Japan, similar flyspeck diseases have occurred on grape, apple, Japanese persimmon, pear, and other cultivated fruit trees. The causal fungus had been identified as *Leptothyrium pomi* (Mont. et Fr.) Sacc. However, the authors'[5] recent investigation revealed that the causal fungus of flyspeck on various fruit trees, as well as that of leaf speckle of banana,[1] greasy blotch of carnation,[4] and flyspeck of apple,[4] was *Z. jamaicensis*.

The disease has been reported to occur over much of the temperate and tropical world.[4] The fungus is widely distributed in Europe, the U.S., and Canada, and also occurs in Australia, New Zealand, and African and Asian countries including Japan.[4,6] When this disease occurs severely on carnation whose stems and leaves are normally covered with wax, the marketable value is reduced greatly.[4] Similarly, the occurrence of this disease on the surfaces of a variety of berries and fruits reduces their commercial value, which is sometimes dependent on their visual appearance.[7] However, cultivated fruit trees and other wild host plants are never killed by this fungus, and moreover, rotting and browning never occur inside the fruits and berries because, as described later in detail, the causal fungus is an ectoparasite that grows on surfaces of host plants. It is likely that the fungus uses waxy bloom crystals as a sole source of nutrition.

II. OVERALL FEATURES OF THE DISEASE ON HOST FRUIT TREES

A. MACROSCOPIC SYMPTOMS AND SIGNS ON A VARIETY OF HOST TREES

The entire surface of various berries and fruits is normally covered (more or less) with a whitish waxy bloom. The incipient symptom of disease that can be observed with the naked eye is speckles, about 1 to 5 mm in diameter, caused by the loss of glaucous wax (Figure 1). When environmental conditions are suitable for symptom development, the speckles enlarge and coalesce, the wax-free area spreads, and finally covers the entire surface of berries and fruits, giving the affected area a shiny appearance. Numerous dark brown to black specks, up to 1 mm in diameter, become visible in the areas free of bloom (Figures 2 to 6). These specks represent sclerotium-like structures that are described later in detail. Hyphae and conidiophores also are formed on host surfaces, but they are not visible without the use of a hand lens. Such symptoms and signs are common for most commercial fruits (Figures 1 to 5).

Similar symptoms and signs also occur on stems and leaves of various host plants, and are particularly prominent on young, greenish shoots on which glaucous, waxy bloom is abundant (Figure 6).

FIGURES 1 to 6. **Figure 1.** The incipient symptom of flyspeck on grape berries. The glaucous bloom has been lost in speckles of 1 to 5 mm diameter (arrows). **Figure 2.** The incipient speckles spread and coalesced with one another, giving large spots free of bloom on the berry surfaces (arrows). Numerous dark-colored specks (arrow heads) appeared in the spots. **Figure 3.** Typical symptom with specks on fruits of Chinese quince. **Figure 4.** A large number of specks produced on an apple fruit. **Figure 5.** Specks produced in an area free of bloom on a plum fruit. **Figure 6.** Specks (arrows) produced on twigs of grapevine.

According to *in vitro* inoculation tests with mycelial inocula of different host origins, i.e., grape berries, Japanese persimmon, and apple fruits,[5] tiny speckles are distinguishable with the naked eye within 4 d postinoculation and, flyspecks appear by 9 to 10 d on all hosts examined regardless of the source of inocula. At least 2 to 3 weeks are required for symptom development in the orchard and vineyard after conidia contact surfaces of fruits and berries.

In several grape-growing areas of Japan, some varieties are cultivated in greenhouses. On grape berries grown under these conditions, only the evanescence of bloom is a visual symptom and the number of flyspecks produced on the berry surface is much less than that on berries grown outdoors. In vineyards where grape berries are severely infected by the causal fungus, the incipient speckles are usually seen on young shoots in late June, about 10 d after the 1-month rainy season starts. By mid- to late July, the pathogen is disseminated to clusters of young berries. In greenhouses, fruit clusters located near the side windows are usually first infected in mid- to late July, and the majority of fruit clusters in the greenhouse may become infected thereafter. Vineyards and greenhouses where this disease occurs severely are frequently surrounded by woods and forests where wild host plants are abundant. Therefore, removal of wild host plants around vineyards, orchards, and green-houses is a very effective way to reduce the primary inoculum and thus to control this disease.

B. HOST RANGE OF THE PATHOGEN

Z. jamaicensis has an extensive host range on flowering plants with a wide geographical distribution. According to the literature survey by Baker et al.,[4] the fungus survives on 78 species in 36 families of flowering plants. Nasu and Kunoh[6] reported 120 species in 44 families as host plants in Japan (Table 1). According to Baker et al.[4] in the U.S. and Messiaen[8] in France, the telemorph of *Z. jamaicensis,* i.e., *Schizothyrium pomi* (Mont. et Fr.) von Arx, has been found on a variety of host plants. In Japan, Katumoto[9] reported *S. disperum* Katumoto sp. nov. on *Quercus salicinae* Bl. and *S. perexigum* (Roberge) Höhnel on *Maesa japonica* (Thunb.) Moritzi, but did not confirm their anamorph stage. Because of the diversity of known hosts, the large pathogen synonymy, the extensive geographical distribution, and the limited studies made, it is certain that the host range is much wider than is known at present.[4] It is most plausible that the fungus is able to survive on all plants whose surfaces are covered with waxy bloom, unless the bloom contains antifungal factors and/or is unsuitable as the nutritional source for the fungus.

III. MORPHOLOGICAL AND PHYSIOLOGICAL CHARACTERISTICS OF THE PATHOGEN

A. FUNGAL STRUCTURES FORMED ON HOST SURFACES

All structures, hyphae, conidiophores, conidia, and sclerotium-like structures of *Z. jamaicensis* are formed on host surfaces at the anamorph stage. Hyaline hyphae with numerous branches are readily visible with a hand lens on host surfaces. The waxy bloom along the hyphae has normally disappeared (Figure 7). This phenomenon is very common on surfaces of almost all host plants examined (Figures 8 to 12). At the advanced stage of symptom development, hyphae are covered with a thin film of the bloom-degenerating product, and thus become invisible in these areas.

Conidiophores that arise from hyphae have a unique shape (Figures 13 and 14). They consist of three distinguishable parts: a tortuous, dark brown sector; an angular, subhyaline, terminal cell; and on its apical sides, two divergent, hyaline, ovate to ampulliform conidiogenous cells that bear a thickened, circular, dark-colored disk[4,10] (Figure 13). When a

TABLE 1
Host Range of *Zygophiala jamaicensis* in Japan

Family	Species
Ginkgoaceae	*Ginkgo biloba* L.
Myricaceae	*Myrica rubra* Sieb. et Zucc.
Betulaceae	*Alnus japonica* (Thunb.) Steud.; *A. traveculosa* Hand.-Mazz.; *Carpinus japonica* Bl.; *C. tschonoskii* Maxim.
Fagaceae	*Castanea crenata* Sieb. et Zucc; *Quercus acutissima* Carr.; *Q. aliena* Bl.; *Q. dentata* Thunb.; *Q. glauca* Thunb.; *Q. mongolica* var. *grosseserrata* (Blume) Rehd. et Wils.; *Q. myrsinaefolia* Bl.; *Q. salicina* Bl.; *Q. serrata* Thunb.; *Q. sessilifolia* Bl.; *Q. variabilis* Bl.
Moraceae	*Morus alba* L.; *M. bombycis* Koidz.
Polygonaceae	*Polygonum cuspidatum* Sieb. et Zucc.
Phytolaccaceae	*Phytolacca americana* L.
Illiciaceae	*Illicium religiosum* Sieb. et Zucc.
Anonaceae	*Asimina triloba* Dunal.
Magnoliaceae	*Magnolia kobus* DC.; *M. liliflora* Desr.; *M. obovata* Thunb.; *M. salicifolia* (Sieb. et Zucc.) Maxim.; *M. virginiana* L.
Cercidiphyllaceae	*Cercidiphyllum japonicum* Sieb. et Zucc.
Ranunculaceae	*Clematis terniflora* DC.
Lardizabalaceae	*Akebia quinata* (Thunb.) Decne.
Menispermaceae	*Cocculus orbiculatus* DC.
Lauraceae	*Actinodaphne lancifolia* (Sieb. et Zucc.) Meisn; *Cinnamomum camphora* (Linn.) Sieb.; *Lindera erythrocarpa* Makino; *L. glauca* (Sieb. et Zucc.) Bl.; *L. obtusiloba* Bl.; *Neolitsea aciculata* (Blume) Koidz.; *Persea thunbergii* (Sieb. et Zucc.) Kosterm.
Saxifragaceae	*Deutzia crenata* Sieb. et Zucc.; *Hydrangea paniculata* Sieb.; *Philadelphus satsumi* Sieb. ex Lindl. et Paxt.
Platanaceae	*Platanus acerifolia* Willd.
Rosaceae	*Chaenomeles lagenaria* Koidz.; *C. sinensis* Koehne; *Kerria japonica* DC.; *Malus pumila* Mill. var. *domestica* Schneider; *Prunus grayana* Maxim.; *P. jamasakura* Sieb. ex Koidz.; *P. mume* Sieb. et Zucc.; *P. persica* Batsch var. *vulgaris* Maxim.; *P. salicina* Lindl.; *Pyrus serotina* Rehd. var. *culta* Rehd.; *P. ussuriensis* Maxim. var. *sinensis* Kikuchi; *Rosa multiflora* Thunb.; *R. sambucina* Koidz.; *Sorbus alnifolia* (Sieb. et Zucc.) Koch; *Spiraea cantoniensis* Lour.
Leguminosae	*Albizzia julibrissin* Durazz.; *Lespedeza bicolor* Turcz. f. *acutifolia* Matsum.; *Millettia japonica* (Sieb. et Zucc.) Gray; *Pueraria lobata* (Willd) Ohwi; *Wisteria brachybotrys* Sieb. et Zucc.; *Sophora tlavescens* Aiton
Rutaceae	*Phellodendron amurense* Rupr.
Euphorbiaceae	*Daphniphyllum macropodum* Miq.; *D. teijsmannii* Zoll.; *Euphorbia adenochlora* Morr. et Decne.; *Sapium japonicum* (Sieb. et Zucc.) Pax et Hoffm.; *S. sebiferum* Roxb.; *Securinega suffruticosa* (Pall.) Rehd. var. *japonica* (Muell. Arg.) Hurusawa
Buxaceae	*Buxus microphylla* var. *japonica* (Muell. Arg.) Rehd. et Wils.
Anacardiaceae	*Rhus javanica* L.
Aquifoliaceae	*Ilex integra* Thunb.; *I. latifolia* Thunb.; *I. pedunculosa* Miq.; *I. rotunda* Thunb.
Celastraceae	*Celastrus orbiculatus* Thunb.; *Euonymus japonicus* Thunb.; *E. sieboldianus* Bl.
Staphyleaceae	*Euscaphis japonica* (Thunb.) Kanitz
Aceraceae	*Acer japonicum* Thunb.; *A. mono* Maxim.; *A. mono* Maxim. var. *mayrii* (Schwer.) Koidz. ex Nemoto; *A. palmatum* Thunb. var. *amoenum* (Carr.) Ohwi; *A. rufinerve* Sieb et Zucc.
Hippocastanaceae	*Aesculus turbinata* Bl.
Vitaceae	*Vitis labrusca* L.; *V. vinifera* L.; *Ampelopsis brevipedunculata* (Maxim.) Trauty.
Tiliaceae	*Tilia japonica* (Miq.) Simonkai
Sterculiaceae	*Firmiana simplex* (L.) Wight et Endl.
Actinidiaceae	*Actinidia chinensis* Planchon
Theaceae	*Camellia japonica* L.; *Stewartia pseudocamellia* Maxim.
Flacourtiaceae	*Idesia polycarpa* Maxim.
Stachyuraceae	*Stachyurus praecox* Sieb. et Zucc.
Myrtaceae	*Eucalyptus globulus* Labill.

TABLE 1 (continued)
Host Range of *Zygophiala jamaicensis* in Japan

Family	Species
Araliaceae	*Acanthopanax spinosus* (Linn. fil.) Miq.; *Dendropanax trifidus* (Thunb.) Makino; *Fatsia japonica* (Thunb.) Decne. et Planch.; *Kalopanax pictus* (Thunb.) Nakai
Cornaceae	*Aucuba japonica* Thunb.; *Cornus brachypoda* Mey.; *C. controversa* Hemsl.; *C. kousa* Buerger ex Hance
Ericaceae	*Vaccinium oldhamii* Miq.; *Vaccinium* spp.
Ebenaceae	*Diospyros kaki* Thunb. var. *domestica* Makino
Oleaceae	*Forsythia suspensa* (Thunb.) Vahl; *F. viridissima* Lindl.; *Ligustrum lucidum* Ait.
Caprifoliaceae	*Lonicera japonica* Thunb.; *L. gracilipes* var. *glabra* Miq.; *Sambucus sieboldiana* Bl. ex Graebn.; *Weigela hortensis* (Sieb. et Zucc.) K. Koch
Gramineae	*Miscanthus sinensis* Anderss.; *Phyllostachys bambusoides* Sieb. et Zucc.; *P. heterocycla* (Carr.) Mitf. var. *pubescens* (Mazel) Ohwi; *Pleioblastus simonii* (Carr.) Nakai
Liliaceae	*Smilax china* L.

conidium is discharged, the disk between conidiophore and conidium splits so that half of it remains on the conidium and half forms the scar of the conidiogenous cell.[4] The conidia are two celled with a slightly granular surface, elliptical to obovate, and constricted at the septum (Figures 13 and 14).

Mycelia produce superficial, thickened, black, circular knots or flyspecks (the incipient sclerotium-like structures) in random clusters[4] (Figure 15). These structures are 100 to 500 μm in diameter and 25 to 60 μm in height, with an irregular margin. At maturity, the entire structure is often cracked (Figure 16), and the central, raised part drops off spontaneously to expose the underlying host surface, leaving only the marginal structure on the host surface.

B. CONIDIOGENESIS

Figures 17 to 24 illustrate conidiogensis of the fungus on a wheat extract agar medium. The initial step of conidiogenesis is the swelling of a hyphal cell (Figure 17). A protrusion subsequently emerges perpendicularly from the swollen cell (Figure 18) and keeps extending. The protrusion is separated from the basal, hyphal cell by a septum to become an original conidiophore cell (Figure 19). Thereafter, the entire cell is tortured and lightly colored. Another septum forms near the apex of the conidiophorous cell to give a terminal, mother cell of the divergent, conidiogenous cells (Figure 19). The conidiogenous cells arise from the mother cell one by one (Figures 20 and 21). A small protrusion is borne on the conidiogenous cell (Figure 22) and initially develops into a single cell (Figure 23). This cell is split later by a septum and constricted there to give a mature form of the two-celled conidium (Figure 23). According to Meredith,[10] exposure of the moist turgid conidial apparatus to dry air caused vertical contraction of the dark sector of the conidiophore, followed by sudden elongation when a gas bubble formed in the sector, forcibly discharging the conidia. This could be repeated several times by altering periods of high and low humidity.[10] Only dark sectors of conidiophores with degenerate conidiogenous cells of several dark scars remain on the medium (Figure 24). However, on host surfaces, conidiogenous cells seem to degenerate soon after conidia are discharged: thus, only the dark sectors are usually visible (Figure 15). It is likely that conidiogenesis occurs in the humid atmosphere during the night, conidia are discharged when the humidity is lowered, and the air temperature is elevated in the morning; at that point, conidiogenous cells shrink.[7]

C. FORMATION OF SCLEROTIUM-LIKE STRUCTURES

The serial step of formation of sclerotium-like structures is readily observed on nutritional agar media. In aged mycelia, numerous septa form in some hyphae at short intervals, resulting

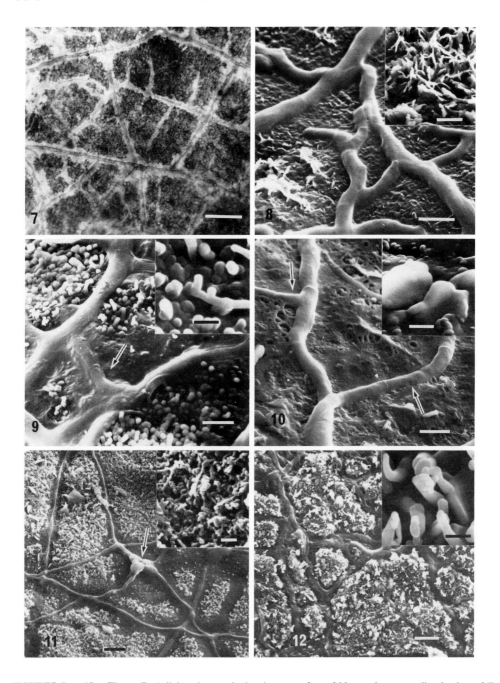

FIGURES 7 to 12. **Figure 7.** A light micrograph showing areas free of bloom along extending hyphae of *Z. jamaicensis* on a grape berry. Bar = 40 μm. **Figure 8.** An area free of bloom along hyphae growing on a Japanese persimmon fruit. The insert shows bloom crystals on a healthy fruit. Bar = 5 μm (insert, 2 μm). **Figure 9.** Hyphae covered with a thin film, probably composed of degenerating product(s) of bloom of Chinese quince. Arrow indicates the most prominent portion of this phenomenon. Enlarged view of bloom crystals on a healthy fruit of Chinese quince are shown in the insert. Bar = 5 μm (insert, 2 μm). **Figure 10.** Hyphae covered with a thin film, similar (arrows) to that of Figure 9, on an apple fruit. Stone-like bloom crystals on a healthy fruit are shown in the insert. Bar = 5 μm (insert, 2 μm). **Figure 11.** Areas free of bloom crystals along hyphae arising from paired, germinated conidia (arrow) on a plum fruit. Bloom crystals on a healthy fruit are shown in the insert. Bar = 10 μm (insert, 2 μm). **Figure 12.** Hyphae growing on surface of a pawpaw fruit. Waxy bloom crystals have disappeared along the hyphae. Bloom crystals with a unique shape are shown in the insert. Bar = 10 μm (insert, 2 μm).

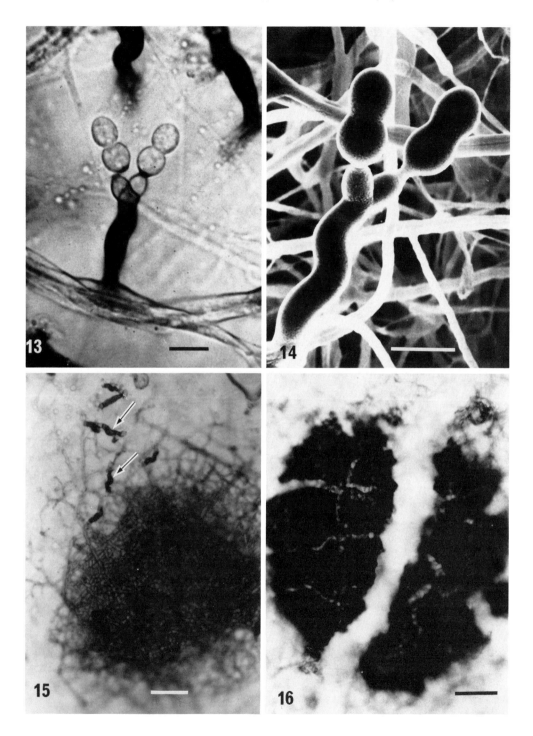

FIGURES 13 to 16. **Figure 13.** A conidiophore consisting of three distinguishable parts: a tortuous, dark brown sector at the base; an angular, subhyaline, terminal cell which bears on its apical sides two divergent, hyaline, ovate to ampulliform conidiogenous cells bearing a two-celled conidium. Bar = 10 μm. **Figure 14.** An SEM of a conidiophore bearing two conidia. Bar = 10 μm. **Figure 15.** An incipient sclerotium-like structure produced on the surface of a grape berry. Arrows indicate conidiophores formed at the marginal area of the sclerotium-like structure. Bar = 50 μm. **Figure 16.** A mature sclerotium-like structure which has cracked at its center. The central, raised part often drops off spontaneously, leaving only the marginal structure on the host surface. Bar = 50 μm.

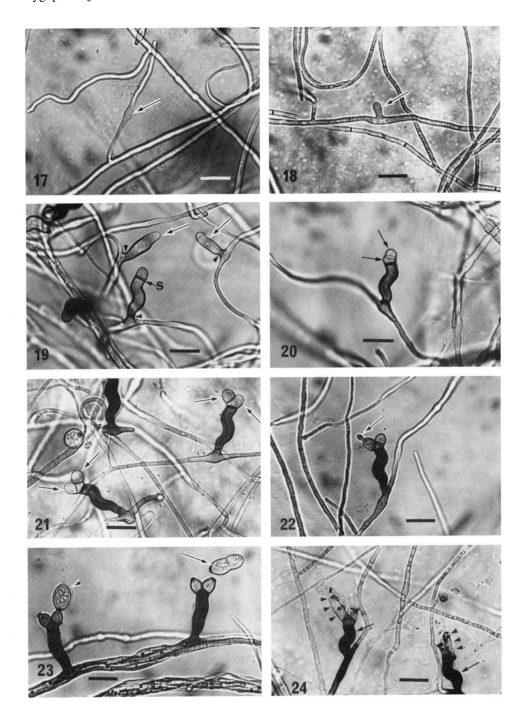

FIGURES 17 to 24. Serial micrographs showing conidiogenesis on an artificial medium. The initial step of conidiogenesis is the swelling of a hyphal cell (**Figure 17**, arrow). A protrusion emerges perpendicularly from the swollen cell (**Figure 18**, arrow) and continues extending (**Figure 19**, arrows). The protrusions are separated from the basal, hyphal cells by septa (Figure 19, arrowheads). Another septum (Figure 19, s) forms near the apex of the conidiophorous cell to give a terminal mother cell of the conidiogenous cells. Two conidiogenous cells arise from the mother cell one by one (**Figures 20 and 21**, arrows). A small protrusion is born on the conidiogenous cell (**Figure 22**, arrow), and initially develops into a single cell (**Figure 23**, arrowhead). This cell is split later by a septum and constricted there to give a mature form of the two-celled conidium (Figure 23, arrow). After conidia have been discharged, only dark sectors of conidiophores with degenerate conidiogenous cells (arrowheads) remain on the medium (**Figure 24**). Bar = 20 μm.

in a large number of small cells. These cells swell individually and diverge to form a cluster of short, relatively light colored cells (Figure 25). The number of such cells increases to build up an initial form of the sclerotium-like structure (Figure 26). As the central part of the cell cluster thickens, the color of the entire cluster becomes darker in color (Figure 27). Finally, the brownish-black sclerotium-like structures form on the medium surface (Figures 28, 29A, and 30). The same structures are also produced in the medium. When the structures are produced in the medium, its surface is raised as indicated by "B" in Figure 29.

D. GERMINATION OF CONIDIA

When conidia are suspended in distilled water, they produce one to three germ tubes.[11] Nearly 90% of conidia produce a single germ tube from each of their cells (Table 2). Although conidia are two-celled under natural conditions, sometimes single-cell conidia are present in the suspension. These conidia probably have been forced to be released from conidiogenous cells prior to septum formation, when the mycelial mat was rubbed with a brush to prepare the suspension. These immature conidia also have a germination ability in distilled water. Nearly 90% of conidia produce two germ tubes on surfaces of grape berries (Table 2).

The optimum temperature for germination of conidia isolated from grapes ranges from 10° to 28°C. Conidia never germinate below 3°C or above 32°C. Germ tubes elongate at 15° to 28°C, most rapidly at 25°C.[7] This temperature range is almost consistent with that for germination of conidia of an isolate from apple.[12]

E. TWO ISOLATES OF THE FUNGUS WITH DIFFERENT PHYSIOLOGICAL NATURES

To date, the presence of physiological races of *Z. jamaicensis* has not been known; however, two isolates with different physiological characteristics, temporarily designated isolates A and B, have been distinguished in Japan.[11] The mycelial colony of both isolates grows very slowly on artificial media. The colony diameter of isolate A reaches a maximum of 4 to 6 cm 20 d postinoculation on a potato sucrose agar (PSA) medium and about 7 cm at 50 d (Figures 31 and 32). The 20-d-old colony is greenish-gray at the center and white to gray at the margin (Figure 31). A large number of conidiophores with conidia are produced throughout the colony. When the incubation is continued for 50 d, the central part of the colony is occupied by a large, dark-colored, sclerotium-like structure. Small specks, which are also the sclerotium-like structures, are scattered at the marginal region (Figure 32). When isolate A is grown on a wheat extract agar medium (WEA), a thin, white colony with small, scattered specks develops (Figure 33). The growth of isolate B is much slower than that of isolate A on both media and reaches, in the largest colonies, 1 to 1.5 cm in diameter 50 d postinoculation (Figure 34). On WEA, isolate B forms only a few sclerotium-like structures that are smaller than those formed by isolate A, but never forms the structures on PSA (Table 3).

Mycelia of isolate A grow actively on PSA at 15° to 28°C and those of isolate B between 10° and 28°C. The latter temperature range was almost consistent with that of an apple isolate in the U.S.[12] Isolate A grows on PSA at a pH range between 4 and 10 and isolate B at a wider range, between pH 4 and 11.[7]

Isolate A has been obtained from grape, Japanese persimmon, and apple, whereas isolate B has been isolated only from the former two hosts. The frequency of isolation of isolate B is higher than that of isolate A in the field, however, inoculation tests revealed that the growth and pathogenicity of both isolates were almost indistinguishable on all host plants examined.[7]

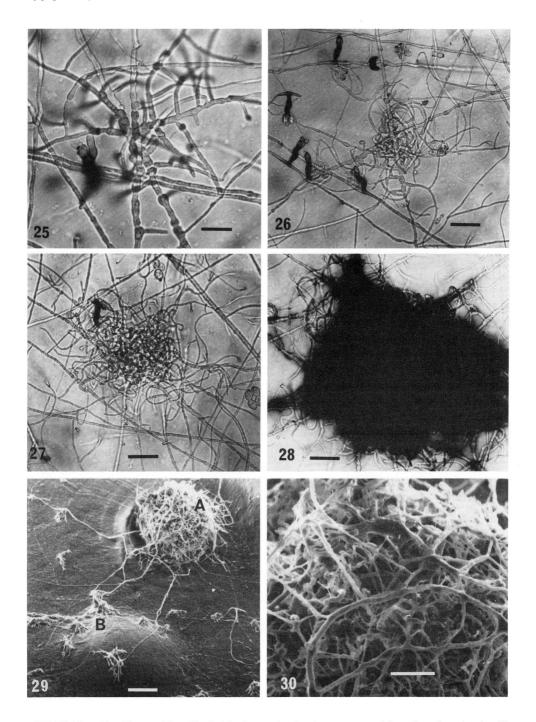

FIGURES 25 to 30. Figures 25 to 28. Serial micrographs showing a process of formation of a sclerotium-like structure on an artificial medium. In aged mycelia, numerous septa form in some hyphae at short intervals, resulting in a large number of small cells. They swell individually and diverge to form a cluster of short, slightly dark cells (**Figure 25**). The number of such cells increase to build up an initial form of the sclerotium-like structure (**Figure 26**). As the central part of the cell cluster thickens, the entire cluster changes to a dark color (**Figure 27**). Finally, a brownish-black sclerotium-like structure forms on the medium surface (**Figure 28**). Bar = 40 μm, except 20 μm in Figure 25. **Figure 29.** An SEM of a sclerotium-like structure formed on the medium surface (A). When the structure is produced in the medium, its surface is raised (B). Bar = 40 μm. **Figure 30.** Enlarged view of hyphal cluster of the sclerotium-like structure shown by A in Figure 29. Bar = 10 μm.

TABLE 2
Germination Types of Conidia of *Z. jamaicensis* in Distilled Water and on Surfaces of Grape Berry

	No. of conidia observed	Germination type							
In distilled water	339	87.0[a]	3.8	0.3	0.6	0.3	0.9	5.6	1.5
On surfaces of berry	73	87.7	0	1.4	0	0	11.0	0	0

[a] $\dfrac{\text{No. of conidia observed}}{\text{Total no. of germinated conidia observed}} \times 100\ (\%).$

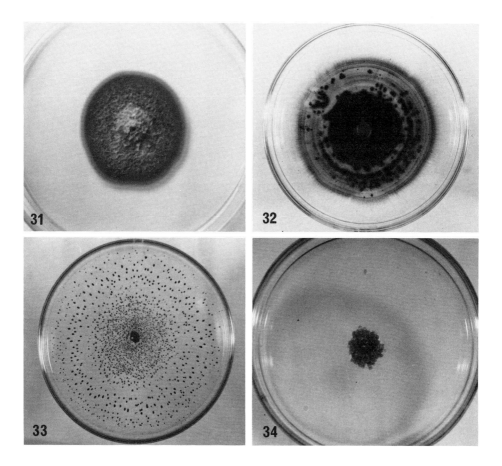

FIGURES 31 to 34. **Figure 31.** A 20-d-old mycelial colony of isolate A with the greenish-gray center and white-gray margin formed on a PSA medium. **Figure 32.** A 50-d-old colony of isolate A with a large sclerotium-like structure at the center and numerous small structures at the margin on PSA. **Figure 33.** A 50-d-old, thin, white colony of isolate A with a large number of scattered, sclerotium-like structures on WEA medium. **Figure 34.** A 50-d-old, small colony of isolate B on PSA.

TABLE 3
Size of Microsclerotium-Like Bodies of Three
Grapevine Isolates of *Z. jamaicencis* **on Media**

Isolate no.	Colony type[a]	Medium[b]	Size, long axis × short axis (μm, av)
1	A	PSA	150–1780 × 150–1180 (830 × 610)
		WEA	230–980 × 190–690 (570 × 430)
2	A	PSA	260–2100 × 220–1760 (820 × 500)
		WEA	220–840 × 200–660 (490 × 390)
22	B	PSA	—
		WEA	120–320 × 90–210 (280 × 140)

[a] Colony type on a PSA medium.
[b] PSA, potato sucrose agar; WEA, wheat extract agar.

IV. ELECTRON MICROSCOPY OF FUNGAL GROWTH ON HOST SURFACES

A. UNIQUE STRUCTURES OF BLOOM CRYSTALS ON SURFACES OF SEVERAL FRUITS

De Bary[13] emphasized that epicuticular wax patterns on plant surfaces varied among species and proposed a classification for types of wax coating as follows: (1) single layers of granules, (2) small rodlets perpendicular to the cuticle, (3) several layers of very small needles of granules, and (4) membrane-like layers of incrustations. In our observations of surfaces of several fruit species, each fruit species had a unique pattern of bloom crystals. Bloom patterns of apple corresponded to type 1 of de Bary's classification, showing scattered, stone-like crystals of various sizes (Figure 10); those of Chinese quince (Figure 9) and pawpaw (Figure 12) were similar to type 2 with a unique rodlet shape; those of pear, grape (Figure 39), Japanese persimmon, and plum (Figures 8 and 11) corresponded to type 3, each having varied sizes and a unique crystalline shape.

B. EVANESCENCE OF BLOOM CAUSED BY HYPHAL ELONGATION

On all infected plants observed, small patches free of waxy bloom are readily visible with a hand lens. Ungerminated, paired conidia often are present at the center of some patches, suggesting that conidia release the bloom-degenerating enzyme prior to germination (Figure 35). Although single conidia are produced independently on the two conidiogenous cells of a conidiophore, they are quite frequently disseminated together and land on host surfaces as paired conidia. Thus, it frequently appears as though the conidia produced four germ tubes on host surfaces, as illustrated in Figures 36 and 37.[11] Each of the paired conidia usually produces two germ tubes elongating toward different directions so as not to cross each other (Figure 37). Bloom crystals around conidia and along hyphae arising from them have degenerated on all fruit surfaces. Such an area free of original bloom crystals is wide near the conidia and tapers toward the hyphal apex (Figures 37 and 38). A higher magnification of electron micrographs indicates that the apical portion of hyphae elongates over bloom crystals that keep the original shapes, but the bloom crystals along the hyphal portion

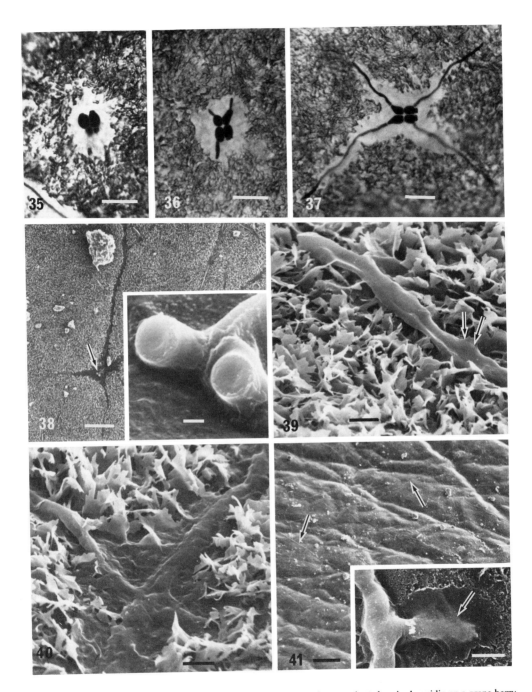

FIGURES 35 to 41. **Figure 35.** An area free of waxy bloom around ungerminated, paired conidia on a grape berry. Bar = 20 μm. **Figure 36.** An area free of waxy bloom around germinated cells of paired conidia on a grape berry. Bar = 20 μm. **Figure 37.** Areas free of waxy bloom along four hyphae arising from paired conidia. Bar = 20 μm. **Figure 38.** An SEM showing areas free of waxy bloom along hyphae arising from paired conidia (arrow) on a grape berry. Enlarged view of scars of the conidia is shown in the insert. Bar = 40 μm (insert, 1 μm). **Figure 39.** A hypha growing on the bloom crystals of a grape berry. Note that crystalline bloom appears unaltered around the hyphal apex, but that along the hyphal part, 10 to 15 μm behind it, it appears to have altered (arrows). Bar = 2 μm. **Figure 40.** A hypha covered with a thin film probably consisting of degenerating product(s) of bloom on a grape berry. Bar = 2 μm. **Figure 41.** A grape berry surface from which bloom has evanesced almost completely. Arrows indicate hyphae growing on the exposed cuticle surface. The insert shows a branching hypha (arrow) covered with a thin film similar to that seen in Figure 40. Bar = 100 μm (insert, 2 μm).

several microns behind the apex have degenerated slightly (Figure 39). Hyphae elongating further from the conidia on the fruit surface are covered with a thin film probably composed of degenerating product(s) of bloom crystals (Figures 40 and 41), although some of the hyphae appear to grow on the degenerating products. Such a pattern of hyphal growth and evanescence of bloom crystals are observed commonly on all plants examined thus far. Assuming that the film consists of the degenerating product(s) of bloom crystals, it might be in a gel-like state after degeneration, and thus the hyphae could grow underneath it. Otherwise, it is hard to account for such a hyphal growth. When networks of growing hyphae expand on fruit surfaces, almost all bloom crystals disappear from the surfaces (Figure 41), giving a shiny appearance to the entire fruit or berry. These observations are supported by Nasu's[14] preliminary data that the mycelia release high concentrations of esterase on the culture medium.

C. FORMATION OF SCLEROTIUM-LIKE STRUCTURES ON HOST SURFACES

At the cleared surface area of fruits and berries free of bloom crystals, some hyphae repeat cell division and branching to form a dense network of short hyphae, although the factor(s) that trigger this phenomenon is unknown (Figure 42). Cell division seems to occur more frequently at the central region of the hyphal network, resulting in a thick mycelial mat where hyphae are tighly intermingled (Figure 43). At the mature stage, the sclerotium-like structure has a plateau-like appearance surrounded by a long plain (Figure 44). The plain consists of a hyphal mass similar to that of the plateau, but the entire thickness is much less than that of the plateau. Hyphae growing in the area surrounding the plain seem to be covered with a thin film when viewed at a high magnification, their contours being visible through the film. Similarly, the entire surface of the plateau appears to be covered with a continuous film, although several hyphae grow on this film. Figure 44 shows a mature stage of the structure formed on a grape berry where the entire structure is covered with a thin film, but individual hyphae are still distinguishable from each other. Figure 45 shows the same structure formed on a pawpaw fruit in which the entire structure is covered with a relatively thick film, making features of individual hyphae indistinguishable.

The cut surface of the plateau of an immature sclerotium-like structure (Figure 47) that has been exposed by a microneedle shows that a great number of hyphae are embedded in a somewhat amorphous substance (Figure 46). Furthermore, micromanipulation revealed that a layer of the bloom-degenerating products is clearly distinguishable at the surface of the plateau of a mature structure (Figures 44 and 47). This filler substance, probably composed of the degenerating product(s) of bloom crystals, appears to be continuous with the thin film covering the plateau surface.

Baker et al.[4] reported that the mycelial structure similar to that shown herein formed ascospores in culture and on apple fruit. Thus, they called this structure a pseudothecium. In contrast, Atkinson[15] regarded the same structrure as a sclerotium. The authors are not aware of a description of ascospore formation on such a structure in literature published in Japan. Moreover, we have never seen ascospore formation on any of the various plant species observed in the field in this country (Japan). Therefore, until ascospore formation is confirmed, the authors tentatively term this a sclerotium-like structure, although we believe that the description of Baker's group may be correct. As mentioned above, tightly intermingled hyphae are embedded in an amorphous material. In this sense, the structure shown herein is unique and differs from the four types of sclerotium which Willetts[16] described, or infection cushion, which consists solely of hyphae and their associated structures.

D. ULTRASTRUCTURAL EVIDENCE OF ECTOPARASITISM

On the infected fruits or berries, the hyphal network covers the entire portion of the affected surfaces. Some hyphae grow over bloom crystals and others are covered with the

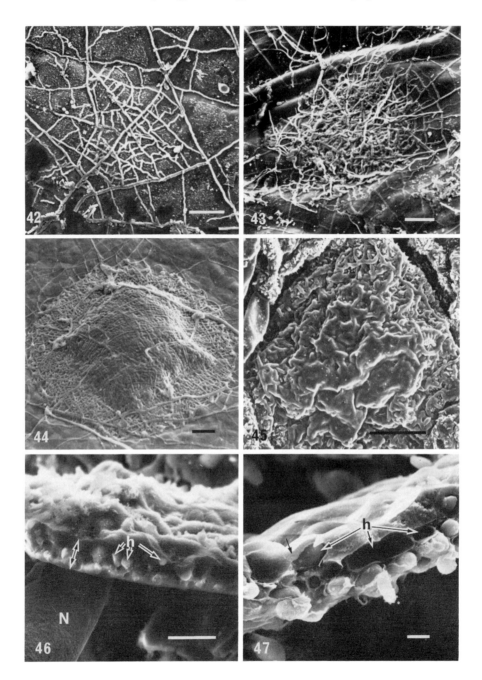

FIGURES 42 to 47. **Figure 42.** A hyphal network showing an initial stage of formation of a sclerotium-like structure on a grape berry. Bar = 40 μm. **Figure 43.** A great number of hyphae have gathered, resulting in a thick mycelial mat at the central region. Bar = 40 μm. **Figure 44.** A mature sclerotium-like structure with a raised center and a marginal plain formed on the exposed cuticle of a grape berry. Bar = 40 μm. **Figure 45.** A mature sclerotium-like structure produced on a pawpaw fruit, the entire surface of which has been covered with a continuous film, thereby making individual hyphae invisible. Bar = 40 μm. **Figure 46.** A cut surface (its thickness indicated by the arrows) of the plateau of an immature sclerotium-like structure on a grape berry, exposed by a microneedle (N). Note the great number of hyphae (h) embedded in the amorphous substance which probably is the degeneration product(s) of bloom crystals. The outer surface is the upper. Bar = 10 μm. **Figure 47.** A cut surface of the plateau of a mature sclerotium-like structure formed on a grape berry in which numerous hyphae (h) are embedded. Arrow indicates the continuous, thin covering. The outer surface is the upper. Bar = 2 μm.

degenerating product(s) of bloom crystals. Micromanipulation of infection sites at the scanning electron microscope (SEM) level revealed that none of the hyphae growing over bloom crystals produced penetration hyphae, nor any of their associated structures, through the cuticle of epicarps of fruits and berries (Figure 48). Similarly, the side view of hyphae covered with a thin film that have been exposed by micromanipulation shows that the film is distinguishable from the cuticle attached tightly to the cell wall of the epicarps (Figure 49). Micromanipulation also indicates no signs of penetration on the cuticle surfaces when the plain and plateau of the sclerotium-like structures have been removed (Figures 50 and 51). All these observations are consistent with light microscopic observations for cross-sections of infected fruits or berries that show no infection hyphae in the host's internal tissues (Figures 52 and 53). These results are supported by the fact that mycelia can develop and form conidia and sclerotium-like structures on isolated waxy bloom[4] and commercially available polyhydrocarbon on glass slides.[7] Thus, *Z. jamaicensis* is an ectoparasite utilizing bloom alone for its survival.

V. CONCLUDING REMARKS

The structures of *Z. jamaicensis* are produced and behave very similarly on all plants examined.[4,7,11,17-19] The plateau of a mature sclerotium-like structure is readily detachable from grape berries in the fields, suggesting that the strucrture may be an overwintering organ.[11] This is supported by the ecological experiment of Nasu,[7] showing that removal of wild trees having this structure on their twigs from the surrounding area of grape orchards effectively reduced the occurrence of flyspecks in the next season.

Although we can find no indication that the cuticle underneath the bloom is penetrated by the fungus, its ability to degenerate bloom structures may have serious physiological consequences for berries and fruits of various infected host plants. Since host plants are never killed by this fungus, it never causes a great loss of the fruit or berry yield. Nevertheless, it reduces the value of marketable plants and brings serious economic consequences to growers of fruits and ornamental plants.

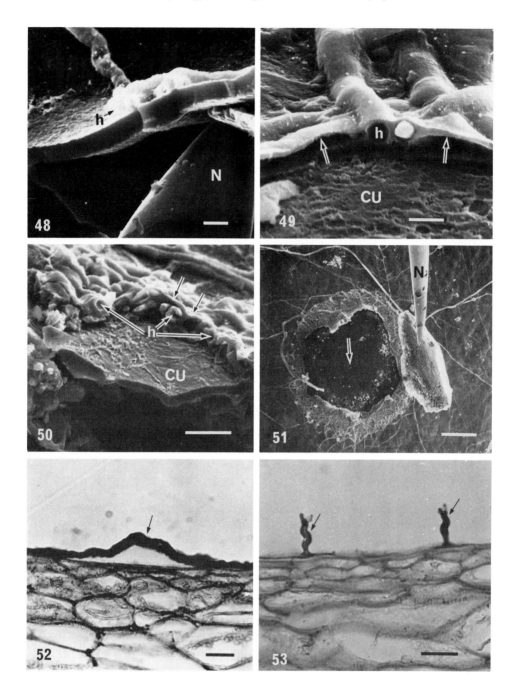

FIGURES 48 to 53. **Figure 48.** A cut surface of the epicarp of a grape berry, exposed by a microneedle (N), which shows no signs of penetration from the superficial hyphae (h). Bar = 2 μm. **Figure 49.** A cut surface of the thin film (arrows) that probably consists of the degeneration product(s) of bloom crystals in which a hypha (h) is embedded. The film is distinguishable from the cuticle (CU) covering the epicarp. Bar = 2 μm. **Figure 50.** A cuticle surface (CU) which was covered with the plain region of a sclerotium-like structure, exposed by a microneedle. Hyphae (h) on the cuticle surface are buried in the degeneration product(s) of bloom crystals (arrows). Bar = 10 μm. **Figure 51.** The central region of a sclerotium-like structure having been opened with a microneedle (N). The arrow indicates a cuticle surface exposed by micromanipulation. Bar = 100 μm. **Figure 52.** A cross-section of a central, raised, empty region of a sclerotium-like structure (arrow) on the epicarp of a grape berry. No hyphae are inside the epicarp. Bar = 40 μm. **Figure 53.** A cross-section showing conidiophores (arrows) formed on the epicarp of a grape berry. No indication has been given of hyphae growing in the epicarp. Bar = 40 μm.

REFERENCES

1. **Martyn, E. B.,** A note on banana leaf speckle in Jamaica and some associated fungi, *Mycol. Pap.,* 13, 1, 1945.
2. **Baker, K. F. and Davis, L. H.,** Greasy blotch of carnation, *Phytopathology,* 43(Abstr.), 585, 1953.
3. **Durbin, R. D. and Snyder, W. C.,** Ecology and hosts of flyspeck of apple in California, *Phytopathology,* 43(Abstr.), 586, 1953.
4. **Baker, K. F., Davis, L. H., Durbin, R. D., and Snyder, W. C.,** Greasy blotch of carnation and flyspeck of apple: diseases caused by *Zygophiala jamaicensis, Phytopathology,* 67, 580, 1977.
5. **Nasu, H., Fujii, S., and Yokoyama, T.,** *Zygophiala jamaicensis* Mason, a causal fungus of flyspeck of grape, Japanese persimmon and apple, *Ann. Phytopathol. Soc. Jpn.,* 51, 536, 1985.
6. **Nasu, H. and Kunoh, H.,** Distribution of *Zygophiala jamaicensis* in Okayama Prefecture, Japan, *Trans. Mycol. Soc. Jpn.,* 28, 209, 1987.
7. **Nasu, H.,** Studies on occurrence and control of flyspeck of grapevine, *Spec. Bull. Okayama Prefect. Agric. Exp. Stn.,* 80, 1, 1990.
8. **Messiaen, C. M.,** personal communication, 1986.
9. **Katumoto, K.,** Notes on some plant-inhabiting Ascomycotina from western Japan, *Trans. Mycol. Soc. Jpn.,* 27, 1, 1986.
10. **Meredith, D. S.,** Violent spore release in some fungi imperfecti, *Ann. Bot.,* 27, 39, 1963.
11. **Nasu, H., Hatamoto, M., and Kunoh, H.,** Behavior of causal fungus and process of lesion formation of flyspeck-affected berries of grape, *Ann. Phytopathol. Soc. Jpn.,* 52, 445, 1985.
12. **Ocamb-Basu, C. M. and Sutton, T. B.,** Effects of temperature and relative humidity on germination, growth, and sporulation of *Zygophiala jamaicensis, Phytopathology,* 78, 100, 1988.
13. **De Bary, A.,** Ueber die Wachsuberzuge der Epidermis, *Bot. Z.,* 29, 145, 1971.
14. **Nasu, H.,** unpublished data, 1988.
15. **Atkinson, J. D.,** Diseases of Tree Fruits in New Zealand, Government Printer, Wellington, New Zealand, 1971, 406.
16. **Willetts, H. J.,** The morphogenesis and possible evolutionary origins of fungal sclerotia, *Biol. Rev.,* 47, 515, 1972.
17. **Nasu, H. and Kunoh, H.,** Scanning electron microscopy of flyspeck of grape, *Ann. Phytopathol. Soc. Jpn.,* 52, 466, 1985.
18. **Nasu, H. and Kunoh, H.,** A unique sclerotium-like structure produced by *Zygophiala jamaicensis* Mason on grape berries, *Trans. Mycol. Soc. Jpn.,* 27, 225, 1986.
19. **Nasu, H. and Kunoh, H.,** Scanning electron microscopy of flyspeck of apple, pear, Japanese persimmon, plum, Chinese quince, and pawpaw, *Plant Dis.,* 71, 361, 1987.

Chapter 7

PATHOLOGICAL ANATOMY OF FIRE BLIGHT CAUSED BY *ERWINIA AMYLOVORA*

Henk J. Schouten

TABLE OF CONTENTS

I. INTRODUCTION

Well over a century ago, Thomas J. Burrill initiated phytobacteriology when he asserted that fire blight in pear was not caused by a fungus, but by a bacterium. In 1878 he wrote that in "the mucilaginous fluid from the browned tissue under our microscope, the field is seen to be alive with moving atoms known in a general way as bacteria. So far as I know, the idea is an entirely new one — that bacteria cause disease in plants — though abundantly proved in the case of animals."[1] About 2000 papers on fire blight, caused by *Erwinia amylovora,* have appeared since the onset of phytobacteriology. These publications reveal that many research workers have searched for a bacterial toxin that would play a key role in fire blight pathogenesis, but in spite of many trials and hypotheses no metabolite from *E. amylovora* has achieved general acceptance as a key role toxin. Cell wall-degrading enzymes (glycanases and glycosidases) have also not been detected.[2-4] Extracellular polysaccharide (EPS) of *E. amylovora* has been regarded by many authors as a virulence factor,[5-12] but it seems to lack direct toxic properties.[12,13] EPS is undoubtedly required for pathogenicity, but many controversial views exist as to its mode of action. Histological studies have appeared to be essential for clarifying the function of EPS and the pathogenicity mechanism of *E. amylovora.* In this chapter, literature on the histology of fire blight is reviewed, beginning after entry of the bacterium into the host via blossom, shoot, or leaf, natural or artificial. The chapter concludes with a view on the degree of intimacy of the relationship between *E. amylovora* and its host.

II. THE BACTERIUM'S OFFENSE

A. THE CAMOUFLAGE PHASE

When *E. amylovora* cells are needle inoculated into a host's shoot, they apparently lie just as they were distributed throughout the puncture for the first 30 to 45 min.[14] Approximately 1 hour after inoculation, the bacteria appear to be embedded in a jelly-like substance that is somewhat irregular in shape.[14,15] This jelly-like mass, secreted by the bacteria, consists mainly of an acidic EPS with galactose as the main component.[16] The structure of EPS, as proposed by Smith et al.,[17] is shown in Figure 1. The virulence of *E. amylovora* seems to be strongly dependent on the quality of this EPS.[10] EPS itself looks inert in intercellular spaces, and is too large to pass through the molecular sieve of the host cell wall matrix.[13] EPS-producing strains show capsules around the bacteria, whereas EPS-deficient bacteria have no capsules and are not virulent.[9,18] The EPS matrix surrounding the bacteria may well serve to confer some protection by camouflaging the presence of the bacteria, thus delaying plant defense responses.[18] Hignett and Roberts[18] showed that an EPS-deficient strain of the pathogen induced immediate defense reactions by the host, resulting in an incompatible interaction. In contrast, EPS from EPS-producing virulent strains interfered with the induction of plant defense responses.[18] The first observable action by the bacteria, therefore, is secretion of the virulence factor EPS, plausibly to preclude recognition by the host.

B. THE FILL PHASE

The bacteria multiply in the jelly-like mass on the margin of the puncture among the lacerated cells, forming a biomass that invades the intercellular spaces of the cortical region, about ten cells proximal to the epidermis.[14] The intercelluar spaces are large in this parenchymal tissue.[19] The bacterial mass fills the intercellular spaces completely, advancing through the shoot in the directions that offer little resistance to their progress.[14,20-24] The biomass migrates most rapidly through the cortical parenchyma in longitudinal and tangential directions, sometimes referred to as the "optimal path" or path of least resistance.[14,24] The

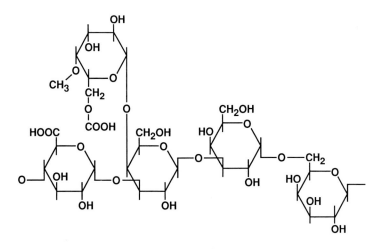

FIGURE 1.   Most likely structure of EPS of *Erwinia amylovora,* according to Smith et al.[17] in their concept, the EPS is composed solely of galactose, glucuronic acid, acetate, and pyruvate in molar ratio 4:1:1:1, together with a small quantity of glucose.

bacteria, embedded in EPS, gorge the intercellular spaces of this ''loose'' inner cortex,[19] replacing the intercellular air, so that the tissue looks as if it is ''water soaked''. From the area of initial infection, the biomass ramifies by formation of new tips and develops an anastomosing system which rapidly fills the connected intercellular spaces.[21,22]

C. THE PRESS PHASE

About 2 d after infection, the bacterial biomass that is a few centimeters behind the invading front in the optimal path begins to exert pressure on the surrounding plant cells.[14,19-24] This pressure may be caused by two mechanisms.[25] One mechanism is via production of new bacteria and new EPS. As long as *E. amylovora* multiplies and produces new biomass in the limited intercellular space, and as long as it cannot escape, the ''multiplication pressure'' increases. Water absorption by the bacteria and subsequent volume increase become more and more difficult due to limited physical space within the host tissue. The bacterial biomass will then suffer lack of water, so that further production of bacterial dry matter will slow down. Consequently, the multiplication pressure has a maximum value. Hypothetically, the maximum value of the pressure caused by production of new bacterial dry matter equals the actual water potential of the tissue minus the water potential at which bacterial dry matter production attains its limit because of drought in the absence of pressure. The multiplication pressure may, according to the theory, attain values as high as 4 MPa (= 40 bars) under humid conditions.[25]

When the multiplication pressure increases, the EPS shrinks by releasing water, thus allowing further dry matter production. Although the dry weight of the EPS then increases, its volume decreases.[26] When the multiplication pressure reaches high values, the extracellular slime is a dense substance around the bacterial cells, with a low water content and with a strong capacity to swell (up to about 400%[27]) by absorbing water when the pressure induces the host tissue to tear.

The other mechanism that may cause pressure is swelling of the bacterial mass without increase of bacterial dry weight. When bacterial mass is accumulated in the intercellular spaces of the host plant and the water potential rises (for instance due to a decrease of the host's transpiration rate at evening), the bacterial mass will tend to absorb water and swell strongly. This swelling without increase of bacterial dry weight may cause a ''swelling pressure'' on the surrounding plant cells. The maximum value of the swelling pressure equals the increase in water potential, which reaches values up to about 2 MPa in fruit tree tissues.[25] The strong swelling capacity of the bacterial mass can be ascribed, in a minor way, to the cells of *E. amylovora* and in a major way to the very hygroscopic EPS.[27,28]

D. THE RUPTURE PHASE

Because of the bacterial pressure, plant cells may be wedged apart, giving rise to schizogenous holes filled with bacterial biomass (Figure 2).[14,19-25] Because of the pressure, less easy to penetrate tissues adjacent to the inner cortex may be penetrated as well. From the optimal path, the bacterial mass may extend radially to the outside until the surface is reached, and to the inside until the pith is invaded. Medullary rays may play an important role in radial migration; however, radial migration proceeds much slower than longitudinal migration.[14,20] The migration through the less-favorable tissues adjacent to the optimal path always seems to be associated with the presence of large masses of bacteria in normal intercellular spaces or in schizogenous cavities. Such migration is slow and intense, covering short distances, but is not limited to one tissue.[14]

Haber[21] showed that *E. amylovora* migrates through leaves in a similar way, gorging intercellular spaces, forming schizogenous cavities, and forcing its way through tissue that offers the least resistance. When bacterial masses penetrate substomatal air chambers of leaves, their progress soon becomes limited. The stomatal opening serves as means for the release of the pathogen as exudate.[21]

When a bacterial mass in a shoot moves radially outward, it may break through the epidermis and come to the surface as a droplet. When the water potential in the host tissue is below -1 MPa, for example, because of a high transpiration rate, the water content of the extracellular bacterial substance is relatively low, so that the bacterial biomass is not fluid but pasty. Incidentally, if temperature is high enough for fast bacterial multiplication, the pasty mass is extruded from the host, and does not appear as droplets but as strands.[29,30] When the air humidity becomes high ($>85\%$), strands change into droplets[29,31] by means of water uptake by the extracellular bacterial slime. Strands may be smooth or beaded, and their length varies from a fraction of a millimeter to several centimeters and the width from 6 to 300 μm. They can easily blow in the wind, are instantly soluble in water, and may therefore play an important role in dissemination of the pathogen. Strands are composed of 80% extracellular slime and of 20% bacterial cells.[30]

On thin, diseased twigs of hawthorn (*Crataegus* spp.) and *Cotoneaster* spp., the bacterial masses sometimes squeeze themselves not out of the twig, but form blisters by lifting the outer cortex. Such a blister, packed with bacterial biomass, can be recognized with the unaided eye as a bump on the twig. Bacterial masses ooze out of the blister when the upper layer of cortex is ruptured mechanically.[32]

Rigid, mature host tissue may be able to resist the pressure by the bacterial masses, thus slowing down further invasion.[25] Evidently, succulence and softness of host tissue are very closely related to susceptibility to fire blight,[33,34] which may be caused partly by the inability of soft tissue to resist mechanical pressure by the bacterial masses.

III. THE PLANT'S DEFENSE

A. THE PLASMOLYSIS PHASE

No reaction of host cells has been observed prior to the "press phase". Plant cells may be surrounded by bacterial biomass and still appear completely healthy.[14,19-24] The first visible effect on host cells is plasmolysis, followed by cellular collapse.[14,19-24] The plasmolysis arises from loss of electrolytes from host cells as indicated by changes in the conductance of ambient solutions.[35,36] Several research workers investigated whether loss of electrolytes

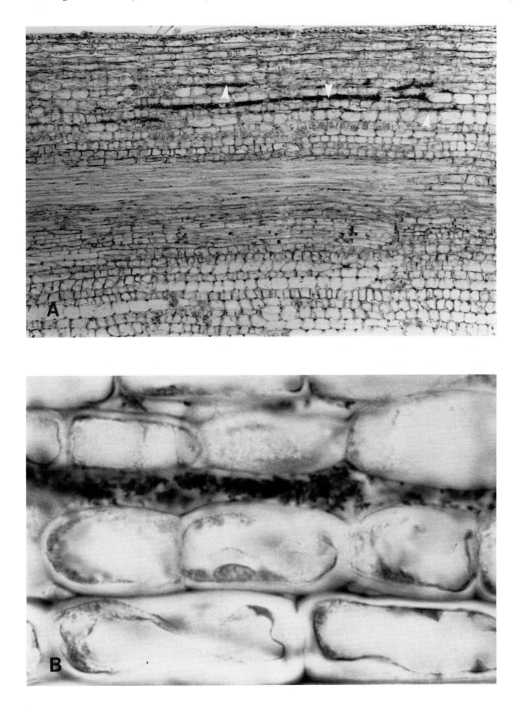

FIGURE 2. (A) Longitudinal section of an apple shoot (cv. M26), 1 cm below the point of inoculation at the shoot tip, 53 h after inoculation with *E. amylovora*. Host cells were separated by the pathogen, giving rise to large, longitudinal, intercellular holes, filled with (arrows) bacterial masses. (Magnification × 160.) (B). As (A). Magnification × 1600. (Photographs were kindly placed at the author's disposal by E. Billing and E. J. Eden-Green, from Horticulture Research International, East Malling, England.)

is caused by a toxin of *E. amylovora;* however, purified EPS from cultures of *E. amylovora* does not cause loss of electrolytes from healthy tissue,[12,13,18] nor do cell-free fluids from cultures of bacteria in artificial media.[13,18] Furthermore, inoculations with sterile filtrates from the interaction between the host and *E. amylovora* do not induce leakage,[18] unless the filtrate has been strongly concentrated.[13] The use of an artificial membrane to separate the bacteria from host tissue showed that contact is required between host and pathogen to induce leakage.[18] These findings indicate that *E. amylovora* does not produce a toxic substance which is responsible for electrolyte leakage and plasmolysis.

Rapidly changing permeability of host cell membranes and subsequent loss of electrolytes often has been regarded as a hypersensitive response to prevent further spread of the pathogen.[37] Klement and Goodman[38] tested the capability of *E. amylovora* to cause the hypersensitive response in leaves of tobacco plants (*Nicotiana tabacum*). They found that neither sterile filtrates from *E. amylovora* broth cultures nor bacterial cell fragments killed bacteria, or that cell-free intercellular fluids from the interaction between living bacteria and tobacco tissue caused the hypersensitive reaction — only living bacterial cells did.[38] Further, it appeared that the hypersensitive response in tobacco coincided with a sharp increase in conductivity of the tissue bathing solution and with deterioration of host membranes.[37] These hypersensitive reactions of the nonhost tobacco are the same as the reactions observed in true hosts of *E. amylovora*. Thus, it seems that plasmolysis and cell collapse are not caused by toxins of *E. amylovora,* but may be merely a general defense reaction of the host cells after contact with the bacteria.

Leaves of resistant pear cultivars show a stronger increase in electrolyte leakage than leaves of susceptible pear cultivars 2 d after infiltration.[36] Further, electrolyte loss usually begins more quickly in incompatible than in compatible combinations.[18,35,36,39] Hignett and Roberts[18] showed that capsules around the bacteria and EPS production interfere with the induction of ion leakage. Capsules around bacteria and EPS might camouflage the bacteria, thus delaying the defense reaction. Furthermore, Hignett and Roberts found a positive correlation between cell leakage and bacterial growth, and they state that host cell leakage is necessary for bacterial growth. They propose that host damage is regulated by *E. amylovora* via changes in the properties of its outer coat in course of pathogenesis.[18] However, Bennett[7] and Billing[9] suggest that cell leakage is induced by another, yet unknown virulence factor of *E. amylovora.*

Mechanical damage, such as derangement of middle lamella and the breaking apart of plasmodesmata by bacterial pressure may also cause leakage and the subsequent defense reaction.

The plasmolysis of host cells may be followed by complete collapse of protoplasts.[12,14,19-24] Host cell walls without the physical support provided by the turgescent protoplast finally yield to the pressure exerted upon them by the bacterial mass, so that schizogenious cavities are enlarged. Because of the plasmolysis, the affected tissue loses its turgor, and wilting occurs.

B. THE DEATH PHASE

The host produces phenolic compounds as part of the nonspecific defense reaction.[35,40] The phenol content increases markedly postinfection, especially in resistant hosts.[35,40,41] Zeller found a simultaneous increase in activities of phenoloxidase and of β-glucosidase in resistant hosts, whereas the compatible reaction was characterized by a strong decrease in the activity of these enzymes.[40] As with the electrolyte leakage, the phenol production occurs sooner in tissue infiltrated with avirulent strains than in tissue infiltrated with virulent strains.[35]

During the terminal stages of the pathogenesis, more than 3 d after invasion of the tissue, bacteria invade host cells through crushed cell walls. Both the host cell's protoplast and cell wall are ultimately destroyed completely, and the schizogenous cavities are transformed into lysigenous cavities.[14-16,23] In dissolved tissue bacterial masses no longer exert pressure.[14] This implies that *E. amylovora* is not able to spread from dead tissue to healthy plant parts. For continuing invasion, the pathogen must advance in front of dead tissue. When lysis and death of host tissue occur in advance of the spreading bacterial masses, further bacterial invasion ceases.

Host cell walls are not dissolved by *E. amylovora*. Seemüller and Beer[3] extensively looked for cell wall polysaccharide degradation, but could not find any. No pectolytic, cellulolytic, or xylolytic enzyme activity was detected *in vivo* or *in vitro,* nor was maceration of infected pear fruit tissue observed.[2,3] As *E. amylovora* seems to lack lysic abilities, the dissolving activities apparently have to be ascribed to the host itself as part of the defense reaction. The lysis after the plasmolysis is probably a continuation of the normal, nonspecific hypersensitive response so that further spread of the pathogen may be prevented.

C. THE ENCLOSE PHASE

Toward the end of the pathogenesis, the host usually forms a defense periderm that isolates diseased tissue from healthy plant parts. Hockenhull[19] studied periderm formation of naturally infected hawthorns (*Crataegus monogyna*) in detail. In healthy cortical parenchyma close to diseased tissue, the formation of a cork barrier was initiated by cells dividing in various planes. The resultant tissue was compact and devoid of intercellular spaces, in contrast to the parenchymal "loose" tissue. Most of these new cells suberized, but a few continued their meristematic activity as phellogen. The phellogen initials produced files of phellem and phelloderm cells in a manner characteristic of periderm tissue (Figure 3). The periderm tissue was maintained compact and free in intercellular space by anticlinal divisions. A complete defense barrier in shoots extended through the cortical and phloem tissue and linked up with the external periderm.[19]

When a diseased region is completely surrounded by a cork barrier, the canker dehydrates, shrinks, and cracks around the edges.[19] Such a completely encapsulated lesion, called a "determinate canker", usually becomes ineffective, no longer playing a role in pathogenesis or contributing to further inoculum production. In such a case, the defense reaction of the plant succeeds. However, the pathogen is still occasionally able to survive a winter in derterminate cankers.[42-44] Lesions are not always delimited by cork barriers. Such indeterminate cankers do not dehydrate and show no cracks. Locating the border between healthy and diseased tissue is then difficult from the exterior. The indeterminate cankers maintain their activity, even after overwintering,[43] appear to be water soaked, and may ooze and extend through the host.[45,46] Anatomical studies revealed that in cases of indeterminate cankers, defense periderm is incompletely formed or halted at an early stage of development. Not even a trace of defense periderm is found in some indeterminate cankers.[19] Beer and Norelli[43] found that inoculation of apple trees during the early growing season produced cankers with determinate margins; later inoculation yielded cankers with indeterminate margins, plausibly due to less activity of the host cortex toward the end of the growing season.

Deep infections extend into the vascular cambium and destroy it; however, no periderm formation is to be found in vascular cambium or in xylem. Instead, vigorous wound healing activities, including the formation of wound callus, are initiated by healthy cambium cells.[19]

IV. OTHER VIEWS ON EPS AND PATHOGENESIS

Goodman and co-workers[5,6,47] pioneered research that revealed that EPS, formerly called amylovorin, is required for pathogenicity. Strikingly, their view on the mode of action of

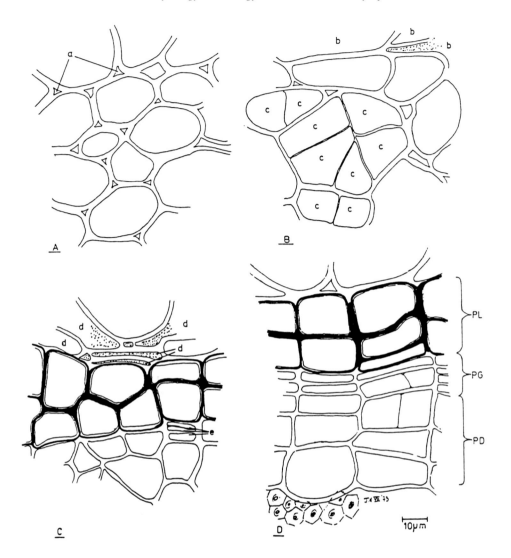

FIGURE 3. Formation of defense periderm in hawthorn (*Crataegus monogyna*) after natural infection by *E. amylovora*.[18] (A) Cross-section of healthy cortical parenchyma, distal to the primary fibers: (a) intercellular spaces. (B) Tissue similar to A, but from a diseased stem. Cells close to the colonized region (b) have divided (c). No suberization is detectable at this stage. (C) Tissue similar to B, but at a later stage of defense barrier development. Bacterial colonization has caused collapse of susceptible cells (d). Suberization of newly divided cells (c in panel B) has occurred, and a compact tissue without intercellular spaces is under development. Early phellogen formation is seen where a cell has begun dividing periclinally (e). (D) Fully developed defense periderm, consisting of phellem (PL) with suberized cell walls, phellogen (PG), and phelloderm (PD). (Drawings courtesy of the Royal Veterinary and Agricultural University, Copenhagen, Denmark.[18])

EPS strongly deviates from the more general belief that EPS functions to delay recognition of the pathogen. In 1976, Goodman and co-workers[47] ascribed plasmolysis and wilting to the toxic activity of EPS itself. They placed the bases of apple shoots in EPS solution (still called amylovorin at that time) and observed plasmolysis and wilting. From such observations Goodman and colleagues suggested a toxic effect of amylovorin.

The following year, Sjulin and Beer[12] concluded from more detailed trials that in Goodman's experiments amylovorin induced plasmolysis and wilt by occlusion of xylem vessels and subsequent restriction of water movement through the xylem, rather than by a direct toxic effect. The water conductivity of the stem was reduced by amylovorin, and amylovorin induced no disruption of membrane integrity or electrolyte leakage. Beer et al.[48] also found no correlation between amylovorin-induced shoot wilting and susceptibility to *E. amylovora*. Indeed, histological studies by Suhayda and Goodman[49] confirmed that EPS, when injected into xylem vessels, caused occlusion and subsequent wilt. Goodman and co-workers[50] verified the idea of occlusion of xylem vessels via EPS by decreasing the viscosity of EPS through the addition of various salts. Decreasing the viscosity of EPS in that way decreased the wilt-inducing capacity of EPS.[50]

However, 2 years later, Goodman[51] recanted his opinion on the viscosity effect of EPS after splitting EPS into smaller molecules, thereby altering its viscosity: "It was readily apparent that viscosity of the polysaccharide preparations played little or no role in wilt development. ... EPS can interact with apple pectin to form a compound that has gel-like properties. As a consequence we suggest that the gel detected ultrastructurally in xylem vessels is due to interaction of bacterial polysaccharide and plant cell wall polymer." Because of vessel occlusion, the xylem parenchyma cells would plasmolyze, giving rise to cavities around xylem vessels. The xylem vessels would then lose physical support by the parenchyma cells and would be ruptured, thus releasing bacteria into the newly formed intercellular space.[51,52] Occlusion of xylem vessels by masses of *E. amylovora* bacteria and by finely granular, gummy material was also reported by Rosen[42] 60 years earlier. In a wider context, Shigo[53,54] pointed out that plugging of xylem vessels is a common response by trees following injury and against advancing wood decay. Plugging would be one of the defense reactions to compartmentalize and enclose injured xylem tissue, according to Shigo.[53,54]

A prerequisite for vessel occlusion, according to the mechanism proposed by Goodman, is that the bacteria enter the xylem vessels. The bacteria are obviously sucked into the xylem vessels when cut shoots are placed in bacterial suspensions and the leaves on the shoots transpire. When shoots are not cut and the leaves transpire, there is a negative pressure in the xylem vessels. As soon as the xylem vessels are damaged, for example, during artificial inoculation or a hailstorm, ambient air or solution is immediately drawn several millimeters into the vessels. The drawn-in air or solution may remain without any further movement, unless it is exuded due to root pressure. It has been reported several times that *E. amylovora* moves quickly into the xylem vessels during artificial inoculation when the vessles are damaged,[20,24,55] but it is still disputable whether these bacteria play a role in the subsequent pathogenesis. In cases of infection through natural openings in blossom or other plant parts, bacterial migration seems to be limited to intercellular spaces. Goodman has adhered to the concept that xylem vessels are sites for early proliferation of *E. amylovora,* and that EPS is needed for occlusion of xylem vessels and thus is responsible for wilt development in host plants.

V. DEGREE OF INTIMACY OF THE PLANT-PATHOGEN RELATIONSHIP

This review of the pathological anatomy of fire blight evokes the idea that the interaction between *E. amylovora* and its host is relatively simple, feasibly based on a camouflaging gel around bacteria and on standard, nonspecific defense reactions by the host. During the early stages of pathogenesis, the bacterium hardly interferes with the metabolism of the host, but only camouflages itself and multiplies. At a later stage, the host reacts to *E. amylovora* in the same way as it does to nonpathogens.[35-38] Apparently, *E. amylovora* does not induce a special host reaction, just a late, nonspecific response. This suggests that the only requirement for successful infection and colonization of the host by *E. amylovora* is the delay in recognition conferred by EPS.

Any effectively camouflaging EPS can probably prevent recognition by many hosts. This supposed broadness of the camouflaging property of EPS would suggest that *E. amylovora* is not specialized to one species or a variety within a species, but that *E. amylovora* has a wide host range. It also arouses the expectation that *E. amylovora* isolates do not differ strongly in host specificity. The host ranges of all pathogenic *E. amylovora* strains should coincide.

Indeed, the host range is wide: besides the genera *Malus* and *Pyrus*, at least 129 species in 37 genera of the family Rosaceae have been reported to be susceptible to fire blight.[56] Furthermore, differences with respect to host specificity of *E. amylovora* isolates have not been clearly demonstrated. Physiological specialization with respect to virulence, which is often the case for intimate partner relationships with gene-for-gene interactions, does not seem to occur in the case of fire blight. The plant-pathogen relationship for fire blight disease seems to be nonspecialized in contrast to intimate relationships in a ''key-and-lock'' situation.

The conclusion from histological studies is that the relationship between *E. amylovora* and its host is nonspecialized rather than intimate, and is supported by the wide host range of *E. amylovora*. Therefore, a gene-for-gene interaction is unlikely for this pathosystem. Within the context of the complex biochemical interactions between host and pathogen in gene-for-gene systems, the outstanding feature of fire blight pathogenesis is its relative simplicity.

REFERENCES

1. **Burrill, T. J.,** Pear-blight. III., *State Hortic. Soc. Trans.,* 11, 114, 1878.
2. **Dye, D. W.,** *Erwinia:* the '*amylovora*' and '*herbicola*' groups, in *Plant Bacterial Diseases,* Fahy, P. C. and Persley, G. J., Eds., Academic Press, New York, 1983, 67.
3. **Seemüller, E. A. and Beer, S. V.,** Absence of cell wall polysaccharide degradation by *Erwinia amylovora, Phytopathology,* 66, 433, 1976.
4. **Geider, K., Falkenstein, H., Bellemann, P., Jahn, N., Schwartz, T., Theiler, R., and Bernhard, F.,** Virulence factors of *Erwinia amylovora, Acta Hortic.,* 273, 227, 1990.
5. **Goodman, R. N., Ayers, A. R., and Politis, D. J.,** The relationship of extracellular polysaccharide (EPS) of *Erwinia amylovora* to virulence, in *Proc. Int. Conf. Plant Pathogenic Bacteria,* Ridé, M. (Ed.), Gibert-Clarey, Tours, France, 1978, 483.
6. **Ayers, A. R., Ayers, S. B., and Goodman, R. N.,** Extracellular polysaccharide of *Erwinia amylovora:* a correlation with virulence, *Appl. Environ. Microbiol.,* 38, 659, 1979.
7. **Bennett, R. A.,** Evidence for two virulence determinants in the fireblight pathogen *Erwinia amylovora, J. Gen. Microbiol.,* 116, 351, 1980.
8. **Bennett, R. A. and Billing, E.,** Origin of the polysaccharide component of ooze from plants infected with *Erwinia amylovora, J. Gen. Microbiol.,* 116, 341, 1980.
9. **Billing, E.,** Studies on avirulent strains of *Erwinia amylovora, Acta Hortic.,* 151, 249, 1984.
10. **Steinberger, E. M. and Beer, S. V.,** Mutants of *Erwinia amylovora* altered in pathogenicity by transposon mutagenesis, *Acta Hortic.,* 217, 167, 1987.
11. **Bellemann, P., Jahn, N., Theiler, R., and Geider, K.,** Transposon mutagenesis of *Erwinia amylovora, Acta Hortic.,* 273, 233, 1990.
12. **Sjulin, T. M. and Beer, S. V.,** Mechanism of wilt induction by amylovorin in *Cotoneaster* shoots and its relation to wilting of shoots infected by *Erwinia amylovora, Physiol. Biochem.,* 68, 89, 1977.
13. **Youle, D. and Cooper, R. M.,** Possible determinants of pathogenicity of *Erwinia amylovora;* evidence for an induced toxin, *Acta Hortic.,* 217, 161, 1987.
14. **Nixon, E. L.,** The migration of *Bacillus amylovorus* in apple tissue and its effects on the host cells, *Pa. Agric. Exp. Stn. Bull.,* 212, 1, 1927.
15. **Miller, P. W.,** Studies of fire blight of apple in Washington, *J. Agric. Res.,* 39, 579, 1929.
16. **Eden-Green, S. J. and Knee, M.,** Bacterial polysaccharide and sorbitol in fireblight exudate, *J. Gen. Microbiol.,* 81, 509, 1974.
17. **Smith, A. R. W., Rastall, R. A., Rees, N. H., Hignett, R. C., and Wait, R.,** Structure of the extracellular polysaccharide of *Erwinia amylovora:* a preliminary report, *Acta Hortic.,* 273, 211, 1990.

18. **Hignett, R. C. and Roberts, A. L.,** A possible regulatory function for bacterial outer surface components in fireblight disease, *Physiol. Plant Pathol.,* 27, 235, 1985.
19. **Hockenhull, J.,** Some anatomical and pathological features of healthy and diseased hawthorn (*Crataegus monogyna*) naturally infected by the fireblight pathogen *Erwinia amylovora, Yearb. R. Vet. Agric. Univ. Copenhagen,* p. 125, 1974.
20. **Bachmann, F. M.,** The migration of *Bacillus amylovorus* in the host tissues, *Phytopathology,* 3, 3, 1913.
21. **Haber, J. M.,** The relationship between *Bacillus amylovorus* and leaf tissues of the apple, *Penn. Agric. Exp. Stn. Bull.,* 228, 1, 1928.
22. **Tullis, E. C.,** Studies on the overwintering and modes of infection of the fire blight organism, *Tech. Bull. Agric. Exp. Stn. Mich.,* 97, 1929.
23. **Wahl, H. A.,** The migration of *Bacillus amylovorus* in the tissue of the quince, *J. Agric. Res.,* 45, 59, 1932.
24. **Eden-Green, S. J.,** Studies in Fireblight Disease of Apple, Pear and Hawthorn, Ph.D. thesis, University of London, London, U.K., 1972.
25. **Schouten, H. J.,** Notes on the role of water potential in the pathogensis of fire blight, caused by *Erwinia amylovora, Neth. J. Plant Pathol.,* 94, 213, 1988.
26. **Schouten, H. J.,** Simulation of pressure in the intercellular space of host tissue caused by multiplication and swelling of *Erwinia amylovora, Neth. J. Plant Pathol.,* 97, 139, 1991.
27. **Schouten, H. J.,** A possible role for the swelling of extracellular slime of *Erwinia amylovora* at increasing water potential, *Neth. J. Plant Pathol.,* 95 (Suppl. 1), 169, 1989.
28. **Politis, D. J. and Goodman, R. N.,** Fine structure of extracellular polysaccharide of *Erwinia amylovora, Appl. Environ, Microbiol.,* 40, 596, 1980.
29. **Eden-Green, S. J. and Billing, E.,** Fireblight: occurrence of bacterial strands on various hosts under glasshouse condition, *Plant Pathol.,* 21, 121, 1972.
30. **Keil, H. L. and van der Zwet, T.,** Aerial strands of *Erwinia amylovora:* structure and enhanced production by pesticide oil, *Phytopathology,* 62, 355, 1972.
31. **Ehrig, F., Ficke, W., and Nachtigall, M.,** Rasterelektronenmikroskopische Untersuchungen zur Entstehung der Bakterienfäden ('strands'') beim Feuerbrand — *Erwinia amylovora* (Burrill) Winslow et al., *Arch. Phytopathol. Pflanzensch.,* 23, 509, 1987.
32. **Schouten, H. J.,** unpublished observations, 1989.
33. **van der Zwet, T. and Keil, H. L.,** Fire Blight: a Bacterial Disease of Rosaceous Plants, U.S. Dept. Agric. Handb. 510, U.S. Government Printing Office, Washington, D.C., 1979, chap. 11.
34. **Tomasik, A. A. and Goodman, R. N.,** Pathogenesis, migration and population dynamcis of *Erwinia amylovora* in susceptible (Jonathan) and resistant (red delicious) apple shoots, in *Proc. Int. Conf. Plant Pathogens and Bacteria,* Civerolo, E. L., Collmer, A., Davis, R. E., and Gillaspie, A. G., Eds., Martinus Nijhoff, Boston, 1987, 672.
35. **Addy, S. K.,** Leakage of electrolytes and phenols from apple leaves caused by virulent and avirulent strains of *Erwinia amylovora, Phytopathology,* 66, 1403, 1976.
36. **Brisset, M.-N. and Paulin, J.-P.,** Study of the interaction between *E. amylovora* and plant tissue through measurement of electrolyte leakage, *Acta Hortic.,* 273, 197, 1990.
37. **Goodman, R. N.,** The hypersensitivity reaction in tobacco: a reflection of changes in host cell permeability, *Phytopathology,* 58, 872, 1968.
38. **Klement, Z. and Goodman, R. N.,** The role of living bacterial cell and induction time in the hypersensitive reaction of the tobacco plant, *Phytopathology,* 57, 322, 1967.
39. **Burkowicz, A. and Goodman, R. N.,** Permeability alterations induces in apple leaves by virulent and avirulent strains of *Erwinia amylovora, Phytopathology,* 59, 314, 1969.
40. **Zeller, W. and Brulez, W.,** Changes in the phenol metabolism of ornamental shrubs (*Cotoneaster species*) infected with *Erwinia amylovora,* in *Proc. Int. Conf. Plant Pathogens and Bacteria,* Civerolo, E. L., Collmer, A., Davis, R. E., and Gillaspie, A. G., Eds., Martinus Nijhoff, Boston, 1987, 686.
41. **Ryugo, K., Okuse, I., and Fujii, Y.,** Correlation between fire blight resistance and phenolic levels in pears, *Acta Hortic.,* 273, 335, 1990.
42. **Rosen, H. R.,** The life history of the fire blight pathogen, *Bacillus amylovorus,* as related to the means of overwintering and dissemination, *Ark. Agric. Exp. Stn. Bull.,* 244, 1929.
43. **Beer, S. V. and Norelli, J. L.,** Fire blight epidemiology: factors affecting release of *Erwinia amylovora* by cankers, *Phytopathology,* 67, 1119, 1977.
44. **Ehrig, F. and Ficke, W.,** Rasterelektronenmikroskopische Untersuchungen zur Überwinterung von *Erwinia amylovora* (Burrill) Winslow et al. im Pflanzengewebe, *Arch. Phytopathol. Pflanzensch.,* 22, 1963.
45. **Van der Zwet, T and Keil, H. L.,** Fire Blight: a Bacterial Disease of Rosaceous Plants, U.S. Dept. Agric. Handb. 510, U.S. Government Printing Office, Washington, D.C., 1979, chap. 4.
46. **Steiner, P. W.,** Predicting canker, shoot and trauma blight phases of apple fire blight epidemics using the MARYBLYT model, *Acta Hortic.,* 273, 149, 1990.

47. **Huang, P. and Goodman, R. N.**, Ultrastructural modifications in apple stems induced by *Erwinia amylovora* and the fire blight toxin, *Phytopathology*, 66, 269, 1976.
48. **Beer, S. V., Sjulin, T. M., and Aldwinckle, H. S.**, Amylovorin-induced shoot wilting: lack of correlation with susceptibility to *Erwinia amylovora*, *Phytopathology*, 73, 1328, 1983.
49. **Suhayda, C. G. and Goodman, R. N.**, Early proliferation and migration and subsequent xylem occlusion by *Erwinia amylovora* and the fate of its extracellular polysaccharide (EPS) in apple shoots, *Phytopathology*, 71, 697, 1981.
50. **Sijam, D., Goodman, R. N., and Karr, R. N.**, The effect of salts on the viscosity and wilt-inducing capacity of the capsular polysaccharide of *Erwinia amylovora*, *Physiol. Plant Pathol.*, 26, 231, 1985.
51. **Goodman, R. N., Butrov, D., and Gidley, M.**, Structure and proposed mode of action for amylovorin, *Acta Hortic.*, 217, 157, 1987.
52. **Goodman, R. N. and White, J. A.**, Xylem parenchyma plasmolysis and vessel wall disorientation caused by *Erwinia amylovora*, *Phytopathology*, 71, 844, 1981.
53. **Shigo, A. L.**, Compartmentalization: a conceptual framework for understanding how trees grow and defend themselves, *Annu. Rev. Phytopathol.*, 22, 189, 1984.
54. **Shigo, A. L. and Marx, H. G.**, Compartmentalization of Decay in Trees (CODIT), Inf. Bull. No. 405, U.S. Department of Agriculture, Washington, D.C., 1977.
55. **Crosse, J. E., Goodman, R. N., and Shaffer, W. H.**, Leaf damage as a predisposing factor in the infection of apple shoots by *Erwinia amylovora*, *Phytopathology*, 62, 176, 1972.
56. **van der Zwet, T. and Keil, H. L.**, Fire Blight: a Bacterial Disease of Rosaceous Plants, U.S. Dept. Agric. Handb. 510, U.S. Government Printing Office, Washington, D.C., 1979, chap. 6.

Chapter 8

PATHOLOGICAL ANATOMY AND HISTOCHEMISTRY OF LEUCOSTOMA CANKER ON STONE FRUITS AND OTHER SELECTED FUNGAL CANKERS OF DECIDUOUS FRUIT TREES

Alan R. Biggs

TABLE OF CONTENTS

0-8493-2939-6/93/$0.00 + $.50

I. INTRODUCTION

There are a great many fungal pathogens of fruit trees that cause twig and stem cankers and that also cause leaf spots and/or fruit rots. There are some fungal pathogens, however, that primarily cause stem cankers and that cause major economic losses in some regions. The latter group of pathogens is the focus of this chapter. Infection of deciduous fruit trees by canker-causing fungi usually involves breaching the host's passive barriers to pathogen ingress via wounding. Wounding in orchard trees can be caused by agents such as insects, humans, lightning, wind, hail, animals, and nutritional and physiological disorders. Alternatively, natural leaf abscission and cold temperature injury also play key roles in creating infection courts for these pathogens. This chapter describes the pathological anatomy and histochemistry of fungal cankers of deciduous fruit trees in focusing on only four diseases: those caused by *Leucostoma* spp. and *Botryosphaeria* spp. on stone fruits, but also found on pome fruits, and *Nectria galligena* and *Valsa ceratosperma*, found primarily on pome fruits. Most of the other canker diseases described in the literature are similar to these examples.

II. LEUCOSTOMA CANKER

Leucostoma canker, also called perennial canker, Cytospora canker, and Valsa canker, is recognized as an important disease of peach in the northern portions of the region suitable for production of temperate fruits, including Canada, and the northeastern U.S.[1] It is mentioned as part of a complex set of factors associated with peach tree short-life syndrome in the southeastern U.S.[2] It is equally important on other stone fruits, including prune and plum in California[3] and Idaho.[4] In Europe, the disease is important on apricot, peach, sweet cherry, and is part of the disease complex of stone fruits called "apoplexy".[5-7] The disease also occurs in South America and Japan.[8,9] The host range of the two causal organisms is similar and they may be found on a wide variety of species within, as well as outside, the family Rosaceae, including apple; apricot; Sitka mountain ash; blackthorn; black, Japanese, flowering, pin, sour, and sweet cherries; chokecherry; nectarine; Russian olive; peach; pear; common, Damson, and wild plums; prune; serviceberry; and golden willow. The pathological anatomy of this disease has been the subject of several investigations.

The fungi that cause Leucostoma cankers on stone fruits, *Leucostoma cincta* (Pers. ex Fr.) Höhn. [= *Valsa cincta* (Pers. ex Fr.) Fr.] and *L. persoonii* (Nits.) Höhn. [= *Valsa leucostoma* (Pers. ex Fr.) Fr.], are ascomycetes in the order Diaporthales, family Valsaceae. The imperfect stages of these fungi, *Leucocytospora cincta* (Sacc.) Höhn. [= *Cytospora cincta* Sacc.] and *Leucocytospora leucostoma* (Pers.) Höhn.) [= *C. leucostoma* Sacc.], respectively, are encountered most commonly in the field.

The appearance of Leucostoma canker varies depending on the part of the tree infected and the specific pathogen involved. Infections of small twigs appear as sunken, discolored areas, often with alternating zonation lines, usually around winter-killed buds or leaf scars from the previous year's foliage (Figure 1). Nodal infections, usually but not always associated with *L. cincta,* are observed 2 to 4 weeks after bud break. The infected tissues become darker with time and an amber-colored gum may ooze from the infected tissue unless the twig is killed entirely. The previous year's shoots that develop in the center of the tree are quite susceptible to nodal infections and, if left untreated, the fungus can invade rapidly into scaffold limbs and large branches. Branch cankers that result from this type of infection will have dead twigs or twig stubs at the center of the canker.

Cankers that form on the main trunk, branch crotches, scaffold limbs, and older branches are the most conspicuous expression of fungal infections. Usually the first external symptom of such cankers is the copious quantity of amber-colored gum. Gum production is a natural response of the tree to irritation (see Chapter 11), but that due to infection by *Leucostoma* spp. is excessive to the point of being detrimental. As the cankers age, the gum becomes dark brown, the infected bark dries out and cracks open, exposing blackened tissue beneath. Cankers appear elliptical along the length of the stem.

Beginning in the late spring and continuing through the summer months, the tree grows rapidly and is able to resist further penetration of the fungus into healthy tissues. During this time, the tree may form a callus ring around the canker, but the fungus usually invades this tissue again in late fall or early spring when the tree is dormant and cannot actively resist penetration. The yearly alternation of callus production and canker extension produces a canker with concentric callus rings. Where tree defenses are compromised by environmental stresses, a callus production may be inhbiited and cankers may appear more diffuse. Branch or twig infections may produce leaf symptoms during the growing season. Leaves on an infected branch often turn yellow, droop down, and may wilt and die. Dead twigs and branches are usually covered with a multitude of pinhead-sized black pycnidia erupting through the dead bark.

A. PATHOLOGICAL ANATOMY OF LEUCOSTOMA CANKERS
1. Peach

In 1-year-old peach shoots exhibiting tip dieback from natural infection in the field by either *L. cincta* or *L. persoonii*,[10] fungal mycelium was present within the cankered area in the cortical tissue, phloem, xylem vessels, and pith. The mycelium was intracellular in the xylem vessels and pith. The pathogens appeared to spread via the pits in these tissues. In the phloem and cortex, fungal growth was intercellular. Mats of fungus mycelium separating adjacent rows of cells often were seen in this region (Figures 2 to 4).

At the transition zone between healthy and diseased tissues, mycelium was detected regularly in the xylem vessels, but was observed infrequently in cortical tissue, phloem, and pith. Mycelium could be observed only in vessels up to 1 cm beyond the visible margin of the canker. The walls of xylem vessels located up to 2 cm beyond the margin of the cankers exhibited a brown discoloration, primarily in the vessels near the pith, but extending throughout the wood with increased proximity to the canker margin (Figures 5 and 6). The lumens of many xylem vessels, both within the cankered region and beyond the canker margin, became plugged with gum that darkened with age.[10]

In 1-year-old peach bark inoculated in the field with mycelium of *L. persoonii*, fungal colonization of cortical and phloem tissues was initially diffuse, although mycelial aggregations formed within 3 weeks.[11] Host responses to inoculation over the initial 1-week period were largely confined to the vascular cambium which gave rise to a barrier zone of gum ducts present in the xylem and forming a distinct band (Figure 7 to 13). Gum ducts continued to enlarge during the 3 to 8 weeks following inoculation, and a second row of ducts was formed in some samples. Gum oozed from the inoculated sites during the 11-month period of study, except during periods of low rainfall and during dormancy. Plugging of vessel elements with gum was sporadic and tyloses were observed only rarely. Diffuse mycelial wedges were present within the gum duct barrier zone in the xylem.

Wound periderm (Figures 10 and 11) was formed in the bark tissues 2 to 3 weeks postinoculation; however, mycelial aggregates accumulated in the necrotic tissue, followed by the eventual penetration of wound periderm tissues by mass action and ingress into living tissue. Formation of a new wound periderm occurred and the proliferation of the fungus was again slowed. Five months after inoculation, the pathogen could be observed penetrating into living tissues by breaching the wound periderm in the bark and along the barrier zone of gum ducts in the xylem. In general, colonization of woody tissue was poor and was

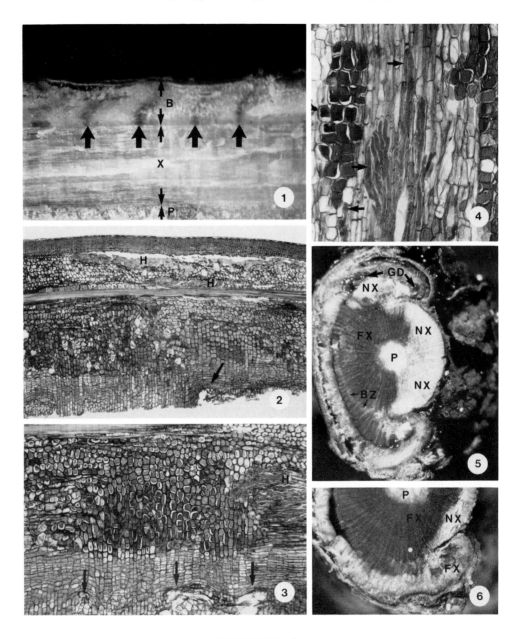

FIGURES 1 to 6.

FIGURES 1 to 6. One- and 2-year-old peach stems inoculated with *L. persoonii*. Figures 1, 5, and 6, stereomicrographs. Figures 2 to 4, light micrographs. **Figure 1.** Longitudinal section through pith (P), xylem (X), and bark (B) (tissues are between oppositely oriented arrows), sampled 7 d postinoculation with *Leucostoma persoonii*, showing 4 polysaccharide-rich reaction zones (large arrows). Colonization from left to right proceeded in the bark through the first zone and was almost to the second. (Magnification × 10.5) (From Biggs, A. R., *Can. J. Bot.*, 62, 2814, 1984. With permission.) **Figure 2.** Longitudinal section of peach bark and vascular cambial tissues 7 d postinoculation with *L. persoonii* showing hyphal aggregations (H) in periderm and cortex regions and alteration of cambial derivatives (arrows). Colonization proceeding from right to left. (Magnification × 294.) **Figure 3.** Longitudinal section of peach bark and vascular cambial tissues 7 d postinoculation with *L. persoonii* showing hyphal aggregation (H) in cortex and alteration of cambial derivatives (arrow). Colonization proceeding from right to left. Note that very little new phloem has been produced. (Magnification × 392.) **Figure 4.** Inter- and intracellular colonization of phloem tissues by hyphae of *L. persoonii* (arrows). Colonization proceeding from bottom to top of micrograph. (Magnification × 476.) **Figures 5 and 6.** Transverse views of peach stems inoculated with *L. persoonii* (Figure 5) or noninfested malt agar (Figure 6). Stems were excised at the base and immersed in stain solution to demonstrate the location of nonfunctional xylem tissues (light areas, NX). Note location of functional xylem (FX) in the new callus tissue in Figure 6 vs. the regions of nonfunctional xylem in callus tissues in Figure 5. Pith (P), gum ducts (GD), and xylem barrier zone (BZ) are present. (Magnification × 5.6.)

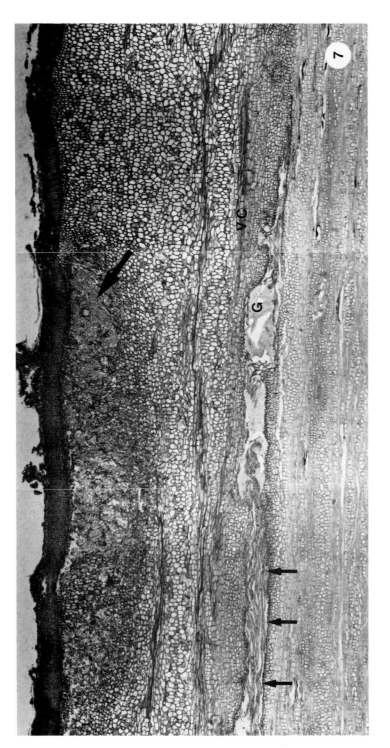

FIGURE 7. Longitudinal mosaic of peach stem inoculated with *L. persoonii*. Note masses of hyphae in periderm and cortical regions (large arrows) and in the gum duct zone in the newly differentiated xylem (small arrows). Colonization proceeding from left to right. Note gum (G), vascular cambium, and dark deposits in older xylem. The vascular cambium in this region will redifferentiate into a suberized periderm-like layer. (Magnification × 378.)

confined to the outer xylem (Figures 14 and 15). Pith tissues remained free from fungal colonization.

Samples taken 11 months after inoculation, at the beginning of the second growing season, showed extensive necrosis of noninvaded bark tissues 2 to 3 mm in advance of fungal colonization.[11] The vascular cambium produced another layer of barrier zone gum ducts. All tissues located outside the previous season's xylem appeared necrotic, suggesting that living tissues had succumbed after the onset of cambial activity. Observations concerning the presence of necrosis in advance of colonization, which occurred twice in the study, were emphasized. The role of a toxin in pathogenesis has been previously reported or suggested for this disease[12-14] and other diseases caused by *Cytospora* spp.[15,16]

The comparative anatomy of 2-year-old peach stems from relatively resistant and susceptible cultivars inoculated with mycelium of *L. cincta* and *L. persoonii* has been described,[17] and these observations have helped resolve some of the discrepancies among earlier studies.[10,11,18] At 7 d postinoculation, the cultivar Candor had longer cankers than the cultivar Sunhaven, following inoculations with either fungus (Figure 16). Colonization around the stem circumference was more extensive for *L. persoonii* on both cultivars relative to *L. cincta* (Figure 17). Regression analyses of the relationship between postinoculation time and canker length showed that 'Candor' was colonized more aggressively by *L. persoonii* than any of the other host/pathogen combinations. In addition, 'Sunhaven' inoculated with *L. persoonii* showed more aggressive colonization than either of the two cultivars inoculated with *L. cincta*. Interestingly, there was a second period of rapid increase in canker length, width, and colonization rate between days 28 and 35 in both cultivars inoculated with *L. persoonii* (Figures 16 and 17).

The extent of fungal invasion, as determined by isolation, varied with time and the particular cultivar/pathogen combination (Table 1). All cultivar/pathogen combinations showed approximately equal invasion of bark and current year's xylem 14 d postinoculation. Fungi were not recovered from 1- or 2-year-old xylem or pith; however, by 28 d postinoculation, the cultivar Candor, inoculated with *L. persoonii* (the more susceptible cultivar paired with the more aggressive fungus), yielded this fungus from 1- and 2-year-old xylem located under the point of inoculation and 1 cm proximal and distal to this point. The extent of xylem colonization in current year's xylem was equal to that found in bark, i.e., the margin of visible symptoms. Notwithstanding, by day 56, *L. persoonii* could be isolated from all tissues, including the pith; the extent of colonization in 2-year-old xylem and pith exceeded the visible canker margin in the bark by 4 cm proximally (the maximum distance examined) and 1 cm distally.

The extent of colonization in the cultivar Sunhaven inoculated with *L. persoonii* and in Candor inoculated with *L. cincta* was similar. Pathogens from both cultivars were isolated infrequently from 2-year-old xylem and, when found, were limited to the region under the point of inoculation. In 1-year-old and current year's xylem, *L. persoonii* was equally advanced relative to bark symptoms in 'Sunhaven', whereas in 'Candor', *L. cincta* was in xylem in advance of visible bark symptoms 3 cm proximally and 1 cm distally. Isolations of *L. cincta* from 'Sunhaven' (the more resistant cultivar paired with the less aggressive fungus) were limited to bark and current year's xylem. Xylem colonization was ahead of that in bark by 1 cm proximal to the visible canker margin but equal to that in bark distally. No pathogens were detected in wounded, noninoculated control tissues.

In mechanically wounded control tissues and in tissues colonized by *L. persoonii* and *L. cincta*, lignified and suberized cells formed a boundary separating living tissues from necrotic tissues and the environment. The location and chemical qualities of these tissues may influence or impart pathogen resistance properties in bark and xylem. In cankered tissues, several discontinuities in bark and xylem boundaries were observed which occurred

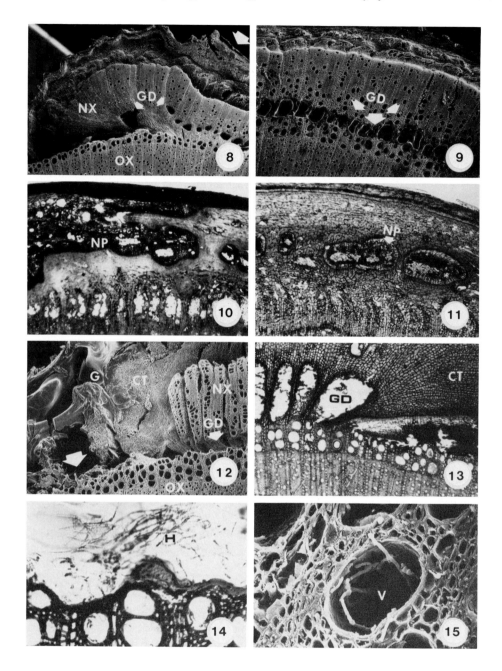

FIGURES 8 to 15.

FIGURES 8 to 15. **Figure 8.** Transverse section of callus of control wound after 4 weeks. Differentiation of callus into zylem and phloem has occurred. Gum ducts have ceased to produce gum. Xylem present at the time of wounding (OX) and xylem formed after wounding (NX) are visible. Necrotic tissue cutoff by NP is visible at upper right (arrow). (SEM; magnification × 18.4.) **Figure 9.** Portion of a stem on side opposite to control wound. Gum ducts (GD) formed in response to inoculated wounds. Gum ducts have begun to coalesce and are still producing gum. (SEM; magnification × 36.8.) **Figure 10.** Transverse section through distal edge of control wound after 4 weeks. Strong positive staining for suberin with Sudan black B is evident. (Light micrograph; magnification × 54.4.) **Figure 11.** Transverse section, approximately 3 mm above control wound. After 4 weeks NP has formed distal finger-like projections surrounding primary phloem fibers. (Light micrograph; magnification × 75.) **Figure 12.** Transverse section from the midportion of an inoculated wound after 4 weeks. Note that large gum ducts and marked lack of differentiation of callus tissue (CT) compared to control wound (Figure 8). Hyphae are visible growing along exposed wood surface (arrow) and tissue is generally embedded in a mass of gum (G). (SEM; magnification × 33.6.) **Figure 13.** Light micrograph of callus tissue similar to that described in Figure 12. Large gum ducts and undifferentiated callus (CT) tissue are visible. (Light micrograph; magnification × 67.2.) **Figure 14.** Transverse section of exposed wood in an inoculated wound after 4 weeks. Mat of hyphae is visible growing in gum deposit along wood surface. (Light micrograph; magnification × 232.) **Figure 15.** Hyphae of *L. persoonii* growing in vessel element that was located in the outermost portion of the xylem present at the time of wounding and was exposed to the environment at the time of sampling. (SEM; magnification × 346.4.) (Figures 8 to 15 from Wisniewski, M., Bogle, A. L., and Wilson, C. L., *Can. J. Bot.,* 62, 2804, 1984. With permission.)

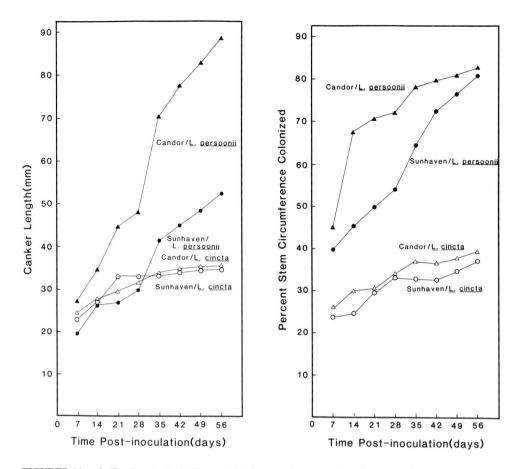

FIGURES 16 and 17. Longitudinal (**Figure 16**, left) and circumferential (**Figure 17**, right) enlargement of cankers on peach cultivars Sunhaven and Candor inocualted with *Leucostoma persoonii* and *L. cincta*. Trees inoculated in the greenhouse and cankers measured weekly for 8 weeks. (From Biggs, A. R., *Phytopathology, 76*, 905, 1986. With permission.)

in the absence or as a result of infection.[17] In noninoculated wounds inflicted with a cork borer, discontinuities in suberin formation occurred at the juncture of the suberized wound periderm with the original periderm and with primary phloem fibers; the centripetal surface of the wound callus as it grew across the exposed xylem; and, in the xylem, the presence of nonsuberized vessels, trachieds, and axial parenchyma. Where suberin was absent in xylem and in bark adjacent to fibers, increased deposition of lignin-like material was observed. The periderm juncture and callus discontinuities appeared time dependent and remained nonsuberized for 2 to 4 weeks, respectively.

In tissues inoculated with *L. persoonii* or *L. cincta,* suberization was less continuous than in wounds. Suberization was observed only infrequently in xylem, although deposition of lignin-like substances occurred in a visible barrier zone. In infected bark, the pattern of suberization was similar to that of wounds, although the shapes of wound phellem cells were altered in some samples. Altered phellem appeared thin-walled and rounded, compared to the relatively thick, rectangular phellem seen at wounds. The junctions of suberized tissues with the original periderm and/or primary phloem fibers often were colonized by wedges of fungus mycelium. In both cultivars inoculated with *L. cincta,* wound periderm formation appeared effective in limiting colonization in bark during the 8-week course of study. With the 'Sunhaven'/*L. cincta* combination, xylem responses (described below) were associated with limited colonization relative to the other cultivar/fungus species combinations.

TABLE 1

**Isolation of *Leucostoma persoonii* and *L. cincta* from Bark,
Xylem, and Pith Tissues Following Inoculation of Peach
Cultivars 'Candor' and 'Sunhaven'[a]**

Cultivar/fungus (days postinoculation)	Bark	Xylem Current	Xylem 1-year-old	Xylem 2-year-old	Pith
'Sunhaven'/*L. cincta*					
14	+	+	–	–	–
28	+	+	–	–	–
56	+	+	–	–	–
'Sunhaven'/*L. persoonii*					
14	+	+	–	–	–
28	+	+	–	–	–
56	+	+	+	+	–
'Candor'/*L. cincta*					
14	+	+	–	–	–
28	+	+	–	–	–
56	+	+	+	+	–
'Candor'/*L. persoonii*					
14	+	+	–	–	–
28	+	+	+	+	–
56	+	+	+	+	+

[a] Trees inoculated in the greenhouse and isolations made by removing stem segments with control wounds or cankers plus 4 cm of nonaffected tissues on either side of the visible canker or wound margin. Stem segments were halved longitudinally and isolations made at 1-cm intervals from the various tissues. Control wound segments were free of pathogens.

The nonsuberized, centripetal region of the callus was readily colonized by both pathogens in the early stages of callus formation. As callus tissues progressed in their development, the suberized boundary comprised of current year's xylem parenchyma was often ruptured from the inside by the coalescing of enlarged gum ducts. The rupture then provided access for the hyphae to the gum duct region, thus facilitating longitudinal invasion in a manner similar to that described by Wisniewski et al.[11] In addition, the rupture allowed fungal access to parenchymatous tissues of the differentiated callus, resulting in rapid circumferential spread of mycelium in the bark. Direct penetration of well-formed wound periderm (at least three cells thick) was observed once in this study. Mycelia were more likely to separate adjacent rows of phellem parallel to the axis of the tissue after access to the tissue was gained via lenticels or growth cracks. Primary ligno-suberized boundary tissues and wound periderms with one to two layers of phellem were penetrated readily by hyphal aggregations produced by both fungi.

In the bark of both 'Sunhaven' and 'Candor' inoculated with *L. persoonii* and *L. cincta*, necrophylactic periderm formed at 21 to 28 d postinoculation and was associated with an effectively delimited lesion in both cultivars inoculated with *L. cincta*, the less pathogenic fungus. For cultivars inoculated with *L. persoonii*, the wound periderm was not as thick, had fewer cells, as was associated with only a slight inhibition of lesion expansion followed by a second rapid increase in canker length as the periderm was breached between 28 to 35 d after inoculation[19] (Figures 16 and 17). In contrast, Hebard et al.[20] observed that periderms in American chestnut were produced by the host and then breached by the chestnut blight fungus, *Chryphonectria parasitica*, at 18 or 28 d after inoculation, depending on the virulence of the isolate. In resistant chestnuts, some virulent isolates continued to expand lesions up to 30 d after inoculation.[21]

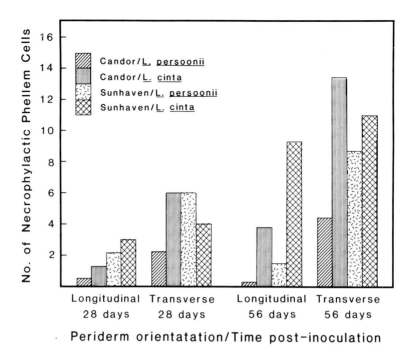

FIGURE 18. Thickness in number of cells of the necrophylactic periderm located at longitudinal and transverse canker margins of peach cultivars Sunhaven and Candor inoculated with *Leucostoma persoonii* and *L. cincta*. (From Biggs, A. R., *Phytopathology,* 76, 905, 1986. With permission.)

The numbers of suberized phellem cells formed following inoculations of peach cultivars Sunhaven and Candor with *L. persoonii* or *L. cincta* differed significantly depending upon fungus species, cultivar, and the orientation of the canker margin (longitudinal or transverse) (Figure 18). Fungus species along accounted for the greatest amount of variation in cell number (29.3%), followed by the cultivar × orientation interaction (14.1%). For example, trees inoculated with *L. cincta* had nearly twice the numbers of necrophylactic periderm cells as those inoculated with *L. persoonii*. At 56 d postinoculation, bark tissues colonized by *L. cincta* possessed significantly more suberized phellem relative to the other fungus species × time combinations. The numbers of phellem cells were significantly less for 'Candor' in the longitudinal orientation (at the proximal and distal canker margins) relative to the other cultivar × orientation combinations. Xylem tissues also became suberized in response to wounding or fungal infection. Suberin was observed in ray parenchyma cells of 1- or 2-year-old xylem present at the time of wounding or inoculation, and in current year's xylem formed after inoculation and at a distance from the point of inoculation. In the latter case, suberization was limited to xylem parenchyma in the gum duct zone, as described by Wisniewski et al.[11]

It was shown via quantitative fluorescence microphotometry that wounded bark of the more resistant cultivar Sunhaven had 73% more suberin relative to the cultivar Candor.[17] Furthermore, the frequency of stem wounds exhibiting suberization at 8 d postwounding was 100% for 'Sunhaven' and only 33% for 'Candor'. Interestingly, amounts of suberin in wounded xylem of 'Candor' were slightly higher than those of 'Sunhaven'. Tissue deposition patterns for lignin and suberin following wounding were independent of the cultivar. For example, 'Sunhaven' formed more suberin in bark and more phloroglucinol-positive material in xylem than 'Candor'. In contrast, 'Candor' formed more lignified material in bark and more intensively suberized xylem ray parenchyma than 'Sunhaven'.

2. Sweet Cherry

Cankers caused by *L. persoonii* on sweet cherry in Germany[7] are initiated in the bark via wounds in the periderm and through leaf scars injured by autumn frosts during the leaf abscission period. Symptoms of the latter type of infection consisted of cankers with dead buds or fruiting spurs in their center. In winter and early spring, the pathogen grows mainly in the bark and less so in the xylem. In summer and fall, the fungus is found growing only in xylem. Because of this alternating, two types of cankers develop on sweet cherry.

The first type of canker is characterized by bark necrosis and wound periderm formation by the host in late spring. Cankers cease to enlarge until the following winter when the fungus, which has survived in cavities around the wound periderm, circumvents the wound periderm and invades healthy bark tissue. In this type of canker, which always originates from bark infections, the fungus is not able to grow from the older xylem or pith into the bark.

In the second type of canker, the fungus can be seen in young xylem immediately beneath the vascular cambium. Goring[7] maintains that the fungus kills the cambial cells by the action of a toxin, resulting in heavy gum production in the bark external to the dead cambium. The pathogen could not be isolated from the necrotic bark tissues in July. Wound periderm production in bark is suppressed in this canker type, also due to the action of the putative toxin. This second canker type is found only on large trees that exhibited extensive necrosis on the trunk and main branches. It was suggested that in the first canker type the current year's xylem shields the vascular cambium from the effects of the fungus which is growing in the older wood.

B. HISTOCHEMICAL ASPECTS OF *LEUCOSTOMA* INFECTION

Few histochemical studies have been conducted on tissues infected with *Leucostoma* spp.[11,17,18] These studies have shown that lignin (and/or lignin-like substances, hereafter referred to collectively as lignin) and suberin are the main constituents of cell walls in the boundary zone that is formed prior to wound periderm formation in bark.[18] The lignin in this ligno-suberized tissue can be detected with phloroglucinol/HCl, but not with the Mäule reaction or the chlorine-sulfite test. The first phloroglucinol-positive reactions were observed 72 h after inoculation or wounding. In inoculated tissues, lignin was detected in crushed phloem sieve elements, sieve element cell walls, and xylem vessel plugs and vessel pits up to 2 cm in advance of hyphae. As the boundary zone developed in the bark, the red phloroglucinol chromophore was first seen in the corners of the compound middle lamellae of adjacent hypertrophied cells. The cell walls in the boundary zone then became uniformly reactive over the next 4 d for wounded tissues. In inoculated tissues, lignification can be delayed up to 21 to 28 d, or in certain locations, it may not develop.[17,18] Results of tests for tissue aldehydes (Schiff reagent prior to periodic acid hydrolysis, orcinol) gave color reactions in the same positions as described for phloroglucinol/HCl.

Tests with azure B as a histochemical reagent for lignin showed that this test was not specific for lignin in peach tissues.[18] Although some green coloration was imparted to lignified cell walls with this reagent, the reaction was weak. Azure B also showed strong affinity for the walls of cells in a zone that appeared amber-yellow without staining, and which appeared strongly reactive with the periodic acid-Schiff (PAS) test for polysaccharide substances. Toluidine blue O also yielded poor color differentiation for lignified tissues.

Nonwounded, healthy periderm exhibited a strong reaction with many of the histochemical reagents for suberin in these studies.[11,17,18] Observations with phosphine GN (chrysaniline yellow), Sudan IV, Sudan black B, and epifluorescence microscopy were the most reliable, whereas results of tests with crystal violet, Sudan III, and methylene blue were variable.[18] Suberized wound periderms appeared pinkish with Sudan IV and examination of sections treated with Sudan IV with blue-violet fluorescence excitation showed bright red fluorescence associated with only the most recently suberized cells. Application of phosphine GN and examination with ultraviolet (UV) fluorescence excitation induced a silvery-white fluorescence of all suberized cells. Nonstained tissues mounted in glycerol and examined with UV fluorescence exhibited violet autofluorescence of suberized tissues, however, in boundary zone tissues this was often masked by the bright blue autofluorescence of lignin. As the boundary zone developed, in both control wounds and inoculated tissues, cells that had become lignified formed a thin suberin lining on the inner sides of the cell walls.[18,19,22,23] These linings were most easily visualized under UV fluorescence in tissues reacted with phloroglucinol/HCl or toluidine blue O (Figure 12). These reagents acted to quench any lignin autofluorescence. Similarly Sudan black B quenched suberin autofluorescence, helping to confirm further the chemical nature of these cell wall layers.[18]

Suberized cells were of two basic types. Suberization occurred in tissues that were present in the wound reaction zone or in advance of colonization. The tissue was comprised of dedifferentiated cells with lignified walls lined with lamellar suberin. Cell division was not involved in the formation of this tissue. Suberization of extant tissue occurred following formation of the impervious boundary zone of the bark in the inner periderm, cortex, and primary and secondary phloem. Secondary suberization in the bark occurred after cell division and resulted in the formation of new phellogen and its derivatives. Secondary suberized cells occurred in the necrophylactic periderm which formed in all wounded or colonized bark tissues.[17]

Callose deposits in infected peach bark and xylem appeared as bluish-green fluorescent flecks when stained with aniline blue and examined under UV excitation. Callose was regularly associated with the amber-yellow, polysaccharide-rich zone in wounded and infected tissues. It was observed was early as 24 h after wounding or inoculation and it occurred at the margins of the wound or infection court, about five to ten cells in from the surface. The intensity and frequency of callose deposition was not influenced by fungal colonization except in necrotic sieve elements observed 7 d postinoculation. In callus tissue, callose was seen in some poorly differentiated phloem sieve elements at 35 d postinoculation.

Degradation of pectin and cellulose was a regular feature of tissues colonized by *L. persoonii*.[18] For pectic substances, the ruthenium red reaction and the iron absorbtion test[18] showed degradation of pectic substances adjacent to mycelial aggregations as early as 7 d postinoculation. Similar results were obtained with histochemical tests for cellulose (birefringence, IKI-H_2SO_4, and zinc-chloriodide).

C. TEMPORAL CHANGES IN THE ANATOMY OF THE INFECTION COURT AND PATHOGEN RESISTANCE

Following wounding, the process of establishing boundaries in the infection court between healthy tissues and the external environment confers resistance to infection in many host/pathogen interactions. Wounds generally become increasingly less susceptible to infection with age.[21,24-26] Resistance to infection in this manner is probably related to a number of nonspecific plant responses leading up to and including formation of primary lignosuberized tissues and secondary wound (or necrophylactic) periderm. The major structural components of these tissues are lignin and suberin.[27] See Chapter 2 for a more thorough discussion of wound anatomy and the interaction of pathogens and wounds.

III. BOTRYOSPHAERIA CANKER OF PEACH

Peach tree fungal gummosis was first observed in central Georgia in 1970[28] and is now reported throughout Alabama, Florida, Georgia, and Louisiana. Three species of *Botryo-*

sphaeria have been associated with the disease and wound inoculations. *B. obtusa* (Schw.) Schoem. and *B. rhodina* (Berk. & Curt.) Arx produced symptoms identical to those produced by *B. dothidea* (Moug. ex Fr.) Ces. & DeNot.[29] The infection court for all three species is pruning wounds, although *B. dothidea* also can penetrate lenticels.[30,31] Penetration of lenticels by the two former species has not been observed, but these species can invade lesions caused by *B. dothidea*. Isolation of *B. obtusa* and *B. dothidea* from naturally infected tissue was more common than *B. rhodina*. *B. dothidea* was more active in the summer, whereas *B. obtusa* predominated in cankers in the winter and spring.[32]

The disease is characterized by numerous sunken necrotic lesions about 7 to 14 mm in diameter and usually located at lenticels. Gum exudation occurs in excess. Shallow, round to oval, brown, gummy lesions, about 12 to 25 mm in diameter, can be seen upon removal of the outer bark. Symptoms of the disease usually occur first on the trunk during the second or third growing season. As the disease progresses, scaffold limbs and twigs may be affected. On young branches lenticels may be swollen, but gumming does not occur. Reproductive stroma of the causal organisms are produced in infected lenticels.

Inoculation of wounded peach bark with spores of *B. obtusa* and *B. dothidea* provided an interesting contrast to the mycelial inoculation studies described above for Leucostoma canker.[33] Initial colonization by the fungus occurred on the wound surface, in necrotic bark tissue on the wound margin, and in necrotic xylem tissue to a depth of about 200 μm. In inoculated tress, few macroscopic symptoms of infection were present during the 8-week course of the study, and normal wound healing responses appeared macroscopically and microscopically to be proceeding similar to the noninoculated wounds. Histologically, no differences in lignin and suberin deposition, periderm formation, and callus formation were observed between noninoculated wounds and wounds inoculated with either fungus after 1 and 2 weeks' incubation. By 28 d after inoculation for both fungi, however, new callus tissue was being colonized by fungal hyphae located on the surface of the xylem beneath the nonsuberized ventral callus surface. The nonsuberized ventral region of callus was primarily parenchymatous with a ligno-suberized outer layer, and was the focal point for fungal pathogenesis. The actions of fungal toxins as irritants to this nonsuberized, highly meristematic region is probably critical to successful pathogenesis. By 8 weeks, the breakdown of tissue in this region was directly associated with fungal hyphae in close ventral proximity. Direct fungal penetration of the callus, although only occasionally observed, was not required to induce gum pockets, cambial alterations, and tissue breakdown. These effects are consistent with the production of a fungal toxin(s), described recently for *B. obtusa*.[34] Intracellular hyphae of both *Botryosphaeria* spp. were observed colonizing cortical and callus parenchyma, and xylem ray parenchyma, vessels, and tracheids. Intercellular hyphae were observed in phloem fibers, necrotic cortical parenchyma external to wound periderm, and necrotic ligno-suberized parenchyma on the callus surface.

IV. VALSA CANKER OF APPLE

Valsa canker, caused by the fungus *Valsa ceratosperma* (Tode ex Fr.) Maire (anamorph *Cytospora sacculus* [Schwein.] Gvritischvilli) is an important disease of apple in the Pacific Rim countries, including Japan, China, and Korea.[35] It is only found occasionally on pear and quince. In northern Japan, the disease is especially severe with 35% of orchards affected to some degree. The disease is manifested as cankers around fruit scars, twig stubs, branch crotches, and sites of winter or mechanical injuries to the bark. The surface of infected bark may appear swollen and water soaked and, when wet, the centers of cankers may appear pinkish. Infections are commonly seen as elongated cankers on the upper sides of the larger limbs of older trees. Most new lesions appear in spring, between March and late April.

Cankers elongate rapidly in the spring and early summer. Canker growth is slow in the remainder of summer and during the winter. The fungus can be isolated from about 5 mm beyond the margin of actively growing cankers. Single hyphae of *Valsa ceratosperma* rapidly invade the cortical and phloem tissues during the spring and early summer. Colonization proceeds similarly, albeit more slowly, during the dormant season. A transition zone of collapsed cells separates infected from noninfected tissue. As temperatures increase in the summer, the host forms a lignified layer in the transition zone, followed by formation of cork layers with thin-walled phellem cells. During this period, the fungus forms mycelial aggregates that penetrate and destroy wound periderm tissues by mass action.[36] However, repeated formation of cork layers eventually results in a decreased rate of lesion development. Wound periderm composed of thick-walled cells is formed by the month of August, however, as temperatures decline in the autumn, healthy tissue is reinvaded by the fungus that grows through defective phellem layers between the periderm and cortex or between the phloem and xylem. The authors concluded that the wound cork layers acted as a temporal barrier to mycelial invasion and that the rate of formation of periderm depended upon host metabolic activity.[36]

V. NECTRIA CANKER OF APPLE, PEAR AND QUNICE

Nectria canker, caused by the fungus *Nectria galligena* Bres. (anamorph *Cylindrocarpon heteronemum* [Berk. & Broome] Wollenweb.) occurs in North America, Chile, Australia, New Zealand, northern continental Europe, the U.K., South Africa, and Japan. In North America, the disease, also known as European apple or crotch canker, occurs in the northeastern U.S. and southeastern Canada and westward to the Pacific Coast. The fungus that causes Nectria canker has a broad host range, including species of aspen, beech, birch, elm, and hickory, among others.[37]

The first visible symptoms of infection are observed at nodes and appear as elliptical, sunken areas. The infected area may appear water soaked and darker than the adjacent, healthy tissue. Infected twigs and branches may be girdled as the cankers enlarge, causing the collapse of portions of the shoot distal to the infection. If shoots are not killed during the first year thay are infected, the host may produce an annual layer of callus surrounding the area colonized by the fungus. The following year, the fungus invades and kills the callus tissue and, when this cycle of healing and reinvasion occurs annually, the cankers develop a concentric or zonate appearance. Gelatinous sporodochia of the fungus are common on cankers during wet weather and the bright red to orange perithecial stage often appears on older cankers in late fall or winter.

Nectria galligena is able to enter the host through wounds inflicted by pruning, frost injury, breakage caused by ice and snow,[38] and small dead brach stubs.[39,40] In apple, *N. galligena* enters the host through pruning wounds, frost injury, infections caused by *Venturia inaequalis* and *Neofabrea malicorticis,* and leaf scars.[41,42] For the fungus to become well established, wounds must reach to the depth of the vascular cambium. The host is able to confine the pathogen to the cortex when wounds are shallow. The fungus is able to colonize all of the tissues it penetrates. In the cortex and phloem, hyphae are intercellular in the early stages of infection, but intracellular hyphae were observed during the later stages. The fungus grows vigorously in the phloem region and enters the xylem via the medullary rays. Although the pathogen is unable to penetrate suberized cell walls,[43] it grows freely in the vessels and tracheids and spreads in the xylem through the pits. Intracellular phenolic compounds, gum in the xylem, and wound lignin at the levels observed in this host/pathogen interaction do not deter fungal colonization.

VI. COMPARATIVE ANATOMY OF PERIDERM FORMATION IN RESPONSE TO WOUNDS AND FUNGAL PATHOGENS

In tree bark, delimitation of fungal cankers ultimately occurs via the formation of ligno-suberized boundary tissue derived from cells present at the time of inoculation and necrophylactic periderm derived from newly differentiated phellogen.[18,19,44-46] The same statement is accurate for "healing" processes which begin immediately after wounding. For some cankers, barriers may be effective for preventing continued colonziation and diseased tissues may be sloughed. For established perennial cankers, bark boundary tissues, which also may contain extensive xylem inclusions,[46] are either directly penetrated or circumvented annually, resulting in a series of concentric callus ridges.

The direct involvement of the ligno-suberized layer and/or wound periderm, or by-products generated during the differentiation of these tissues, in the resistance of woody plants to fungi has never been demonstrated conclusively. Several lines of evidence indicate that periderms function to prevent desiccation of internal layers, prevent inward movement of pathogen toxins or enzymes, and serve as physical barriers to colonization. The process of generating new tissues also may result in lignification of pathogen hyphae[17] or in the production of fungistatic compounds which act as phytoalexins.[44]

Several recent papers have described the formation of necrophylactic periderm and associated tissues in response to wounding and pathogen inoculation of peach trees.[17,19,33] Some notable differences were observed. For example, pathogens affect the position or location of periderm and related tissues in wounded bark relative to their location in the absence of pathogens (Figures 19 and 20). In wounds, new tissues are regenerated usually within 1 to 2 mm of the wound surface.[18] When wounds are colonized by aggressive pathogens, the formation of new tissues may be several centimeters from the infection court. While this may seem obvious and perhaps insignificant, the location of natural defensive barriers in trees is of paramount importance for the proper management of canker diseases.

Pathogens affect the morphology and differentiation of new tissues and may retard or inhibit the formation of periderm and/or the deposition of suberin in wound periderm and xylem barrier zones (Figures 21 and 22).[7,11,19,33,36] Such alterations could influence the effectiveness of the periderm in limiting further spread of the pathogen within the host.

Few hosts are able to wall out the initial invasion of a pathogen with the formation of a single ligno-suberized boundary zone and necrophylactic periderm, as usually occurs at wounds. Instead, the tree often forms a series of periderms, each exhibiting varying degrees of development, within a single season in response to the advancing fungus. The fungus will continue to spread in the host tissue until conditions become more favorable for the host to form a complete periderm and less favorable for the pathogen to penetrate further. This is the reason many canker lesions on juvenile stems have vislble concentric areas of discoloration during the initial establishment phase. For perennating cankers, the tree also produces annual rings of callus tissue which represent the successful formation of necrophylactic periderm and callus tissue for that year, but gives no clue as to the number of attempts during that particular year.

VII. CANKER EXPANSION AND ROUTES OF REENTRY FOR PATHOGENS

Very few studies on hardwood cankers in general, or fruit tree cankers in particular, have presented evidence to show conclusively how fungi invade the phellogen layer to reinfect the phloem after the canker has stopped expanding for the season. For many canker-causing fungi, the main mode of reinfection is by direct hyphal penetration of the wound

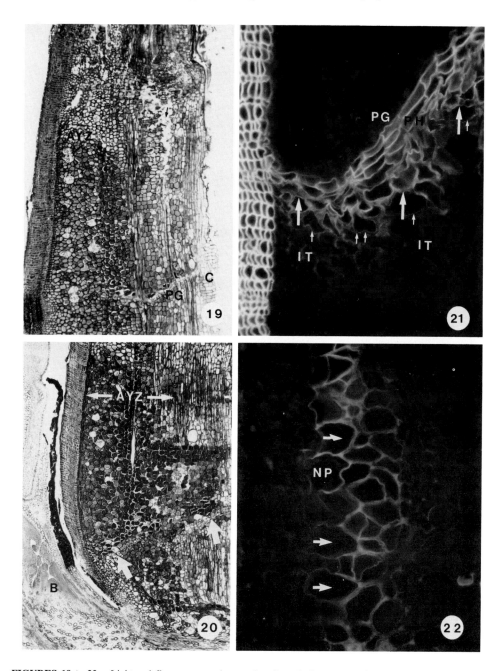

FIGURES 19 to 22. Light and fluorescence micrographs of peach tissues wounded (Figure 21) or inoculated with *L. persoonii* (Figures 19, 20, and 22). **Figure 19.** Longitudinal section through a portion of colonized bark showing the relative position of polysaccharide-rich zone (AYZ), ligno-suberized tissue (IT), and newly generated phellogen (PG). Note hyphae (arrow) and small callus lip (C). (Magnification × 448.) **Figure 20.** Longitudinal section through bark and lateral bud (B) showing location of polysaccharide-rich zone (AYZ and small arrows) and dedifferentiating tissue leading to generation of ligno-suberized tissue and new phellogen (large arrows). (Magnification × 144.) **Figure 21.** Transverse section of bark tissue stained with phosphine and examined under UV excitation, showing differentiated phellem (PH) and cells of the ligno-suberized zone (IT, its internal boundary delimited by arrow). (Magnification × 640.) **Figure 22.** Transverse section of a portion of the ligno-suberized boundary zone and new periderm (NP) adjacent to peach bark tissue colonized by *L. persoonii* (from right) showing altered phellem cell shape (arrows). Note contrast between altered tissues in this figure and normal periderm in Figure 21. (Magnification × 640.) (Figures 19 to 21 from Biggs, A. R., *Can. J. Bot.,* 62, 2814, 1984. With permission. Figure 22 from Biggs, A. R., *Phytopathology,* 76, 905, 1986. With permission.)

periderm. Penetration may be either direct via mass action of hyphae or indirect via hyphal penetration of fibers or sclereids that occasionally bridge wound periderms and connect diseased areas with healthy tissue. Penetration of periderms by mass action has been described for Leucostoma canker of peach,[11] Nectria canker of apple,[42] and Valsa canker of apple.[36] In contrast, *Cryphonectria parasitica,* which forms diffuse cankers on American chestnut, grows under wound periderm before it forms,[48] and thus can rapidly girdle stems. In addition, Hebard et al.[20] showed that mycelial fans of this fungus also could directly penetrate developing wound periderms.

Other avenues of reinfection occur in natural and pathogen-induced discontinuities along the wound periderm and new callus tissue produced by the host in response to invasion. For example, *L. persoonii* can circumvent wound periderm by colonizing the nonsuberized ventral callus region or by growing between the original periderm and wound periderm external to the cortex, or between the wound periderm and phloem fibers.[18,19] Similar routes of reinfection were described for Valsa canker of apple.[36]

The lumens of phloem fibers may also provide a means of entry for fungi into healthy tissues delimited by wound periderm. Apple bark colonized by *N. galligena*[42] and poplar bark colonized by *Cytospora chrysosperma*[15] may be reinvaded in this manner. Similarly, *Eutypella parasitica* may reinvade maple bark through sclereids or other discontinuities in the wound periderm.[49] Colonization of phloem fibers also has been described for Leucostoma canker of peach[18] and illustrated for Botryosphaeria canker of peach.[33]

Ashcroft[50] proposed that fungi such as *N. galligena* which colonize the xylem may reinvade the bark by growing through the medullary rays and into the wound wood (CODIT wall 4) and ray parenchyma into healthy bark that was not delimited by wound periderm. This has never been demonstrated, although French[49] mentions it as a possibility of *E. parasitica.*

It is likely that most fungal canker pathogens can reinvade the host by any or all of the above routes. Reinvasion is accomplished by single hyphae or fungal mass action in combination with toxin production, and may be facilitated by environmental factors, such as cold injury to the new phellogen and host physiological status.

VIII. CONCLUSIONS

Histological studies have provided a rudimentary understanding of the host/pathogen interaction between woody hosts and canker-causing fungi. We known that resistance to canker pathogens is largely nonspecific and therefore we must pay increasing attention to the host if progress in understanding these diseases is going to continue. Elucidating the factors that trigger periderm formation in woody plants could lead to improved control of these diseases via manipulation of the host reaction following pruning. The relative role of resistance in bark and xylem tissues is not understood and more research is needed to elucidate the relative importance of the many and varied contributions of partial resistance in woody plants. We know little about the importance and extent of pathogen variation for these diseases. We do know that environment plays a major role in the interaction between host and pathogen, although few studies have attempted to quantify environmental influences. Because of this, we lack the information required to accurately predict the occurrence of canker diseases. More importantly, we lack the knowledge required to formulate reliable plant health criteria to take advantage of current levels of resistance.

REFERENCES

1. **Biggs, A. R.,** Integrated control of Leucostoma canker of peach in Ontario, *Plant Dis.,* 73, 869, 1989.
2. **Ritchie, D. F. and Clayton, C. N.,** Peach tree short life: a complex of interacting factors, *Plant Dis.,* 65, 642, 1981.
3. **DeVay, J. E., Gerdts, M., English, H., and Lukezic, F. L.,** Controlling Cytospora canker in President plum orchards in California, *Calif. Agric.,* 28, 12, 1974.
4. **Helton, A. W. and Konicek, D. E.,** Effects of selected *Cytospora* isolates from stone fruits on certain stone fruit varieties, *Phytopathology,* 51, 152, 1961.
5. **Babos, K., Rozsnyay, Zs. D., and Kelment, Z.,** Apoplexy of apricots. V. Pathological and histological investigations of the apoplexy of apricots, *Acta Phytopathol. Acad. Sci. Hung.,* 11, 71, 1976.
6. **Portilla, M. T. and Tuset, J. J.,** Chancros en fruitales de hueso: *Leucocytospora cincta* (Sacc.) Höhnel un coelomycete activo en el melocotonero, *Inv. Agrar. Prod. Prot. Veg.,* 1, 259, 1986.
7. **Goring, M. -C.,** Zur Ätiologie der Valsa-Krankheit an Süsskirsche. Histologische Untersuchungen über Eintrittspforten und Ausbreitung von Leucostoma persoonii (Nits.) Hohnel, *Mitt. Biol. Bundesanst. Land. Forstwirtsch. Berlin-Dahlem,* Vol. 162, 1975.
8. **Biggs, A. R.,** Leucostoma canker, in *Compendium of Stone Fruit Diseases,* Ogawa, J. M. and Zehr, E. I., Eds., American Phytopathological Society, St. Paul, MN, 1992, 20.
9. **Togashi, K.,** Studies on the pathology of peach canker, *Bull. Imp. Coll. Agric. For.,* Morioka, Japan, 16, 1931.
10. **Tekauz, A. and Patrick, Z. A.,** The role of twig infections on the incidence of perennial canker of peach, *Phytopathology,* 64, 683, 1974.
11. **Wisniewski, M., Bogle, A. L., and Wilson, C. L.,** Histopathology of canker development on peach trees after inoculation with *Cytospora leucostoma, Can. J. Bot.,* 62, 2804, 1984.
12. **Rozsnyay, D. S. and Barna, B.,** Apoplexy of apricots. IV. Studies on the toxin production of *Cytospora (Valsa) cincta* Sacc., *Acta Phytopathol. Acad. Sci. Hung.,* 9, 301, 1974.
13. **Tsakadze, T. A.,** The action of the toxin of *Cytospora leucostoma* on the plant cell, *Bull. Cent. Bot. Gdn. Moscow* 35, 75, 1959 (in Russian).
14. **Svircev, A. M., Biggs, A. R., and Miles, N. W.,** Isolation and partial purification of phytotoxins from liquid cultures of *Leucostoma cincta* and *L. persoonii, Can. J. Bot.,* 69, 1998, 1991.
15. **Biggs, A. R., Davis, D. D., and Merrill, W.,** Histology of cankers on *Populus* caused by *Cytospora chrysosperma, Can. J. Bot.,* 61,.563, 1983.
16. **Koganezawa, H. and Sakuma, T.,** Possible role of breakdown products of phloridzin in symptom development by *Valsa ceratosperma, Ann. Phytopathol. Soc. Jpn.,* 48, 521, 1982.
17. **Biggs, A. R.,** Comparative anatomy and host response of two peach cultivars inoculated with *Leucostoma cincta* and *L. persoonii, Phytopathology,* 76, 905, 1986.
18. **Biggs, A. R.,** Boundary zone formation in peach bark in response to wounds and *Cytospora leucostoma* infection, *Can. J. Bot.,* 62, 2814, 1984.
19. **Biggs, A. R.,** Intracellular suberin: occurrence and detection in tree bark, *Int. Assoc. Wood Anat. Bull.,* 5, 243, 1984.
20. **Hebard, F. V., Griffin, G. J., and Elkins, J. R.,** Developmental histopathology of cankers incited by hypovirulent and virulent isolates of *Endothia parasitica* on susceptible and resistant chestnut trees, *Phytopathology,* 74, 140, 1984.
21. **Russin, J. S. and Shain, L.,** Initiation and development of cankers caused by virulent and cytoplasmic hypovirulent isolates of the chestnut blight fungus, *Can. J. Bot.,* 62, 2660, 1984.
22. **Biggs, A. R.,** Detection of impervious tissue in tree bark with selective histochemistry and fluorescence microscopy, *Stain Technol.,* 60, 299, 1985.
23. **Biggs, A. R. and Stobbs, L. W.,** Fine structure of the suberized cell walls in the boundary zone and necrophylactic periderm in wounded peach bark, *Can. J. Bot.,* 64, 1606, 1986.
24. **Butin, H.,** Über den Einfluss des Wassergehaltes der Pappel auf ihre Resistenz gegenüber *Cytospora chrysosperma* (Pers.) Fr., *Phytopathol. Z.,* 24, 245, 1955.
25. **Cline, M. N. and Neely, D.,** Wound-healing process in geranium cuttings in relationship to basal stem rot caused by *Phythium ultimum, Plant Dis.,* 67, 636, 1983.
26. **Bostock, R. M. and Middleton, G. E.,** Relationship of wound periderm formation in resistance to *Ceratocystis fimbriata* in almond bark, *Phytopathology,* 77, 1174, 1987.
27. **Kolattukudy, P. E.,** Biochemistry and function of cutin and suberin, *Can. J. Bot.,* 62, 2819, 1984.
28. **Weaver, D. J.,** A gummosis disease of peach trees caused by *Botryosphaeria dothidea, Phytopathology,* 64, 1429, 1974.
29. **Britton, K. O. and Hendrix, F. F.,** Three species of *Botryosphaeria* cause peach gummosis in Georgia, *Plant Dis.,* 66, 1120, 1982.

30. **Weaver, D. J.**, Role of conidia of *Botryosphaeria dothidea* in the natural spread of peach tree gummosis, *Phytopathology*, 69, 330, 1979.

31. **Pusey, P. L.**, Influence of water stress on susceptibility of nonwounded peach bark to *Botryosphaeria dothidea*, *Plant Dis.*, 73, 1000, 1989.

32. **Britton, K. O. and Hendrix, F. F.**, Population dynamics of *Botryosphaeria* spp. in peach gummosis cankers, *Plant Dis.*, 70, 134, 1986.

33. **Biggs, A. R. and Britton, K. O.**, Presymptom histopatholgoy of peach trees inoculated with *Botryosphaeria obtusa* and *Botryosphaeria dothidea*, *Phytopathology*, 78, 1109, 1988.

34. **Venkatasubbaiah, P., Sutton, T. B., and Chilton, W. S.**, Effect of phytotoxins produced by *Botryosphaeria obtusa*, the cause of black rot of apple fruit and frogeye leaf spot, *Phytopathology*, 81, 243, 1991.

35. **Sakuma, T.**, Valsa canker, in *Compendium of Apple and Pear Diseases,* Jones, A. L. and Aldwinckle, H. S., Eds., American Phytopathological Society, St. Paul, MN, 1990, 39.

36. **Tamura, O. and Saito, I.**, Histopathological changes of apple bark infected by *Valsa ceratosperma* (Tode ex Fr.) Maire during dormant and growing periods, *Ann. Phytopathol. Soc. Jpn.*, 48, 490, 1982.

37. **Grove, G. G.**, Nectria canker, in *Compendium of Apple and Pear Diseases,* Jones, A. L. and Aldwinckle, H. S., Eds., American Phytopathological Society, St. Paul, MN, 1990, 35.

38. **Lortie, M.**, Pathogenesis in cankers caused by *Nectria galligena, Phytopathology*, 54, 261, 1964.

39. **Grant, R. J. and Childs, T. W.**, Nectria canker of northeastern hardwoods in relation to stand improvement, *J. For.*, 38, 797, 1940.

40. **Zalasky, H.**, Penetration and initial establishment of *Nectria galligena* in aspen and peachleaf willow, *Can. J. Bot.*, 46, 57, 1968.

41. **Wiltshire, S. P.**, Studies on the apple canker fungus. I. Leaf scar infection, *Ann. Appl. Biol.*, 8, 182, 1921.

42. **Crowdy, S. H.**, Observations on apple canker. III. The anatomy of the stem canker, *Ann. Appl. Biol.*, 36, 483, 1949.

43. **Krähmer, H.**, Wound reactions of apple trees and their influence on infections with *Nectria galligena, J. Plant Dis. Protect.*, 87, 97, 1980.

44. **Mullick, D. B.**, The non-specific nature of defense in bark and wood during wounding, insect and pathogen attack, *Recent Adv. Phytochem.*, 11, 395, 1977.

45. **Biggs, A. R., Merrill, W., and Davis, D. D.**, Discussion: response of bark tissues to injury and infection, *Can. J. For. Res.*, 14, 351, 1984.

46. **Biggs, A. R.**, Anatomical and physiological responses of bark tissues to mechanical injury, in *Defence Mechanisms of Woody Plants Against Fungi,* Blanchette, R. A. and Biggs, A. R., Eds., Springer-Verlag, Berlin, 1992, chap. 2.

44. **Biggs, A. R.**, Responses of angiosperm bark tissues to canker and canker rot fungi, in *Defence Mechanisms of Woody Plants Against Fungi,* Blanchette, R. A. and Biggs, A. R., Eds., Springer-Verlag, Berlin, 1992, chap. 3.

48. **Bramble, W. C.**, Reaction of chestnut bark to invasion by *Endothia parasitica, Am. J. Bot.*, 23, 89, 1934.

49. **French, W. J.**, Eutypella Canker on *Acer* in New York, *State Univ. Coll. For. Tech. Publ.*, No. 94, Syracuse, NY, 1969.

50. **Ashcroft, J. M.**, European canker of black walnut and other trees, W. Va. Agric. *Exp. Stn. Bull.*, 261, 1934.

Chapter 9

THE DEVELOPMENT AND ULTRASTRUCTURE OF GUM DUCTS IN CITRUS IN RESPONSE TO PATHOGENS

Esther Shedletzky

TABLE OF CONTENTS

I. INTRODUCTION

The phenomenon known as gummosis is defined as the secretion of gum as a result of injuries such as wounding, mechanical pressure, attack of microorganisms or insects, or physiological disturbances (water stress, etc.). Gummosis can be seen in many diseased plants and is abundant in several groups, including the order Leguminosae and the families Anacardiaceae, Meliaceae, Rosaceae, and Rutaceae.[1] Gum is present in special cavities, known as gum ducts; it fills the lumen of tracheary elements as well as parenchymatous cells;[2-10] or it can be found in different plant organs including shoot, root, and fruit.

Since many Rosaceae species, are important fruit trees, gummosis in this family has been investigated extensively and will not be reviewed here (see Chapter 8 and References 11 and 12 and papers cited therein). The description of gum duct structure and formation presented in this chapter focuses on *Citrus*, where gum secretion is the common result of many fungal and viral bark diseases.[13-15]

II. LOCATION AND GENERAL STRUCTURE OF GUM DUCTS

The first sign of brown rot gummosis caused by *Phytophthora citrophthora* in *Citrus* trees, is often the secretion of gum.[14] The fungus usually attacks healthy and well-developed trees and can cause rapid death. The main threat is the death of large areas of the bark and girdling of the tree. The fungus penetrates the stem through cracks in the bark or cortex, and kills and causes browning of the tissues in which it has spread. It also induces the formation of gum ducts below and above the lesion. The gum, an acidic polysaccharide which can absorb a large volume of water and expand, is present in the gum ducts under positive pressure. It exudes through cracks in the bark or cortex killed by the fungus. The gum itself does not seem to have any effect on the tree, and apparently serves as a mechanical barrier, since it dries up on the surface of the tree and blocks the penetration of secondary pathogens. It also prevents water loss through the cracked bark.

When young stems of *Citrus aurantium* L. and *C. volkamariana* Pasquale were infected artificially with the fungus *P. citrophthora*[16] (Figure 1), it was found that epithelial cells of a longitudinal gum duct were formed by the cambium initials toward the xylem. The rate of formation and size of gum ducts depended on the cambial activity at the time of infection of the stems. During periods of high cambial activity, gum ducts reached an average distance of 3 to 5 cm above and below the lesion; when cambial activity was low, gum duct length reached only a few millimeters.

Gum ducts are round in cross sections (Figure 2); they are elongated, but are observed to fuse above and below xylem rays forming a net-like pattern in tangential sections. Cross sections in young stems in the immediate vicinity of the lesion caused by the spreading of the fungus reveal the largest gum ducts. In addition to gum found in the ducts in this region, gum can also be found inside vessel elements and parenchyma cells of the cortex and phloem, and sometimes in very young stems even inside the parenchyma cells of the pith. Both the size and number of the ducts decrease with increasing distance from the lesion.

In mature trees in the orchard, where infection occurs naturally, the area covered with gum ducts is much larger than that found in young stems that were artificially infected. This may be due to the fact that in the orchard conditions may favor a continuous growth and spreading of the fungus in the bark, and thus a continuous stimulus for gum duct formation.

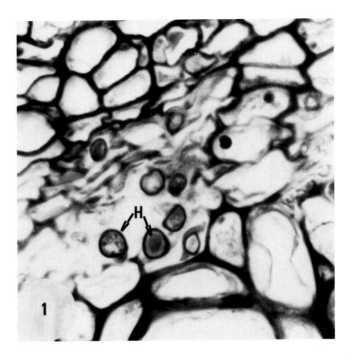

FIGURE 1. Light micrograph of the cambial zone where all cells are dissolved and penetrated by the hyphae of the fungus *Phytophthora citrophthora*. H, hyphae. (Magnification × 950.)

III. THE STRUCTURE AND ULTRASTRUCTURE OF DEVELOPING GUM DUCTS FORMED IN RESPONSE TO PATHOGENS

Early stages of gum duct formation were described by different authors as being either lysigenous or schizogenous. Gum ducts formed lysigenously are the result of the dissolution of cell walls at the cambial zone where xylem mother cells form, as in cherry shoots.[18] Schizogenously formed gum ducts are the result of separation of cells, followed by the differentiation of polysaccharide-secreting cells. Ducts formed via the latter method were found to develop in a very similar manner at the cambial zone in *Citrus* stems[16] and in vascular bundles of almond fruit.[12]

In *Citrus*, the first signs of gum duct formation, both after artificial infection with the fungus *P. citrophthora* (the causal agent of brown rot gummosis) and the application of the ethylene-releasing compound, ethrel, are the swelling and dissolution of the middle lamella between xylem mother cells (Figures 3 and 4) and the formation of a cavity (Figure 5). These changes are seen 1 or 2 d after artificial infection of stems; the result is a schizogenous cavity. In almond fruits, in which a similar separation of vascular parenchyma cells occurs[12] an increase in the activity of the enzyme polygalacturonase in the extracellular space of the mesocarp tissue was found prior to the onset of the schizogenous stage. It was suggested that this increase corresponds to the degradation of the middle lamella, which results in the separation of cells.[19]

The duct lumen continues to increase due to the schizogenous activity 2 to 3 d after gum duct induction, and the cells surrounding the formed cavity differentiate into secretory epithelial cells (Figure 6). These cells are round in cross sections of the stem and elongated when seen in longitudinal sections. They have dense cytoplasm filled with Golgi bodies, cisternae of rough endoplasmic reticulum (ER), and numerous mitochondria (Figure 7), all

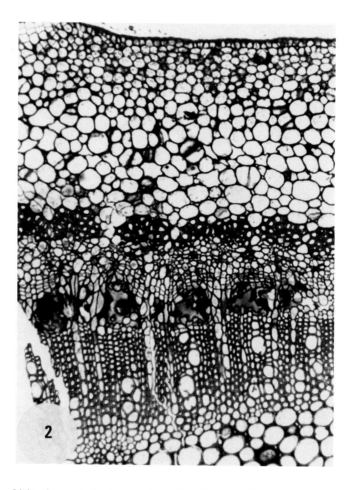

FIGURE 2. Light micrograph showing round gum ducts between xylem rays. (Magnification × 160.)

of which are signs of high metabolic activity. These signs of activity become more obvious 3 to 5 d postinfection, when numerous vesicles and small vacuoles containing fibrillar material are seen associated with the Golgi bodies (Figure 8). Some of these vesicles and vacuoles are associated with the area of the plasmalemma facing the duct lumen, and they appear to fuse with the plasmalemma and discharge their contents to the outside of the cytoplasm (Figure 8). The fibrillar material seen in these vesicles and vacuoles, as well as the usually thicker wall facing the duct lumen of the epithelial cell, stains strongly for polysaccharides with the PATAg method[20,21] (Figure 8). This indicates a directional transport of polysaccharides formed by the Golgi apparatus and moved out of the cell into the gum duct lumen. Synthesis of gum polysaccharides and their transport into the gum ducts is probably also the case in the development of the traumatic gum ducts in branches of cherry trees. Here, Stösser[22] demonstrated the incorporation of ^{14}C-labeled sucrose and glucose into the duct initials and the exuded gum.

As the epithelial cells of the *Citrus* ducts pass their stage of active synthesis and transport of polysaccharides, their cytoplasm gradually darkens, shrinks, and withdraws from the cell wall facing the duct lumen (Figure 9). Gum accumulates in the space formed between the shrinking cytoplasm and the cell walls as well as in the wall itself. The cambium resumes its regular pattern of divisions and forms new xylem elements (Figure 10). This results in

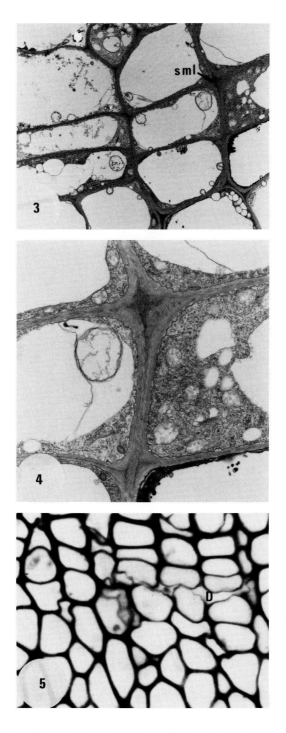

FIGURES 3 to 5. **Figure 3**: Electron micrograph of the cambial zone showing the beginning of swelling of the middle lamella (sml) in the corners of the cells. (Magnification × 2920.) **Figure 4**. Enlargement of the area indicated by an arrow in Figure 3. (Magnification × 10,600.) **Figure 5.** Light micrograph showing the separation of cells and the beginning of the formation of the gum duct (D) lumen. (Magnification × 1600.)

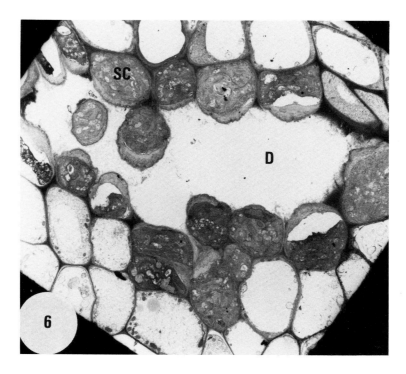

FIGURE 6.   Electron micrograph of a gum duct (D) with secretory cells (SC) at their most synthetically active stage. (Magnification × 2400.)

the embedding of the gum ducts in the xylem tissue (Figure 11). The ducts are then seen in cross sections as large, round, elliptical, or irregular in shape. They are lined by thin-walled parenchyma cells in which remnants of gum and dead cytoplasm are seen. Cells with broken walls filled with polysaccharides are also present as are remnants of the cell wall inside the duct cavities. This indicates a lysigenous stage, which follows the early schizogenous stage. The cavity is first formed by separation of cells, followed by active secretion of polysaccharides, and ending in breakdown of some of the gum-filled cells.

It should be mentioned that the above pattern of duct formation is seen in areas that are further away from the cite of fungal infection or ethylene application. The closer the microscopical sections are to the site of gum duct induction, the less organization one can see in them; many dead and broken cells and much larger ducts and larger quantities of gum are seen.

IV. PHYSICOCHEMICAL PROPERTIES OF GUM IN RELATION TO FUNCTION IN WOUND SEALING

Gum is a branched polysaccharide usually containing uronic acids (galacturonic and/or glucuronic acid) and two or more neutral sugars such as arabinose, galactose, xylose, rhamnose, etc.[23,24] Gum composition differs for different species.[25] *Prunus* gums possess a backbone of glucuronic acid and mannose residues to which branched side chains containing residues of galactose and arabinose[11] are attached. *Acacia* gums, on the other hand, have a backbone of galactose residues and branches consisting of galactose, arabinose, and rhamnose residues.[11] *Citrus* gum also has a backbone of galactose residues, which is highly branched and has side chains consisting of arabinose and glucuronic acid residues.[23-29]

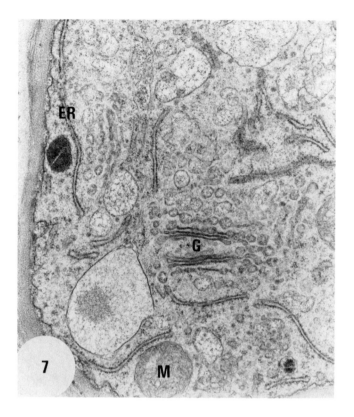

FIGURE 7. Electron micrograph showing a secretory cell at the stage of gum production. Golgi bodies (G) and numerous associated vesicles are seen, as well as small and larger vacuoles filled with fibrillar material (ER, endoplasmic reticulum; M, mitochondrion). (Magnification × 22,500.)

Gums show molecular weight dispersion.[25,30,31] They contain a water-soluble fraction and a gel fraction. The molecular weight of soluble gum from different *Citrus* species ranges between 200,000 to 300,000,[32] while gum remaining as gel can reach a molecular weight of 2 million.[28]

A high content of uronic acids makes the gum a polyanion. The carboxylic groups on different chains within a macromolecule cause the formation of repulsion forces that maintain the polysaccharide macromolecules in a swollen gel and increase gum solution viscosity as gum concentration increases, as well as in very low gum concentrations.[32,33]

A high content of calcium ions is also present in the gum (e.g., 4.1 and 4.3 mmol/g dry gum in *C. volkamariana* Pasquale and *C. limetta* Risso, respectively[32]). These ions interact with the carboxylic groups to from cross-linkages between molecules and contribute to gel formation.[29,34,35]

Evidence that *Citrus* gum does not inhibit growth of the fungus *P. citrophthora*, when nutrient amounts are not limited, is presented in Table 1. This is also supported by Kardoŝová et al.,[36] who found that fungi can grow on and utilize substrate-containing gum.

The above-mentioned properties cause gum gelation and swelling when hydrated, and upon drying, the gum forms a hard layer. As mentioned earlier, the accepted view is that gum, which oozes through wounds and cracks in the cortex or bark, serves to block them. The viscosity of the fluid in tracheary elements increases via the penetration of the soluble fraction of the gum; as gum concentration increases and a gel forms, water movement in the elements slows down and can be blocked completely. These effects may prevent the

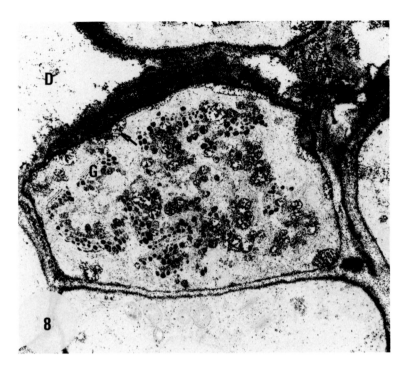

FIGURE 8. Electron micrograph of a secretory cell stained for polysaccharides with the PATAg technique. Intense staining is seen in Golbi bodies (G) and associated vesicles as well as in small vacuoles, some of which seem to fuse with the plasmalemma (arrow) facing the duct (D) lumen. The wall at that side also shows intense staining, and apparently is impregnated with polysaccharides. (Magnification × 12,000.)

penetration and spreading of pathogens as well as water loss through the wounded and infected areas in the tree.[13,29,37-41] It seems that the gum itself has no toxic effect on the growth of fungi.

The role of gum production in peach stems has been the subject of considerable discussion.[11] Although gumming probably represents a general response to injury and infection, the proven effectiveness of the response as a protective barrier has yet to be documented. In fact, the production of excessive gum requires significant metabolic activity and thus may represent a significant loss of carbohydrates from the tree that would otherwise be available for vegetative and reproductive growth. One could speculate that the production of copious amounts of gum, apparently without much benefit, may represent a high energy expenditure, thus leading to depletion of reserves and poor growth, which eventually leads to decline. If true, such a hypothesis would cast doubt on the role of a strong gumming response as an effective protective barrier. However, such speculation warrants further research before relegating a strong gumming response to a list of unfavorable breeding characteristics for peach. It also indicates that evidence is needed to demonstrate this trait as an operative mechanism in plant defense.

V. PATHOGENS, GROWTH SUBSTANCES, AND GUM DUCT FORMATION

The induction of gummosis by growth substances is well documented in the literature. Gummosis was induced by naphthalene acetic acid (NAA), which was used for sprout control in peach[42] and for the control of pruning in peach and nectarine,[43] as well as by the ethylene-

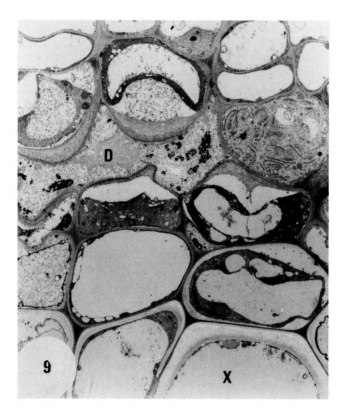

FIGURE 9. Electron micrograph of a gum duct (D) with secretory cells at their final stage of secretion. Their cytoplasm is dark and withdrawn from the cell wall (X, xylem). (Magnification × 2900.)

releasing compound, ethrel (ethephon, CEPA, or 2-chloroethane-phosphonic acid), which was applied to different *Prunus* sp.[18,41,44]

NAA, ethrel, and ACC (1-amino-cyclopropane-1-carboxylic acid), the immediate precursor of ethylene in higher plants, were found to induce gum duct formation in *Citrus* stems. The growth substance with the most rapid and strongest effect in *Citrus* was ethylene, which always caused the formation of gum ducts except in midwinter, when cambial activity was the lowest. Very regular ducts were formed beginning with the dissolution of the middle lamella and the differentiation of three or four layers of epithelial cells that have very little (if any) dead cells. The pattern of gum duct formation was found to be very similar to that described in stems infected with the fungus *P. citrophthora*.[16]

The formation of ethylene by diseased plants is a well-known phenomenon[45-48] which is attributed to the host and/or the pathogen. A sharp peak in ethylene release from *Citrus* stems infected with *P. citrophthora* followed by a sharp decrease was observed 2 d after artificial infection (Figure 12). This corresponds with the development of the disease symptoms (fungal growth stopped 2 to 3 d after infection) and the beginning of gum duct formation, which was observed in the cambial zone 2 to 3 d after infection. It seems that when the rate of ethylene production in the infected stems reached a high enough level, it triggered a change in the differentiation pattern of the cambium from the production of normal xylem elements to the production of gum-secreting cells. Gum ducts induced by ethylene were formed earlier than those induced by other growth substances. This may suggest that ethylene is a primary trigger of differentiation. The fact that ACC, which is the immediate precursor of ethylene in higher plants, but not in fungi, caused the formation of gum ducts, supports the idea that the released ethylene originated from the host.

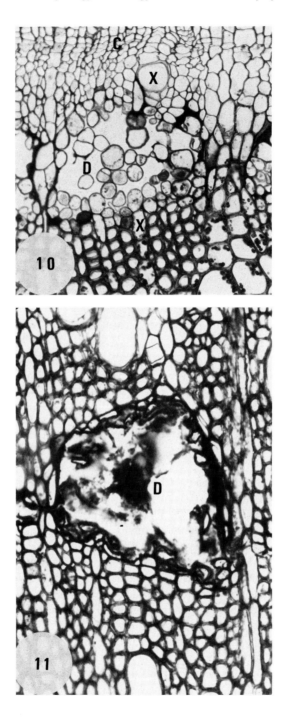

FIGURES 10 and 11. **Figure 10**. Light micrograph showing the resumption of cambial (C) activity and xylem (X) differentiation after gum ducts (D) formed and the surrounding cells passed their secretory stage. (Magnification × 300.) **Figure 11**. Light micrograph of a gum duct (D) embedded in the xylem. (Magnification × 530.)

TABLE 1
Growth of *Phytophthora citrophthora* Mycelium on Potato
Dextrose Agar (PDA) in Presence or Absence of
1% (w/v) Gum of *C. limon* (L.) Burm.[a]

Days after mycelium transfer	Area of colony on PDA	Area of colony on PDA + gum	Growth (%)[a]
1	1.21 ± 0.02[b]	1.46 ± 0.02	120
2	4.21 ± 0.04	3.55 ± 0.03	119
5	16.87 ± 0.08	15.76 ± 0.03	97

[a] Percentages shown in this column are relative to the area of the colony growing on PDA, which was assumed to be 100%.
[b] Standard deviation for n = 10.

From Gedalovich, E., Gummosis in *Citrus* Trees — Structure, Ultrastructure and the Development of the Traumatic Tissue that Produces Gum in *Citrus* Trees, Ph.D. thesis, The Hebrew University of Jerusalem, Jerusalem, Israel, 1985.

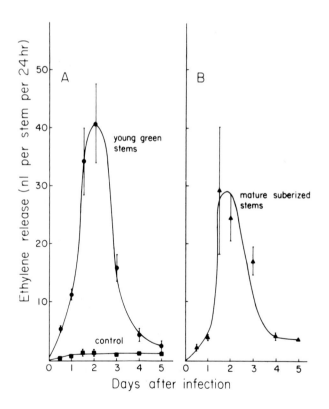

FIGURE 12.   Ethylene release from *Citrus* stems artificially infected by the fungus *Phytophthora citrophthora*. (A) Young green stems (●); control: wounded uninfected stems (■). (B) Mature suberized stems (▲). A sharp peak in ethylene release is seen 2 to 3 d after artificial infection, which corresponds to the onset of gum duct formation.

In contrast to the extensive gum production caused by auxin in *Prunus* sp., its effect on gum duct formation in *Citrus* was local. Also, in the case of auxin, gum duct production, as in the case for artificial infection and the application of other growth substances, was correlated to the activity of the cambium. The formation of gum ducts as a result of auxin treatment can be attributed to an auxin-induced ethylene formation.[49-52]

It is possible that the direct effect of ethylene is the induction of the synthesis of hydrolytic enzymes. These enzymes subsequently dissolve the middle lamella between xylem mother cells, forming the duct lumen. This was found to occur in the formation of aerenchyma in waterlogged and ethylene-treated plants[53,54] and in specific cells of abscission zones.[55] However, this does not explain the synthesis of new gum polysaccharide material in the vessel lumen, as described above. Directly or indirectly, ethylene must also induce the energy-requiring biochemical and histological processes of specific gum polysaccharide synthesis.

VI. CONCLUDING REMARKS

Gummosis, or the formation of gum ducts, in plants is a common phenomenon known to occur in many fruit trees. Most of the more recent work done on the development of gum ducts suggests a process of differentiation of a special tissue that actively synthesizes and exudes polysaccharides from its cells. This process seems to be induced by ethylene, which is synthesized by the plant in response to infection. A variety of factors, including environmental stresses, wounding, and pathogens such as fungi and viruses, are apparently capable of causing ethylene formation at rates sufficient to induce gummosis. More research on the role of gum in plant defense and repair processes is required in order to clarify its contribution to the pathological condition.

ACKNOWLEDGMENTS

The author wishes to thank Dr. Ella Werker and Prof. A. Fahn, Department of Botany, The Hebrew University of Jerusalem, Israel, for critical reading of the manuscript. During the period of manuscript preparation, the author was supported by grant number I-1386-87 from the U.S.-Israel Binational Agricultural Research and Development Foundation (BARD).

REFERENCES

1. **Howes, F. N.**, *Vegetative Gums and Resins*, Chronica Britannica, Waltham, MA, 1949.
2. **Bell, A. A.**, Biochemical mechanisms of disease resistance, *Annu. Rev. Plant Physiol.*, 32, 21, 1981.
3. **Bevington, K. B.**, Development of union abnormalities in grafts between lemon (*Citrus limon*) and *Poncirus trifoliata*, *Aust. J. Agric. Res.*, 27, 661, 1976.
4. **Butler, O.**, A study of gummosis of *Prunus* and *Citrus*, with observations on squamosis and exanthema of the *Citrus*, *Ann. Bot.*, 25, 107, 1911.
5. **Catesson, A.-M., Czaninski, Y., Péresse, M., and Moreau, M.**, Sécrétions intravasculaires de substances "gommeuses" par les cellules associées aux vaisseaux en réaction à une attaque parasitaire, *Bull. Soc. Bot. Fr.*, 123, 93, 1976.
6. **Chattaway, M. M.**, The development of tyloses and secretion of gum in heartwood formation, *Aust. J. Sci. Res. Ser. B.*, 2, 227, 1949.
7. **Magnano di San Lio, G., Davino, M., and Catara, A.**, Osservazioni istologiche su una "gommosi" del bergamotto, *Riv. Patol. Veg. (Pavia)*, 14, 99, 1978.
8. **Matarese Palmieri, R., Tomasello, D., and Magnano di San Lio, G.**, Origine e caratterizzazione istochimica delle gomme in piante di agrumi con sindromi a diversa eziologia, *Riv. Patol. Veg. (Pavia)*, 15, 43, 1979.

9. **Moreou, M., Catesson, A.-M., Péresse, M. and Czaninski, Y.**, Dyamique comparée des réaction cytologiques du xylème de l'oeilet en présence de parasites vasculaires, *Phytopathol. Z.*, 91, 289, 1978.

10. **Schneider, H.**, Deposition of wound gum, callose, and suberin as response to diseases and wounding of *Citrus, Bull. Soc. Bot. Fr.*, 127, 143, 1980.

11. **Boothby, D.**, Gummosis of stone-fruit trees and their fruit, *J. Sci. Food Agric.*, 57, 603, 1983.

12. **Morrison, J. C. and Polito, V. S.**, Gum duct development in almond fruit, *Prunus dulcis* (Mill) D. A. Webb, *Bot. Gaz.*, 146, 15, 1985.

13. **Fawcett, H. S.**, *Citrus Diseases and Their Control*, 2nd ed., McGraw-Hill, New York, 1936.

14. **Klotz, L. J.**, Fungal, bacterial and nonparasitic diseases and injuries originating in the seedbed, nursery and orchard, in *The Citrus Industry*, rev. ed., Vol. 4, Reuther, W., Calavan, E. C., and Carman, G. E., Eds., Division of Agriculture, University of California, Berkeley, 1978, 1.

15. **Wallace, J. B.**, Virus and viruslike diseases, in *The Citrus Industry*, rev. ed., Vol. 4, Reuther, W., Calavan, E. C., and Carman, G. E., Eds., Division of Agriculture, University of California, Berkeley, 1978, 67.

16. **Gedalovich, E. and Fahn, A.**, The development and ultrastructure of gum ducts in *Citrus* plants formed as a result of brown-rot gummosis, *Protoplasma*, 127, 73, 1985.

17. **Gedalovich, E. and Fahn, A.**, Ethylene and gum duct formation in *Citrus, Ann. Bot.*, 56, 571, 1985.

18. **Stösser, R.**, Investigations on gum duct formation in cherries using a plastic embedding-medium, *Sci. Hort.*, 11, 247, 1979.

19. **Morrison, J. C., Greve, L. C., and Labavitch, J. M.**, The role of cell wall-degrading enzymes in the formation of gum ducts in almond fruit, *J. Am. Soc. Hortic. Sci.*, 112, 367, 1987.

20. **Roland, J. C.**, General preparation and staining of thin sections, in *Electron Microscopy and Cytochemistry of Plant Cells*, Hall, J. L., Ed., Elsevier/North Holland Biomedical Press, Amsterdam, 1978, 1.

21. **Thiéry, J. P. and Rambourg, A.**, Cytochemie des polysaccharides, *J. Microsc. (Paris)*, 21, 225, 1974.

22. **Stösser, R.**, Die autoradiographische Lokalisierung der ^{14}C-Aktivität nach Applikation markierter Zuckr beider Gummibildung von Süsskirschen, *Gartenbauwissenschaft*, 43, 231, 1978.

23. **Aspinall, G. O.**, Gums and mucilages, *Adv. Carbohydr. Chem. Biochem.*, 24, 333, 1969.

24. **Aspinall, G. O.**, Pectins, plant gums, and other plant polysaccharides, in *The Carbohydrates*, Vol. 2B, Pigman, W. and Horton, D., Eds., Academic Press, new York, 1970, 515.

25. **Smith, F. and Montgomery, R.**, *The Chemistry of Plant Gums and Mucilages*, Reinhold, New York, 1959.

26. **Anderson, E., Russell, F. H., and Seigel, L. W.**, The gum from lemon trees, *J. Biol. Chem.*, 113, 683, 1936.

27. **Connell, J. J., Hainsworth, R. M., Hirst, E. L., and Jones, J. K. N.**, Grapefruit and lemon gums. I. The ratio of sugars present in the gums and the structure of the aldobinic acid (4-D-glucuronosido-D-galactose) isolated by graded hydrolysis of the polysaccharides, *J. Chem. Soc.*, p. 1696, 1950.

28. **Stoddart, J. F. and Jones, J. K. N.**, Some structural features of *Citrus limonia* gum (lemon gum), *Carbohydr. Res.*, 8, 29, 1968.

29. **Towle, G. A. and Whistler, R. L.**, Hemicelluloses and gums, in *Phytochemistry*, Vol. 1, Van Nostrand Reinhold, New York, 1973, 198.

30. **Stoddart, J. F.**, *Stereochemistry of Carbohydrates*, Wiley-Interscience, New York, 1971.

31. **Clarke, A. E., Anderson, R. L., and Stone, B. A.**, Form and function of arabinogalactans and arabinogalactan-proteins, *Phytochemistry*, 18, 521, 1979.

32. **Gedalovich, E.**, Gummosis in *Citrus* Trees — Structure, Ultrastructure and the Development of the Traumatic Tissue that Produces Gum in *Citrus* Trees, Ph.D. thesis, The Hebrew University of Jerusalem, Jerusalem, Israel, 1985.

33. **Pasika, W. M.**, Polysaccharides polyelectrolytes, in *Extracellular Microbial Polysaccharides*, Sandford, P. A. and Laskin, A., Eds., *Am. Cement Soc. Symp. Ser.*, 45, 128, 1977.

34. **Rees, D. A.**, Structure, conformation, and mechanism in the formation of polysaccharide gels and networks, *Adv. Carbohydr. Chem. Biochem.*, 24, 267, 1969.

35. **Smidsrød, O. and Hang, A.**, Dependence upone the gel-sol state of the ion-exchange properties of alginates, *Acta Chem. Scand.*, 26, 2063, 1972.

36. **Kardošová, A., Rosík, J., and Kubala, J.**, Utilization of degraded peach gum polysaccharide by *Aspergillus flavus, Phytochemistry*, 14, 341, 1975.

37. **Jones, J. K. N. and Smith, F.**, Plant gums and mucilages, *Adv. Carbohydr. Chem.*, 4, 243, 1949.

38. **Hough, L. and Pridham, J. B.**, The composition of plum gums, *Biochem. J.*, 73, 550, 1959.

39. **VanderMolen, G. E., Beckman, C. H., and Rodehorst, E.**, Vascular gelation: a general response phenomenon following infection, *Physiol. Plant Pathol.*, 11, 95, 1977.

40. **Moore, K. E.**, Barrier-zone formation in wounded stems of sweetgum, *Can. J. For. Res.*, 8, 389, 1978.

41. **Olien, W. C. and Bukovac, M. J.**, Ethephon-induced gummosis in sour cherry (*Prunus cerasus* L.). I. Effect on xylem function and shoot water status, *Plant Physiol.*, 70, 547, 1982.

42. **Couvillon, G. A., Bass, S., Joslin, B. W., Odom, R. E., Robertson, J. E., Sheppard, D., and Tanner, R.,** NAA-induced sprout control and gummosis in peach, *HortScience,* 12, 123, 1977.

43. **Blanco, A. and Gomez-Aparisi, J.,** Containment pruning and auxin paint effects on peach and nectarine, *Acta Hortic.,* 160, 177, 1986.

44. **Wilde, M. H. and Edgerton, L. J.,** Histology of ethephon injury on 'Montmorency' cherry branches, *HortScience,* 10, 79, 1975.

45. **Achilea, O., Chlutz, E., Fuchs, E., and Rot, I.,** Ethylene biosynthesis and related physiological changes in *Penicillium digitatum*-infected grapefruit *(Citrus paradisi), Physiol. Plant Pathol.,* 26, 125, 1985.

46. **Achilea, O., Fuchs, E., Chlutz, E., and Rot, I.,** The contribution of host and pathogen to ethylene biosynthesis in *Penicillium digitatum*-infected citrus fruit, *Physiol. Plant Pathol.,* 27, 55, 1985.

47. **Archer, S. A. and Hislop, E. C.,** Ethylene in host-pathogen relationships, *Ann. Appl. Biol.,* 81, 121, 1975.

48. **Yang, S. F. and Pratt, H. K.,** The physiology of ethylene in wounded plant tissues, in *Biochemistry of Wounded Plant Tissues,* Kahl, G., Ed., Walter de Gruyter, Berlin, 1978, 595.

49. **Imaseki, H., Yoshii, H., and Todaka, I.,** Regulation of auxin-induced ethylene biosynthesis in plants, in *Plant Growth Substances 1982,* Wareing, P. F., Ed., Academic Press, London, 1982, 259.

50. **Jones, J. F. and Kende, H.,** Auxin-induced ethylene biosynthesis in subapical stem sections of etiolated seedlings of *Pisum sativum* L. *Planta,* 146, 649, 1979.

51. **Yang, S. F., Adams, D. O., Lizada, C., Yu, Y., Bradford, K. J., Cameron, A. C., and Hoffman, N. E.,** Mechanism of ethylene biosynthesis, in *Plant Growth Substances 1979,* Skoog, F., Ed., Springer-Verlag, Berlin, 1982, 219.

52. **Yu, Y.-B. and Yang, S. F.,** Auxin-induced production of ethylene and its inhibition by aminoethoxyvinylglycine and cobalt ion, *Plant Physiol.,* p. 1074, 1979.

53. **Kawase, M.,** Role of cellulase in aerenchyma development in sunflower, *Am. J. Bot.,* 66, 183, 1979.

54. **Kawase, M.,** Effect of ethylene on aerenchyma development, *Am. J. Bot.,* 68, 651, 1981.

55. **Osborne, D. J.,** The ethylene regulation of cell growth in specific target tissues of plants, in *Plant Growth Substances 1982,* Wareing, P. F., Ed., Academic Press, London, 1982, 279.

Chapter 10

PATHOLOGICAL ANATOMY OF ROOT DISEASES CAUSED BY *PHYTOPHTHORA* SPECIES

Darren P. Phillips

TABLE OF CONTENTS

I. INTRODUCTION

This chapter is concerned with the pathological anatomy of root diseases caused by *Phytophthora* species, with particular emphasis on fruit tree diseases. The genus *Phytophthora* (Gr. *phyton*: plant; *phthora*: destructan) contains some of the most destructive fungal pathogens in agriculture and consists of more than 45 species.[1-3] The group of fungi, commonly known as water molds, belong to the oomycetes* which reproduce asexually by motile spores and zoospores, and sexually by heterogametangial contact to produce thick-walled, nonmotile oospores.[6,7] Some species are homothallic while others are heterothallic.[8,9] Pathogenic species include:

1. Host-specific pathogens of one or several related hosts such as *P. infestans* (Mont (de Bary) on potato[10,11] and tomato;[12,13] *P. megasperma* Drechs. f.sp. *glycinea* (Kuan & Erwin) or *P. megasperma* Drechs. var. *sojae* Hildeb. on soybean,[14-16] and *P. vignae* Purss. on cowpeas[17,18]
2. Nonspecific pathogens with a broad host range such as *P. cinnamomi* Rands pathogenic to over 1000 plant species throughout temperate and tropical climates, including avocados[19] and English walnuts;[20,21] *P. cactorum* (Leb. & Cohn) Schroet., pathogenic to hundreds of host species, including apple,[22-24] pears,[25] and peaches;[26] *P. citrophthora* (Sm. & Sm.) Leonian, a citrus pathogen;[27-31] *P. syringae* (Kleb.) Kleb., pathogenic to both stone fruits, such as apricots[32] and almonds;[33] and *P. palmivora* (Butl.) Butler, pathogenic to cocoa and pawpaw, among over 700 other hosts.[34,35]

Other important fruit tree pathogenic species of *Phytophthora* include *P. citricola* Sawada,[21] *P. cryptogea* Pethyb. & Laff.,[20,36] *P. cambivora* (Petri) Buism.,[22,32,37] *P. megasperma* Drechs.[32,38] and *P. parasitica* Dastur.[30,39] The *Phytophthora* spp. that cause root diseases of fruit trees are nonspecific pathogens with relatively broad host ranges.

Phytophthora species usually experience a soilborne phase during part or all of their life cycle,[40,41] causing decay in plant roots, and they may form cankers around the base of trunks (Figure 1A to D).[33,42-51] The life cycle of *P. cinnamomi*, one of the most successful members of the genus, is representative of heterothallic soilborne forms (Figure 2).[52] *P. cinnamomi*, like most other soilborne *Phytophthora* spp., penetrate plant roots preferentially in the zone of elongation behind the root tip and wounds (Figures 3 and 4A to D).[19,45,53-58] After infection, mycelium may produce necrotic lesions and spread throughout the root system and collar, destroying tissue as it progresses (Figures 4E and 5B and 5D).[20,42,59] These are primary disease symptoms. Disease caused by *P. cinnamomi* is also characterized by the development of secondary symptoms such as leaf chlorosis, dieback of twigs and outer branches of the plant, and leaf microphylly. Secondary symptoms such as these are often the first signs observed when *Phytophthora* spp. attack root and crown tissues, whether it be *P. cinnamomi* infecting avocados (Figure 1D)[42,48] or Australian native plant communities,[60,61] several different *Phytophthora* spp. attacking walnuts (Figures 5A and C),[20] or *P. cactorum* attacking apples.[62]

Concomitant with the invasion of root tissue by hyphae are the many changes in host anatomical structure and physiological processes.[63-66] The following discussion centers on such postinvasive changes produced by plant hosts in response to infection by *Phytophthora* spp. The interaction between *P. cinnamomi* and the commercially important fruit tree, avocado pear (*Persea americana* var. *drymifolia* Mill.), is used as a case study of the

---

* However, recent work that groups Oomycetes with the kingdom Protoctista rather than the kingdom Fungi[4] is now being accepted by some researchers.[5]

FIGURE 1. (A) A cocoa tree which has undergone sudden dieback due to severe attack by *Phytophthora palmivora*. Note extensive leaf senescence and absence of cocoa pods on the branches. (B) Inset of (A). Canker (arrow) has circumvented the whole trunk, destroying all soft outer tissues and cambium. (Photographs taken by author in capacity as research scientist based at the Cocoa Black Pod Research Trust of Papua New Guinea (PNG), Kar Kar Island, PNG.) (C) A *P. cinnamomi*-infected peach tree showing extensive trunk canker (arrow). (Photograph courtesy of Dr. F. Greenhalgh, Plant Research Institute, Burnley, Victoria, Australia.) (D) An avocado tree attacked by *P. cinnamomi* in an orchard at Mt. Tamborine, Queensland, Australia. Note the dieback of the outer branches, wilted leaves (common secondary symptoms of the disease caused by *P. cinnamomi*), and the exposed remaining fruit (arrow). (From Phillips, D. P., Responses of Susceptible and Resistant Avocado Cultivars to Infection by *Phytophthora cinnamomi*, Ph.D. thesis, University of Melbourne, Melbourne, Australia, 1989.)

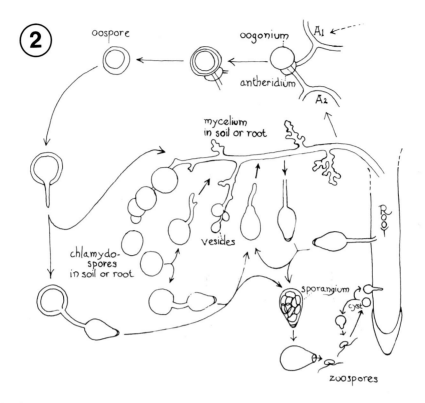

FIGURE 2.   The life cycle of *Phytophthora cinnamomi*, illustrating alternative stages in the life cycle by which the fungus adapts to environmental changes. (Redrawn from Weste, G., in *Plant Diseases: Infection Damage and Loss*, Blackwell Scientific, Oxford, 1984, 273. Original drawing by H. J. Swart. With permission.)

pathological anatomy of root diseases caused by *Phytophthora* spp. throughout the chapter, because little information is available for most other interactions involving *Phytophthora* spp. and fruit trees.

II. INFECTION PROCESS IN SUSCEPTIBLE HOSTS

Where hosts have co-evolved with fungal pathogens, susceptibility to disease is the exception rather than the rule.[67] Plants are continually exposed to attack by a vast range of air- and soilborne fungi, viruses, insects, nematodes, and bacteria. Yet, as is commonly noted in reviews of host defense mechanisms, the majority of plants remain healthy and grow vigorously most of the time.[67,68] This phenomenon may be an indirect reflection on the capacity of either:

1.   The plant's first line of defense — the cuticle of leaves, cork of stem, and exodermis of roots — to resist infection
2.   The ability of the pathogen to first reach, germinate on, and penetrate the host

If infection is successful, plants may also resist subsequent invasion of a pathogen by presenting a large range of both nonspecific and specific defense reactions which, when fully differentiated, may be a combination of physiological, biochemical, or physical responses. The defense mechanisms themselves are not considered to bear any particular degree of specificity, but this may be denoted by the manner in which they are induced. Preformed

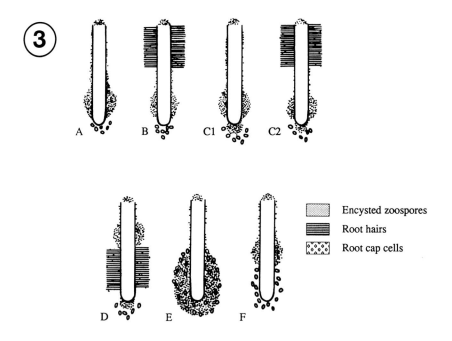

FIGURE 3.   Variation in encystment pattern, diagrammatic representation. (Redrawn from Hinch, J. and Weste, G., *Aust. J. Bot.*, 27, 679, 1979. With permission.)

(Pattern A) The cells at the tip of the root represent the root cap cells that were present on practically all roots studied. These cells are sloughed off as the plant grows. This represents a common type of encystment pattern; most zoospores are attracted to the region immediately behind the root tip, forming several layers. The rest of the root, except the cut end, attracted only small numbers of zoospores; most zoospores accumulated near the center. Examples: *Themeda triandra* (syn. *T. australis*), *Acacia* spp. (without sheath).

(Pattern B) This represents the most common type of pattern found, with most zoospores accumulating in the region behind the root tip. Few zoospores were found on the root hairs. Examples: *Isopogon ceratophyllus*, *Leptospermum juniperinum*, *Banksia* spp. (except *B. spinulosa*), *Eucalyptus* spp., *Pittosporum undulatum*.

(Pattern C1) This represents the accumulation of zoospores in the tip-most region of the root and in the region behind the root tip; the rest of the root attracted few zoospores. Examples: *Melaleuca* spp., *Hakea* spp., *Banksia spinulosa*.

(Pattern C2) This is a similar pattern to C1, but with few zoospores attracted to the root hair region. Example: *B. spinulosa*.

(Pattern D) In cases in which the root hair region started immediately behind the root tip, most zoospores accumulated in the region of maturation. Once again, few zoospores accumulated in the root hair region. Example: *Hakea* spp.

(Pattern E) With this type, characterized by *Xanthorrhoea australis*, *X. resinosa*, and *X. minor*, large numbers of zoospores encysted in the extensive, highly pigmented root cap sheath.

(Pattern F) In some plants with root cap sheaths, mostly *Acacia* spp., most zoospores encysted in the maturation region immediately behind the sheath.

fungitoxic compounds, secondary metabolites such as oxidized phenolics (quinones), and phytoalexins are examples of physiological responses. Preexisting or induced structural barriers such as tyloses and gums, tissue lignification and or suberization, papillae or cell wall appositions, hyphal encapsulation, and meristematic barriers such as wound periderms and cork tissue are examples of physical barriers.[67,69-72] Overt disease expression (susceptibility) only occurs when the pathogen circumvents a number of the host defense responses, or the host fails or it too slow in expression of defense reactions.

The histology of pre- and early postpenetration events by *Phytophthora* spp. has been reviewed in recent years.[55,73-76] Such events are summarized for *P. cinnamomi* (Table 1).

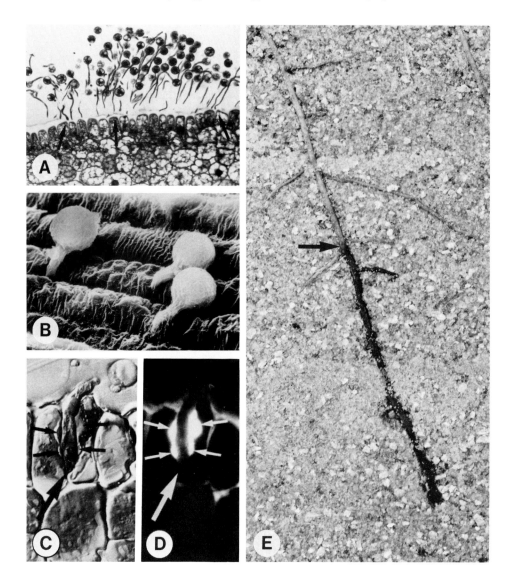

FIGURE 4. (A) Light micrograph of a section of *Zea mays* root taken in the zone of elongation 90 min postinfection. The fungal cysts have germinated and hyphae are penetrating the root mucilage layer and the epidermal cell walls (arrows). (Magnification × 140.) (A, C, and D from Hinch, J. M. et al., *Protoplasma*, 126, 178, 1985. With permission.) (B) Scanning electron micrograph of zoospores of *P. cinnamomi* that have encysted on the root surface of a *Lupinus angustifolius* L. seedling. The germ tube of each cyst penetrates the epidermal cells inter-cellularly. (Photograph courtesy of Dr. G. Weste, School of Botany, University of Melbourne, Parkville, and Dr. J. Hinch, Plant Research Institute, Burnley, Victoria Australia.) (C, D) Light and fluorescence micrographs of serial sections, taken 2 h postinfection, showing a single hypha penetrating the host epidermal layer through the middle lamella. Callose deposits of varying thickness (small arrows) have formed behind the hyphal tip (large arrow). (Magnification × 512.) The sections in the light and fluorescence micrographs are stained in toluidine blue and 0.1% decolorized aniline blue, respectively. (A, C, and D from Hinch, J. M. et al., *Protoplasma*, 126, 178, 1985. With permission.) (E) A lesioned root in avocado clonal plant, rootstock 'Duke 7' (moderately resistant), 12 d postinoculation at its tip with a droplet containing 80 zoospores of *P. cinnamomi*. The lesion is 6.2 cm long and stopped extending (arrow) on day 10 postinoculation. The fungus was readily reisolated from necrotic tissue at the time of harvest and the plant was held at a soil-root temperature of 25°C. (Part E from Phillipa, D. P., Responses of Susceptible and Resistant Avocado Cultivars to Infection by *Phytophthora cinnamomi*, Ph.D. thesis, University of Melbourne, Melbourne, Australia, 1989.)

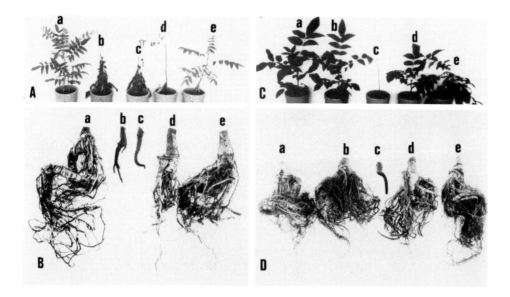

FIGURE 5. Tops and roots of 6-month-old seedlings of Northern California black walnut (*Juglans hindsii*) (A and B) and Paradox (*J. hindsii* × English walnut (*J. regia*)), (C and D) grown for 3 months in noninfested soil (a) and artificially infested soil with *Phytophthora cactorum* (b), *P. cinnamomi* (c), *Phytophthora* sp. (isolate 1029) (d), and *P. megasperma* (e). Note that all four *Phytophthora* spp. (b, c, d, and e) caused visible damage to *J. hindsii* (A and B), whereas only *P. cinnamomi* (c) severely affected 'Paradox' seedlings (C and D). (From Mircetich, S. M. and Matheron, M. E., *Phytopathology*, 73, 1481, 1983. With permission.)

Attraction to and penetration of roots by *P. cinnamomi* and other *Phytophthora* spp. is rapid and similar irrespective of the susceptibility or compatibility of the host. This has been clearly shown in *P. cinnamomi* disease of avocado[45] and eucalypt roots,[77,90] *P. megasperma* var. *sojae* attack of soybean hypocotyls,[91] *P. megasperma* f.sp. *glycinea* attack of soybean roots,[14] *P. infestans* attack of potato tubers,[10] and *P. fragariae* Hickman infection of roots of both susceptible and resistant strawberry cultivars.[92]

A difference in attraction of zoospores to root tips was reported between rabbiteye (resistant) and highbush (susceptible) blueberries infected with *P. cinnamomi*,[93] but not by Hinch and Weste[53] for Australian native species (Figure 3). However, it was also noted that young rabbiteye rootlets were equally susceptible to penetration and subsequent infection compared with highbush rootlets.[93]

Primary roots are most frequently invaded by *P. cinnamomi*[19,48,73] and other *Phytophthora* spp. in the zone of elongation, and this region provides little or no mechanical resistance to penetration (Figure 4A). The epidermis is young and thin walled and the exodermis is undifferentiated and endodermis unthickened,[55,73] compared with more mature root tissue further back from the root tip. However, species such as *P. cinnamomi* are known to also infect major roots possessing tissue that has been secondarily thickened.[50,94,95]

Germ tubes from encysted zoospores preferentially penetrate the outer epidermal layers of primary roots intercellular, i.e., between the anticlinal cell walls (Figures 4B to D).[54,55] In one study of *P. megasperma* f.sp. *glycinea* on soybean, up to 94% of germ tube penetrations were reported to be through anticlinal cell walls, with only 5% forming swollen germ tubes closely appressed to the root surface.[14] Appressoria do not appear to be a feature of germ tube penetration for zoospores of *Phytophthora* spp. Evidence exists of the middle lamella of epidermal cells being partially hydrolyzed by invading germ tubes and hyphae.[55,86] *P. cinnamomi* has been found to demonstrate some capacity to produce enzymes that dephenolize

TABLE 1
Pre- and Early Postpenetration Events by *P. cinnamomi*

Time after inoculation (h)	Event	Ref.
0	Zoospore release and preferential attraction to root tip and/or cut or exposed wounds	19, 53
	Attraction is nonspecific for hosts, and irrespective of their susceptibility	45, 55, 77
	Attraction can be chemotaxic for sugars, amino acids, and ethanol, and/or electrotaxic	19, 78, 79
0.5	Zoospore encystment at root surface	55, 77
	Adheres, recognition being mediated by L-fucose receptors on zoospore and possibly pectin-based material on root	80–82
	Cyst germination	
	Prealignment of emerging germ tube	83
0.5–1	Penetration of epidermis, preferentially intercellular along anticlinal cell walls through middle lamellae	45, 55, 64
	Evidence of both enzymic and mechanical penetration	54, 77
2–4	Infection established in cortex; hyphae are sealed off from empty cysts by amorphous plugs at cyst bases; mainly intercellular development in outer cortex	55
	Microsporangia present	45
16	Hyphal penetration of endodermis and stele	73
18–24	Lesion readily apparent and necrosis is widespread	45, 54, 73
	Sporangia on root surface	73
	Hyphae in stele, concentrated in phloem and xylem lumen; largely intracellular	84
	Papillae/wall appositions may occur in cortex	54, 85
	Extensive cell wall hydrolysis in root tip	55, 86
4–6 d	Hyphal ramification extensive throughout stele and cortex; hyphal vesicles and swellings abundant in cortex	45, 73, 87–89
	Small haustorial-like structures/side branches from intercellular hyphae in cortex	84, 85

lignin,[96,97] and partially disintegrate walls of xylem vessels with secondary thickening in the susceptible avocado rootstock 'Topa Topa',[42] as does *P. capsici* in infected tissue of pepper roots.[98]

A. HOST-SPECIFIC RESPONSES

Once infection hyphae have gained entry into the outer layers of host tissue, whether inter- or intracellulary, host-specific *Phytophthora* spp. characteristically form haustoria, which penetrate the cell walls into the cytoplasm, but not the vacuole. This is true for infections caused by *P. infestans*,[10] *P. parasitica* var. *nicotianae* (Dastur) Waterh.,[74] and *P. megasperma* var. *sojae* on soybean hypocotyl,[16,91] but not for *P. megasperma* f.sp. *glycinea*-infected soybean root tissue.[14] Haustorial size may change. However, they all have a cell wall continuous with that of the intercellular hyphae and are generally larger than those formed in resistant hosts (Figures 6A and B).[10] The haustoria may also be surrounded by an extrahaustorial matrix.[10] Besides the extensive development of haustoria, cell wall apposition development in association with them may[16,98] or may not[10] be commonplace.

It is reported that no discernible differences exist in the ultrastructure of both inter- and intracellular hyphae of *P. infestans* when invading tubers of both susceptible and resistant cultivars of potato.[10] Furthermore, in *P. parasitica* var. *nicotianae* infection of tobacco roots,

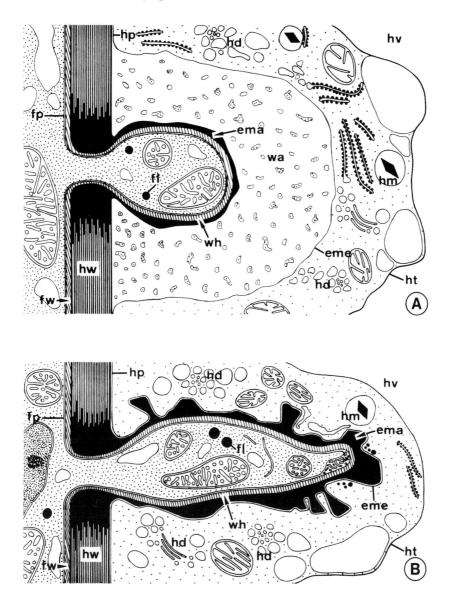

FIGURE 6. Summary diagram of haustorial apparatus of *Phytophthora infestans* in a resistant (A) and a susceptible (B) potato cultivar. Extrahaustorial matrix (ema), extrahaustorial membrane (eme), fungal lipid bodies (fl), fungal plasma membrane, (fp), fungal cell wall, (fw), host cell dictyosome (hd), host cell microbody (hm), host cell plasma membrane (hp), host cell tonoplast (ht), host cell vacuole (hv), host cell wall (hw), wall apposition (wa), wall of haustorium (wh). (From Hohl, H. H. and Stössel, P., *Can. J. Bot.*, 54, 900, 1976. With permission.)

little or no change in cell ultrastructure was observed in cells of the susceptible host-pathogen race combinations in advance of the pathogen,[74] unlike that which occurred in the resistant host/pathogen race combination (see Section III.A). The lack of host cell necrosis until 24 to 48 h postinfection in suspects has also been noted for *P. infestans* on potato.[75] Cell necrosis eventually followed, but did not prevent continued growth and development of fungal mycelium or subsequent sporulation as would have occurred in biotrophs such as rust. Mycelial development was usually severe and systemic throughout host tissue.

B. NONSPECIFIC HOST RESPONSES

The process of infection by a nonspecific pathogen such as *P. cinnamomi* in a susceptible host is summarized in Table 1. In a moderately resistant rootstock such as the avocado host 'Duke 7', lesions progresssed into well-differentiated primary host tissue containing vascular cambium and lignified stelar elements, but was then arrested (Figure 4E).[42,89] In the susceptible rootstock 'Topa Topa', the pathogen was not arrested and infection was extensive.[42] The histology of lesioned root in both rootstocks revealed three zones of tissue with distinctive anatomy in association with the lesion boundary.[42,89] The changes that resulted from infection are summarized in Table 2 (see Section III.B). In most cases, the infection of a susceptible host is characterized by extensive and intensive inter- and intracellular hyphal development and ramification throughout both cortical and stelar tissues, as depicted in light micrographs of lesioned tissue of the moderately resistant rootstock 'Duke 7' (Figures 7A to C).[89] Cell lysis and degradation is often extensive and accompanied by accumulation of material that stains deep blue-black and/or blue-green in toluidine blue O, indicating the presence of polyphenols and tannins (Figure 7C).[55,73,76,89,99] This is confirmed by strong autofluorescence of infected tissue.[73,89] Small haustorial-like structures, or side branches, are sometimes observed such as for hosts infected by *P. cinnamomi*[42,73] and *P. capsici*.[98]

III. POSTINFECTION CHANGES IN THE ANATOMY OF RESISTANT HOSTS

Plant defenses include a large range of nonspecific mechanisms such as cell wall modification, which appear to function as physical barriers and may be closely correlated with the prevention or restriction of disease. Whether such responses play a general role in successful host defense is controversial, and each case must be carefully considered individually. Some examples of physical defense mechanisms resulting in anatomical changes are discussed below.

A. HOST-SPECIFIC RESPONSES

In resistant hosts of host-specific pathogens haustorial structures are also formed upon infection of host tissue, but differ from those formed in susceptible hosts. They are often much smaller and surrounded by a thinner extrahaustorial matrix with well-developed appositions typically encasing the entire haustorium (Figure 6A).[10,11] The wall appositions are usually characterized by electron-translucent material embedded with small electron-dense patches.[10]

A vital response normally observed in resistant hosts is the hypersensitive response. Hypersensitivity, the rapid collapse and death of host cells, is generally considered to be a specific defense response that is a critical factor in preventing the establishment of a biotroph early on in infection in many resistant hosts.[70] The rapid collapse of host cells leads to cessation of growth and development of the invading pathogen if this should be a biotroph, i.e., a parasite requiring living host tissue for growth. The response is formed in epidermal and outer cortical cells immediately adjacent to invading hyphae within a few hours of infection, such as in roots of resistant lines of tobacco infected by *P. parasitica* var. *nicotianae*[74] and soybean hypocotyl infected by *P. megasperma* var. *sojae*.[16] In infection of resistant lines of strawberry by *P. fragariae*, hyphal penetration also did not extend beyond the outer layer of root tip cells.[92] Postinvasive sporulation in such interactions is usually reported as being either greatly reduced or altogether absent.[75]

A second major host defense response associated with host-specific resistance is physiological — the production of phytoalexins. Phytoalexins can be defined as "antimicrobial compounds of low molecular weight synthesized by and accumulated in plants after exposure to microorganisms".[100] These compounds produced quickly enough, and in sufficiently high

TABLE 2
Summary of Histological Observation of Uninfected Root Tissue and *P. cinnamomi*-Infected Tissue of 'Duke 7' (moderately resistant) and 'Topa Topa' (susceptible) Clonal Rootstock Material[42]

Character	Infected 'Duke 7'	'Topa Topa'	Uninfected controls (both rootstocks)
Zone 1/Lesion (infected tissue)			
Hyphae	Inter- and intracellular	Inter- and intracellular	Absent
Cell	Necrotic	Necrotic	Non-necrotic
Walls	Lysed/fragmented	Lysed/fragmented	Intact
Contents	Plasmolyzed	Plasmolyzed	Turgid
Stain with toluidine blue O	Blue-black/dark blue-green	Blue-black/blue-green	Red-purple/blue
Phenolics/tannins	Accumulated/widespread	Accumulated/widespread	Confined to few cells, scattered
Autofluorescence	Strong	Strong	Weak
Xylem			
Discoloration	Absent	Absent	Absent
Vessel Walls	Lignification reduced	Lignification reduced	Lignification complete
% Blockage by tyloses/detritus	Up to 40–45%	Up to 40%	<10–15%
Starch grains	Absent	Absent	Present in central xylem, pericycle, and outer cortex
Lesion	Restricted	Extensive/unrestricted	Absent
Zone 2/Lesion Boundary or Sham-Inoculation Point in Control Root			
Cortex			
Lesion area	Restricted	Unrestricted	Absent
Cell necrosis	Restricted	Restricted	Absent
Hyphae	Inter- and intracellular, but rapidly restricted in advance of lesion	Inter- and intracellular rapidly declining in advance of lesion	Absent
Cell walls	Thickened in regions of hyphal disappearance	Unthickened	Intact, unthickened
Cell wall thickening autofluorescence	Present/lignin-based material	Absent	Absent
Wound periderm	Present/dividing infected and uninfected tissue (necrophylactic)	Absent, except for localized sites in exodermis	Absent, except for localized sites in exodermis
Papillae/callose plugs	Absent	Absent	Wounding absent
Starch grains	Absent	Absent	Present
Stele			
Xylem discoloration	Present	Present	Absent
Xylem blockage/tyloses	Up to 50% max	Up to 40–45%	< 10–15%
% Lignified elements	Increases sharply within 0.3 cm advance of lesion boundary	Increases within 0.3 cm advance of lesion front	Increases directly with increased distance from root tip
Phloem bundles	Lignified/whorled by cells	Lignified, but no cell whorling	Nonlignified and not whorled
Zone 3/Uninfected Root			
Cells	Non-necrotic	Non-necrotic	Non-necrotic
Cell contents	Turgid	Turgid	Turgid

TABLE 2 (continued)
Summary of Histological Observation of Uninfected Root Tissue and
***P. cinnamomi*-Infected Tissue of 'Duke 7' (moderately resistant) and**
'Topa Topa' (susceptible) Clonal Rootstock Material[42]

Character	Infected 'Duke 7'	'Topa Topa'	Uninfected controls (both rootstocks)
	Zone 3/Uninfected Root		
Starch grains	Present	Present	Present
Phloem bundles	Nonlignified (normal)	Normal	Normal
Wound periderm	Localized regions in exodermis	Localized regions in exodermis	Localized regions in exodermis
Xylem discoloration	Absent	Absent	Absent
% Xylem blockage	< 25%	< 15%	< 10–15%
Hyphae	Absent	Absent	Absent

concentrations at the site of infection, inhibit the growth of both fungi and bacteria in many higher plants.[101-103] This is well documented for members of the Leguminales, particularly in the subfamily Papilionaceae,[104] and in the Solanaceae.[105]

It is generally accepted that a strong functional association exists between phytoalexin accumulation at the infection site and the expression of disease resistance in cultivars of soybean resistant to infection by *P. megasperma* Drechs f.sp. *glycinea* Kuan & Erwin, the phytoalexin glyceollin III being the most important.[106,107] The phytoalexin rishitin is produced in resistant cultivars of potato in response to infection by various races of *P. infestans*, although the extent to which it determines the expression of resistance is uncertain.[63]

Thus, in host-specific pathogen interactions such as those described, it is phytoalexin production and hypersensitivity which are the two common mechanisms by which host resistance is believed to be conferred:[70] by reducing penetration of the host, restricting growth of the fungus within, and reducing the pathogen's capacity to sporulate.[108,109]

B. NONSPECIFIC HOST RESPONSES

In interactions involving species of *Phytophthora* with wide host ranges, a multitude of postinvasive anatomical changes occur in response to infection. Cytoplasmic aggregation, haloes, papillae, and tyloses are often formed rapidly at the cellular level,[54,85,110] and new tissues such as meristematic barriers may develop as infection proceeds.[89,95,111]

Papillae, composed of heterogeneous materials deposited between the plasma membrane and cell wall, are formed at sites of perturbation elicited by fungal attack or wounding.[112] Most papillae contain callose, phenolic derivatives, cellulose, and silica. Detailed studies have been made on the significance of papillae in host resistance in a number of host-pathogen interactions, including resistance exhibited to *Phytophthora* spp.:

1. *P. cinnamomi* by the nonhost, *Zea mays* L. (Figures 4C and D),[54] 13 Australian native plant species of varying susceptibility,[85] and the resistant species, *Acacia pulchella* R. Br.[113]
2. *P. infestans* by potatoes[10]

It is difficult to ascertain whether papillae play a significant role in mechanical or physical resistance to invasion, as the majority of reports are based on visual, qualitative associations rather than quantitative assessment and direct measurements of the mechanical forces that may be involved. Host/pathogen interactions involving *Phytophthora* spp. are no exception.

The effectiveness of anatomical changes such as the formation of wound periderms and

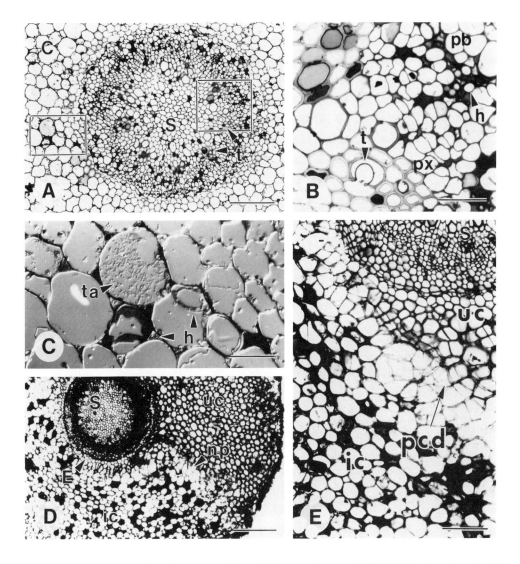

FIGURE 7. Transverse sections from roots of plants of the avocado rootstock, 'Duke 7' (moderately resistant), infected with *Phytophthora cinnamomi*. (A to C) Thin microtome section stained with toluidine blue and counterstained with *p*-aminosalicylic acid, 0.3 cm behind the lesion boundary and 5.9 cm from the root tip of a plant grown at 15°C. Scale bar = 100 μm. (A) Note widespread inter- and intracellular hyphal development, xylem blockage by tyloses (t). S, stele; C, cortex. (B) Inset from right side of A. Hyphae (h) concentrated in phloem bundle (pb). px, protoxylem. Bar = 25μm. (C) Inset from left side of A. Accumulated tannins (ta) and inter- and intracellular hyphae in the cortex. Nomarski optics was used. Bar = 25 μm. (D and E) Hand transverse section stained with toluidine blue, 0.2 to 0.4 cm in advance of a lesion boundary, 8.7 to 8.9 cm from root tip of a plant grown at 20°C (Zone 2). Bar = 250 μm. (D) Necrophylactic periderm (np) extending from the endodermis (E) to the epidermis between infected (ic) and uninfected (uc) cortical tissue. (E) Periclinal cell wall division (pcd) acquiring the appearance of exophylactic periderm. Note the translucent nature of dividing cells in toluidine blue (From Phillips, D. et al., *Phytopathology,* 77, 691, 1987. With permission.)

cork in sealing off uninfected tissue from diseased, macerated tissue has been well documented in woody hosts for:

1. *P. cinnamomi* attack on eucalypts (Figures 8A to D)[95,114] and a moderately resistant avocado rootstock, 'Duke 7'[42,89]
2. *P. cactorum* attack of apples[44,115]

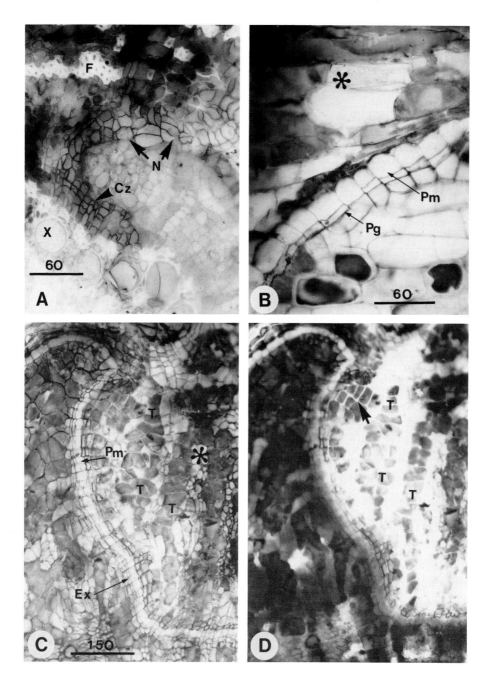

FIGURE 8. Necro- and exophylactic periderm formation in roots of jarrah, *Eucalyptus marginata* (susceptible), in response to invasion by *Phytophthora cinnamomi* following artificial inoculation. (A) A zone of typical necrophylactic periderm (N) and suberization of cells in the cambial zone (Cz) (Sudan black stain). X, xylem; F, phloem fibers. All units on scale bars represent microns. (B) Exophylactic periderm, the type formed deep in the phloem, causing cylindrical isolation of necrotic tissue. Characteristic phellem (Pm) cells closely aligned with the cells of the phellogen (Pg). Exophylactic periderm always has such an ordered appearance compared with necrophylactic periderm. Necrotic tissue on the side of the periderm is marked with an asterisk. (C) Exophylactic periderm (Ex) well differentiated along the radial edge of the lesion. Lesion to right of micrograph (asterisk). Condensed tannins in vacuoles of some parenchyma. (Toluidine blue stain, T.S. root.) T, tanniferous parenchyma. (D) Strong autofluorescence of tissue zone at the edge of the lesion. Same section as shown in C. Autofluorescence of cell walls (arrow) and vacuolar contents of parenchyma. Cells that did not obviously contain phenolics fluoresced most intensely. (From Tippett, J. T. and Hill, T. C., *Eur. J. For. Pathol.,* 14, 431, 1984. With permission.)

3. Various pathogens or physical stimuli artificially applied to white pines,[116] slash pines,[117,118] chestnut trees,[119,120] firs and hemlock,[121,122] and peaches[123,124]

Tippett and Hill[95] noted that once a pathogen such as *P. cinnamomi* progressed into *Eucalyptus* roots with a secondary structure, the host had the potential to mobilize a greater variety of resistance mechanisms than those available in primary root tissue. This was due to the presence of meristematic tissue such as cambium and phellogen, which is continuous along such roots (Figures 8A to D). In the susceptible *Eucalyptus marginata* Sm., the form of the lesion in secondary roots reflected and was controlled by the pattern of postinoculation periderm development.[95]

However, such meristematic barriers can be circumvented by the pathogen if they are incomplete, either in their extension between lesioned and nonlesioned tissue or in lignification and/or suberization.[125] In some cases the pathogen may breach the phloem lesion boundaries directly.[114] In roots of the grass, *Themeda triandra* Forsskal (syn. *T. australis* [R. Br.] Stapf.), *P. cinnamomi* circumvented lignified barriers by growing up through the vessels in the stele where meristematic tissue could not be formed.[73]

Meristematic tissue in the form of stimulated cell division, hyperplasia, has also been associated with *P. cinnamomi* attack of a moderately resistant avocado rootstock, 'Duke 7',[89] and with prevention of disease in young tomato plants by three vascular wilt pathogens.[126]

Tyloses and gums may be formed in xylem vessels of woody dicots as a result of infection or wounding. They have been reported as a mechanism for resistance to vascular wilt pathogens in resistant species of cotton, tomatoes, and bananas. The rapid occlusion of xylem vessels by tyloses is believed to prevent or reduce the systemic spread of fungal spores.[111,126-129] Tyloses are formed by axial or ray parenchyma cells growing out through the pit cavities into lumina of vessels they adjoin.[130] Tyloses are formed as a response to invasion by *P. cinnamomi* in some susceptible eucalypts[95,131] and xylem of roots in both resistant and susceptible avocado rootstocks.[42,89] It is not clear whether tyloses do limit the spread of *P. cinnamomi* in resistant plants such as other *Persea* spp. used for rootstocks and certain eucalypt species.

Suggested mechanisms by which all the anatomical modifications described may impede or restrict infection include mechanical blockage and permeability barriers between pathogen and host. It is suggested that permeability barriers may act to prevent the exchange of phytotoxins, cell wall degrading enzymes, suppressors, and the conductance of nutrients and water.[112] However, it is generally accepted that the anatomical modifications described contribute toward and act in concert with biochemical responses in limiting disease in many host-pathogen interactions.[112]

In recent years, the development and screening for resistance in rootstocks of avocado has resulted in the selection of several cultivars or rootstocks, Martin Grande (G755 selections), Thomas, Barr Duke, Duke 7, G1033, and G6, which exhibit field resistance of varying degree to root rot induced by *P. cinnamomi*.[19,132-135] The level of resistance shown by various avocado rootstocks had been compared,[132-135] but the basis of the resistance has not been defined for many of the selections. Anatomical responses that were considered to be an important component of the resistance exhibited by 'Duke 7' have been defined.[89] Table 3 gives the basic histology of uninfected plants. However, the responses established for the moderately resistant rootstock 'Duke 7' were yet to be compared with the pathological anatomy of a susceptible rootstock (Table 2).

When inoculated with *P. cinnamomi*, roots of 'Duke 7' formed limited lesions, while roots of the susceptible rootstock 'Topa Topa' formed lesions that were more extensive. Histological examination of the tissue changes revealed two distinct anatomical responses

TABLE 3
Histological Features of Uninfected (control) Roots of 'Duke 7' and 'Topa Topa'
Avocado Rootstocks[42]

Distance from root tip (cm)	Histological features
0.0	Root cap cells possess phenolic-based material in cytoplasm and/or vacuoles
	Large vacuolate cells with probable secretory function throughout root cap and region of elongation
0.3–0.6	Number of protophloem sieve elements equals future number of xylem rays
	Files of tannin-rich cells in provascular tissue of root tip
4.5–5.0	Vascular cambium present
	Well-differentiated primary root; stelar tissue possesses xylem elements with secondary lignification
	Starch grains, if present, scattered in outer cortex
>10.0–12.0	Xylem rays enlarged to form continuous, concentric ring
	Starch grains in outer cortex and/or pericycle and central pith
>20.0–25.0	Secondary root possessing secondary xylem development
	Cork cambium, periderm present
	Starch grains and amyloplasts containing starch grains concentrated in secondary xylem

at the lesion boundary ("Zone 2" in Table 2) in 'Duke 7': the formation of necrophylactic periderm in the cortex and whorls of cells walling off infected phloem bundles in the stele. Necrophylactic periderm separated necrotic infected tissue from nonnecrotic uninfected cortical tissue (Figures 7D and E), as occurs for *P. cinnamomi*-infected eucalypt roots (Figure 8A).[95] This zone extended from the epidermis through to the endodermis and always coincided with the very rapid disappearance of infected cortical tissue (Figure 7D). The response was not observed in 'Topa Topa' clones. Thus, the primary differences in host anatomical defense response between the moderately resistant and susceptible avocado rootstocks lie in "Zone 2".

In one 6- to 12-month-old 'Duke 7' clone, an infected secondary order root was sealed off from the main tap root of the plant by a zone of exophylactic periderm in which the cell walls stained positively for the presence of lignin-like material in 1% phloroglucinol in 20% conc. HCl (Figures 9A and B).[42]

Phloem bundles, which were the site of intensive hyphal colonization in lesioned tissue, were devoid of hyphae after the formation of whorls of cells and cell wall thickening at the lesion boundary in 'Duke 7' (Figures 10A to C).[42,89] Cell whorling was rarely observed in 'Topa Topa', although cell wall thickening of some phloem bundles was observed in some plants at the lesion boundary (Figures 11A and B).[42]

It was also noted that both 'Duke 7' (Figures 12A to D)[42] and 'Topa Topa' (Figures 13A to D)[42] could form localized regions of necrophylactic periderm in response to wounds of the outer cortex and exodermis in uninfected, healthy tissue. Components common to the formation of necrophylactic periderm, whether extended or localized, were

1. The deposition of a lignin-type material along the outer cell walls and intercellular spaces of the first non-necrotic cell layer immediately adjacent to the site of cell necrosis or tissue degradation, as determined by both toluidine blue O and phloroglucinol-HCl staining (Figures 12C and D, 13B and D).
2. The presence of non-toluidine blue-staining cork-like layers of cells for the first one to two cell layers immediately adjacent to the wound (Figures 12B to D)
3. The formation of disordered peri- and anticlinal cell division for two to three cell layers behind the cork-like cell layers in which their contents had remained unstained by toluidine blue (Figures 12B and C; 13B).

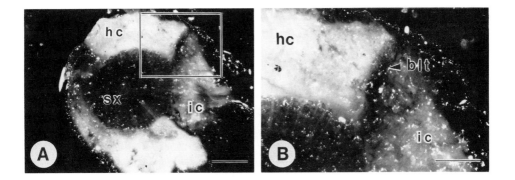

FIGURE 9. An avocado clonal plant, rootstock Duke 7 (moderately resistant), infected with *Phytophthora cinnamomi*. Hand sections of root tissue possessing secondary growth in which a lateral root infected with the fungus (ic) has been sealed off from healthy cortical (hc) tissues of the main root by bands of lignified tissue (blt). The stain is acidified phloroglucinol, which stains cell walls with secondary wall thickenings (lignification) red. Note core of secondary xylem (sx) in main root. Scale bars for A and B are 1 and 0.5 mm, respectively. B is an inset of A. (From Phillips, D. P., Responses of Susceptible and Resistant Avocado Cultivars to Infection by *Phytophthora cinnamomi*, Ph.D. thesis, University of Melbourne, Melbourne, Australia, 1989.)

Tylose formation in xylem vessels (Figures 7B) was prevalent in lesioned tissue of both rootstocks, with the most frequent occurrence being observed immediately behind the lesion boundary or front. This is depicted for both rootstocks in Figures 14 A and B.[42] Whether tyloses played a significant role in the disappearance of hyphae from xylem element/vessels at the lesion boundary has not been established, but it is suggested that tyloses may play a secondary role in freeing xylem of hyphal development by either:

1. Providing a continuous physical barrier.
2. Limiting the uptake of water, thereby causing a localized reduction in tissue water potential, which has been shown to both slow the rate of *P. cinnamomi* infections[42,136] and lead to droughting and dieback symptoms in the host;[65,137] in both control roots at the original point of inoculation and uninfected tissue >3.0 cm in advance of lesion boundaries, the percentage of xylem elements containing tyloses was <15%.

Vessel walls were also heavily discolored yellow-brown or orange-red in the region of the lesion boundary. Again, the discoloration decreased rapidly with increased distance in advance of the lesion boundary. The strong yellow-brown to orange-red discoloration of xylem vessels walls at the lesion boundary may also play a functional role in host defense. Beckman[111] noted that hosts responding to invasion by vascular wilt fungi can utilize phenolic compounds such as dopamine in defense, releasing such compounds into the immediate vicinity of the pathogen.

The observation that wall appositions or papillae were not seen in the interaction between *P. cinnamomi* and avocados, such as rootstocks 'Duke 7' and 'Topa Topa',[42] has also been reported elsewhere.[45] However, in an ultrastructural study of *Phytophthora* root rot-resistant and -susceptible species of *Persea*, Dugger and Zentmyer[138] noted that cell wall appositions were formed in the susceptible species examined, but did not elaborate on the nature of the association of these structures with hyphal invasion. Hypersensitive response was not expressed in avocados infected by *P. cinnamomi*, unlike interactions involving host-specific *Phytophthora* spp.[10] *P. cinnamomi* can invade dead host cells and the collapse of host cells does not impede its growth.

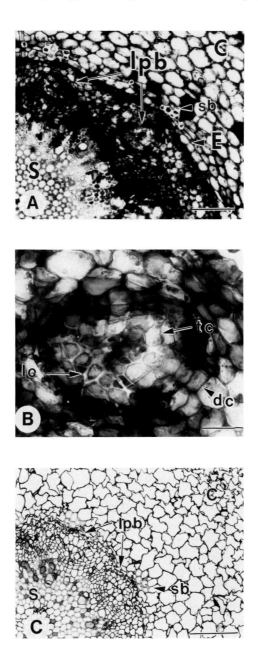

FIGURE 10. Hand and microtome transverse sections of root tissue, 0.2 to 0.5 cm in advance of the lesion boundary, 5.9 cm back from the root tip, in tissue of the avocado rootstock 'Duke 7' (moderately resistant) infected by *Phytophthora cinnamomi*. Plant held at 30°C. (A) Lesioned phloem bundles (lpb) surrounded by whorls of periclinally dividing cells. Metaxylem stained yellow-brown, toluidine blue O staining, and counterstained with decolorized aniline blue. Scale bar = 100 μm. sb, sclereid bundles; E, endodermis; S, stele; C, cortex. (B) Inset of A showing central cell walls and intercellular spaces of lesioned phloem bundles thickened with a luminescent yellow-brown compound (lc) which does not stain in toluidine blue O. Note translucent inner cells (tc) and outer ring of dividing cells (dc). Bar = 25 μm. (A and B from Phillips et al., *Phytopathology*, 77, 691, 1987. With permission.) (C) Microtome thin section of lesioned phloem bundle (lpb) stained with toluidine blue O and counterstained with potassium iodide solution. Bar = 100 μm. (From Phillips, D. P., Responses of Susceptible and Resistant Avocado Cultivars to Infection by *Phytophthora cinnamomi* Ph.D. thesis, University of Melbourne, Melbourne, Australia, 1989.)

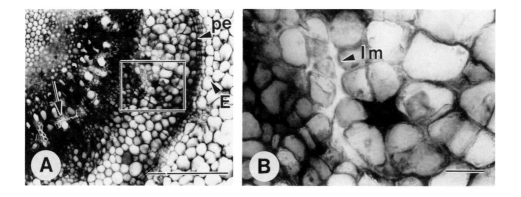

FIGURE 11.   Hand transverse sections of root, 0.2 to 0.5 cm behind the lesion boundary, 19.5 cm back from the root tip, in tissue of the avocado rootstock 'Topa Topa' (susceptible) infected by *Phytophthora cinnamomi*. (A) Section showing yellow-brown staining of xylem vessel walls (arrows) and partial cell whorling near a phloem bundle. Scale bar = 250 μm. Toluidine blue staining. E, endodermis; pe, pericycle. (B) Inset of A showing luminescent yellow-brown material (lm), nonstaining in toluidine blue, deposited along intercellular spaces and cell walls of cells in region of phloem bundle. Bar = 25 μm. (From Phillips, D. P., Responses of Susceptible and Resistant Avocado Cultivars to Infection by *Phytophthora cinnamomi*, Ph.D. thesis, University of Melbourne, Melbourne, Australia, 1989.)

In spite of the defense reactions produced by 'Duke 7' and 'Topa Topa', summarized in Table 2, the pathogen remained viable at or slightly in advance of the lesion boundary. It was therefore concluded that in addition to the anatomical changes exhibited by 'Duke 7' in response to *P. cinnamomi* invasion, other biochemical factors were probably involved in lesion arrest. Whether a similar range of postinvasive responses exist for interactions involving fruit trees and other *Phytophthora* spp., besides that shown for *P. cactorum* attack of apple,[44,115] remains unanswered.

However, with respect to stone fruits infected by *Phytophthora* spp., evidence of root rot lesions is apparently much less common than the presence of trunk cankers,[181] even though there are some reports of root lesion development in association with trunk canker formation.[26,139,140] This may in part explain the absence of information on pathological anatomy of root diseases for such fruits, but as noted by Wilcox,[182] does not account for the relative lack of information on the pathological anatomy of trunk and crown infections. Much work remains to be done on the pathological anatomy of root disease for many different host/pathogen combinations involving fruit trees and *Phytophthora* spp.

IV. ALTERATIONS IN HOST DEFENSE RESPONSE DUE TO EXTERNAL INFLUENCES

A. STRESS PREDISPOSITION

The capacity of a host plant to respond to the presence of a pathogen is determined genetically; however, its disposition to disease, both before and during infection, can be shifted to increased or decreased susceptibility by environmental factors.[141] Of the five major environmental variables, light, air purity, temperature, water, and nutrients, the latter four have the most pronounced effects, alone or in concert.[142] Thus, environmental stress can affect host plant response, increasing susceptibility to disease by directly or indirectly limiting the expression and translation of gene products,[143] or it may help actively exclude pathogens.[142,144]

Classic examples of "stress predisposition" affecting host-pathogen interactions, such as those involving *Phytophthora* spp., lie in studies on the effects of osmotic (saline) and

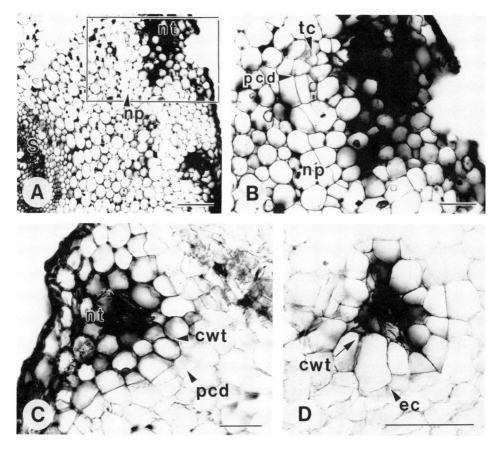

FIGURE 12. Hand transverse sections of localized regions of necrophylactic periderm that line superficial wounds of the outer cortex in healthy root tissues of the avocado rootstock, 'Duke 7'. (A) Localized region of necrophylactic periderm (np) lining necrotic tissue (nt). S, stele. Scale bar = 100 μm. (Toluidine blue stain.) (B) Inset of A. One to two cell layers immediately adjacent to site of cell necrosis appeared translucent (tc) and did not stain in toluidine blue. Cell layers, 2 to 3 cells deep, usually had undergone stimulated periclinal cell division (pcd). (Bar = 50 μm; toluidine blue stain.) (C) Periclinal cell division (pcd) around a small region of necrosis in the outer root cortex. Some cells in the innermost layer appear enlarged. The cell wall thickening (cwt) reacted positively, i.e., stained red, in the presence of acidified phloroglucinol. nt, necrotic tissue. Bar = 50 μm. (D) Peridermal tissue ringing site of cell necrosis. Note cell wall thickening (cwt) and enlarged cells (ec). (Acidified phloroglucinol stain; bar = 50 μm.) (From Phillips, D. P., Responses of Susceptible and Resistant Avocado Cultivars to Infection by *Phytophthora cinnamomi*, Ph.D. thesis, University of Melbourne, Melbourne, Australia, 1989.)

oxygen stress. In chrysanthemum roots exposed to salinity stress and infected by *P. cryptogea*, host defense response was altered from one of resistance to one of susceptibility.[145] In nonstressed roots, infection hyphae were confined to the outer three to four cell layers, particularly in the older portions of the roots. Cell wall appositions were common at points of contact or penetration of host cells by hyphae and resistance was expressed. In contrast, in salinity-stressed roots, hyphae colonized tissue much more rapidly and extensively, and cell wall appositions and organelle migration were reduced or lacking.[145] Root rot severity increased significantly with increased salt stress before or after inoculation of tomato seedlings with zoospores of *P. parasitica*.[146]

Although the previous two examples are not associated with fruit trees, conditions of high soil salinity are also found to render roots and stems of citrus seedlings more susceptible

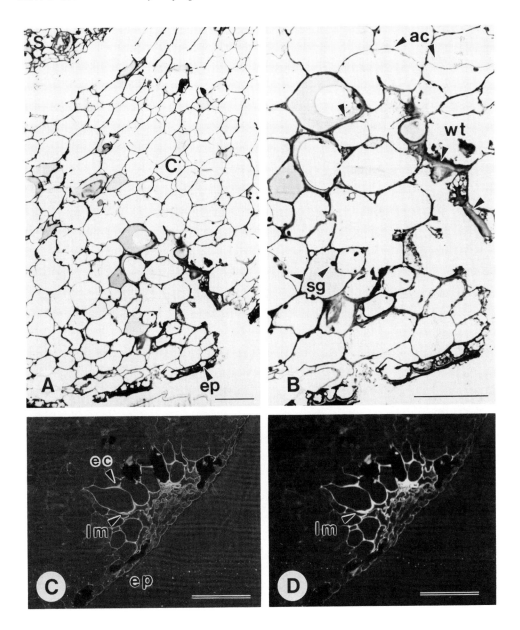

FIGURE 13.   Thin transverse microtome sections of localized regions of necrophylactic periderm lining superficial wounds of the outer core in healthy root tissue of the avocado rootstock, 'Topa Topa' (scale bar = 50 μm). A–D. (A) Root cortical tissue, 3.0 cm in advance of the lesion boundary of a *Phytophthora cinnamomi*-infected 'Topa Topa' clonal plant. Note breach of epidermis (ep) in the outer cortex (C). S, stele. (B) Inset of (A). Outer cell walls thickened (wt) with material staining luminescent blue-green (arrows) immediately adjacent to ruptured cells. Note anticlinal cell division (ac) and starch grains (sg). A and B stained with toluidine blue and counterstained with potassium iodide solution. (C) Well-formed region of necrophylactic periderm in outer cortex showing autofluorescence. Cells immediately adjacent to site of wounding are enlarged (ec) and their outer walls are thickened with a material which is both highly autofluorescent and fluorescent (see Figure 12D). lm, luminescent material. (D) Serial section of that shown in C under decolorized aniline blue fluorescence. (From Phillips, D. P., *Responses of Susceptible and Resistant Avocado Cultivars to Infection by* Phytophthora cinnamomi, Ph.D. thesis, University of Melbourne, Melbourne, Australia, 1989.)

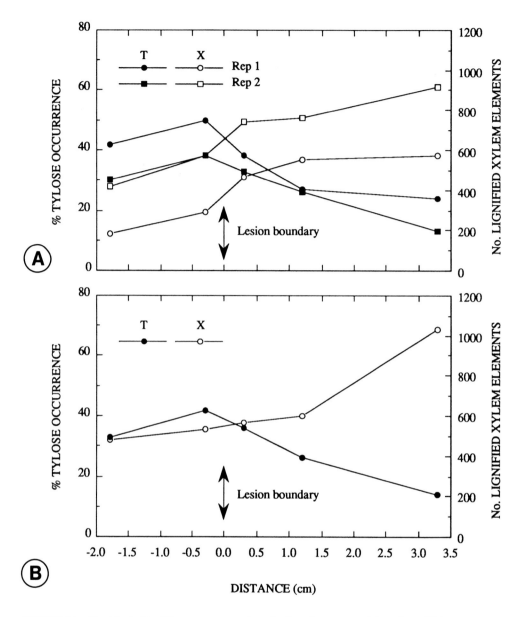

FIGURE 14. Combined plot of the percentage number of xylem elements containing tyloses (T) in a transverse section of the root, and the number of xylem elements with secondary wall thickening (X), in the region of the lesion boundary of *Phytophthora cinnamomi*-infected roots of two avocado rootstocks, 'Duke 7' (moderately resistant) and 'Topa Topa' (susceptible). (A) Two clonal replicates of the rootstock 'Duke 7'. (B) A clone of the rootstock 'Topa Topa'. (From Phillips, D. P., Responses of Susceptible and Resistant Avocado Cultivars to Infection by *Phytophthora cinnamomi*, Ph.D. thesis, University of Melbourne, Melbourne, Australia, 1989.)

to infection to *P. citrophthora*.[147] Sulistyowati and Keane[147] attributed the basis to physiology, whereby high soil salinity reduced accumulation of an antifungal substance 6,7-dimethoxycoumarin (DMC). DMC is reported to be a phytoalexin that is accumulated more rapidly in the resistant species, troyer citrange, than in the susceptible species, rough lemon.[183] *Phytophthora* root rot of citrus was previously reported to be more severe under saline conditions by Blaker and MacDonald.[148] This field of research may become increasingly important, with major problems developing in association with increasing salinization of ground soils in many regions of agricultural production around the world, such as the *Citrus* and avocado production areas of the Murray-Darling River Basin (Australia) and the Coachella Valley of California (U.S.).

Furthermore, water stress, either in the form of very high or low extremes in water status, is also reported to predispose many different hosts to attack by *Phytophthora* spp.[149,150] The factors involved in mediating or affecting host defense responses may differ, depending on whether dealing with waterlogged or saturated (high water status) soils or dry (low water status) conditions, and are not well understood. Compared with the limited number of histological and/or cytological studies that exist on the effects of salinity stress predisposition, no studies exist to help demonstrate or clarify the basis of the effects of water saturation or depletion on host/pathogen interactions involving *Phytophthora* spp.

For many different host/pathogen combinations involving *Phytophthora* spp., saturation of soil prior to inoculation or infection predisposes roots and/or crowns of plants to greater root and/or crown rot, such as with alfalfa,[151] cherry seedlings,[36,37] normally resistant cultivars of rhododendron,[152] and shortleaf and loblolly pines.[153] The same can also be said of avocados[154] and 'Delicious' apples subject to extended periodic flooding after planting in artificially infested soil.[155] The saturation of the soil is associated with low oxygen availability, thus the basis of the stress predisposition may rest with oxygen-mediated defense responses. However, Kuan and Erwin[151] demonstrated for alfalfa seedlings, with scanning electron micrographs (SEM), that water saturation also caused an increased in root damage via cracking of the root surface epidermal layers. This, in conjunction with increased exudation of compounds chemotaxically attractive to zoospores of *P. megasperma* Drechs. f.sp. *medicaginis* Kuan & Erwin, are believed to partly account for the increase in root rot severity. Unfortunately, the work did not extend to histological studies of both control and infected root tissue to help provide more information on the nature of the interaction.

Oxygen depletion is said to inhibit postinvasive host defense response in a number of ways, the resultant effects being both physiological and anatomical, i.e., possible suppression of lignin biosynthesis,[156] which is an important component of host defense response, as highlighted earlier. Consequently, there is again a great need for much more investigation into the histological and cytological nature of host/pathogen interactions involving *Phytophthora* spp. under water stress conditions, whether it be osmotic or matric.

B. SYSTEMIC FUNGICIDES

In recent years the alkyl phosphonates, such as the systemic fungicide Fosetyl-Al, aluminium ethyl phosphonate, have, with certain exceptions, been used widely to control oomycete pathogens, including several species of *Phytophthora* in a range of host/pathogen systems.[157-168] The active component was found to be phosphorous acid,[169] usually in the form of ''phosphonic acid'' or phosphonate (the salts or esters of phosphonic acid) according to IUPAC rules.[170,171] However, the precise mode of action of phosphonate is still somewhat uncertain, with evidence for both direct[157] and indirect modes of action.[110] The fungicide may improve resistance of susceptible hosts by modifying both host defense response and infective fungal mycelia physiologically[172-175] and anatomically.[110,176-178]

Susceptible lines of tobacco seedlings, when pretreated with Fosetyl-Al, became resistant to infection by *P. nicotianae* var. *nicotianae* zoospores.[110] Pretreatment of susceptible tobacco lines induced changes that included rapid cytoplasmic aggregation and deposition of papillae around and migration of host nuclei toward sites of hyphal penetration. The point of infection in 86% of cases was situated at anticlinal cell wall junctions of the epidermis. Hypersensitive-like cell death around invading intercellular hyphae was also induced within 24 h of inoculation. These changes were absent from untreated seedlings. In untreated seedlings, some papillae were deposited adjacent to penetration pegs, penetration was not prevented, and mycelium was formed throughout the cortex, along with haustoria and sporangia.

The mode of action of the phosphonate anion, phosphonic acid, in altering the host/ pathogen interaction is discussed by Guest and Grant,[5] who suggest it is complex. Phosphonate, by direct fungistatic action, can slow the growth of the pathogen and inhibit its sporulation, which in turn allows the host defense more time to develop and kill the invading pathogen. Yet, at phosphonate levels sufficiently low to not be toxic to the pathogen, the interaction may also be indirectly influenced by either increasing the production or activity of elicitors produced by the pathogen in the form of stress metabolites and/or inhibiting the production or activity of suppressors produced by the pathogen, molecules which are capable of suppressing the defenses of the host.[5] Whatever the combination, the result is one of improved recognition of the host to invasion by the pathogen and subsequent elicitation of normal host defense mechanisms.

V. CONCLUSIONS

The pathological anatomy of *Phytophthora*-infected avocado root tissue shows several deviations from uninfected root. These include the formation of barriers comprising extended zones of periderm and the walling of infected phloem elements. Similar changes may occur but have not been described for other fruit trees infected by *Phytophthora* species, with the exception of apples infected by *P. cactorum*. Much work remains to be done in this field of research for *Phytophthora* diseases of citrus, pears, walnuts, and (particularly) stone fruits. Further research is also urgently required into the effects of stress predisposition, such as osmotic or salinity stress, on host anatomical defense response. A greater understanding of the interaction from a histological perspective could assist in attempts to control its impact on host resistance, particularly as soil salinization is becoming an increasing problem in many areas of fruit tree production.

The way in which systemic fungicides, such as Fosetyl-Al or phosphonate, can modify host anatomical defense in what are considered to be normally susceptible host lines,[110] again raises the question of the degree or form of elicitation of host defense response between closely related cultivars or rootstocks of various host/pathogen systems. In diseases caused by *Phytophthora* spp., the question has been extended to interactions involving both so-called host-specific[16,110] and nonspecific[42,73] pathogens (even though no examples of host-specific interactions exist for fruit trees as such), suggesting that the question extends beyond matters of host specificity.

The differential degree and rate of response have been considered to form the crux of the interaction between *P. cinnamomi* and the closely related avocado rootstocks 'Duke 7' and 'Topa Topa'.[42] This phenomenon appears to have also been recently demonstrated for the interaction between *P. cinnamomi* and the eucalypts, *Eucalyptus marginata* Sm. (susceptible) and *E. calophylla* R. Br. (field resistant).[179] ''Resistant'' clonal lines of *E. marginata,* taken from remnant field trees that survived *P. cinnamomi* attack, produce postinvasive resistance mechanisms in response to infection by *P. cinnamomi* which are similar to that of the resistant *E. calophylla*.[179,180] In several different aspects of the avocado-*P. cinnamomi* interaction it was apparent that the defense mechanisms or expression of resistance instigated as an immediate response by 'Duke 7' were also initiated by the susceptible rootstock 'Topa Topa', but more slowly or with a lower magnitude.[42] Hence, it is more likely a question of degree or form of elicitation of defense in such closely related lines, rather than a total lack of a specific response that determined the susceptibility of 'Topa Topa.' This is particularly significant when considering the role of anatomical defense mechanisms such as periderm in host defense, as these mechanisms are sensitive to external constraints.

Thus, a gradient exists from susceptibility to resistance within nonspecific hosts such as the various lines of avocado and *P. cinnamomi,* and this corresponds with the gradient in both rate and magnitude of defense response.

ACKNOWLEDGMENTS

I thank Drs. Gretna Weste, Jill Hinch, M. Chandrashekar, Michele Barson, and Wayne Wilcox for critical readings of the manuscript. I particularly thank Drs. Hans Hohl, John Mircetich, Joanna Young (Tippett), Jill Hinch, Gretna Weste, and Harry Swart, and Mr. Ken Pegg, Mr. Frank Greenhalgh, and Mr. Peter Taylor for permission to use and provision of copies of figures used in the compilation of this chapter. The advice and assistance of Dr. David Guest and staff in the Bureau, particularly Peta Michielsen and Bob Georgeson's expertise with photographic plates, was also much appreciated. The photographs illustrated in Figures 1A and B were taken while the author was working for the Cocoa Black Pod Research Trust of Papua New Guinea (CBPRT of PNG), Kar Kar Island, Papua New Guinea. Thanks also go to the Bureau of Rural Resources, Department of Primary Industries and Energy, Canberra, Australia, for the use of facilities.

REFERENCES

1. **Zentmyer, G.A.,** The world of Phytophthora, in *Phytophthora: Its Biology, Taxonomy, Ecology and Pathology,* Erwin, D. C., Bartnicki-Garcia, S., and Tsao, P. H., Eds., American Phytopatholigical Society, St. Paul, MN, 1983, 1.
2. **Waterhouse, G. M., Newhook, F. J., and Stamps, J. D.,** Present criteria for classification of *Phytophthora,* in *Phytophthora: Its Biology, Taxonomy, Ecology and Pathology,* Erwin, D. C., Bartnicki-Garcia, S., and Tsao, P. H., Eds., American Phytopathological Society, St. Paul, MN, 1983, 139.
3. **Taylor, P. A., Pascoe, I. G., and Greenhalgh, F. C.,** *Phytophthora clandestina* sp. nov. in roots of subterranean clover, *Mycotaxon,* 22, 77, 1985.
4. **Margulis, L. and Schwartz, K.,** *Five Kingdoms. An Illustrated Guide to the Phyla of Life on Earth,* W. H. Freeman, New York, 1982, 376.
5. **Guest, D. and Grant, B.,** The complex action of phosphonates as antifungal agents, *Biol. Rev.,* 66, 159, 1991.
6. **Ribeiro, O. K.,** *A Source Book of the Genus Phytophthora,* J. Cramer, Vaduz, Liechtenstein, 1978, 417.
7. **Erwin, D. C., Bartnicki-Garcia, S., and Tsao, P. H., Eds.,** *Phytophthora: Its Biology, Taxonomy, Ecology and Pathology,* American Phytopathological Society, St. Paul, MN, 1983, 391.
8. **Erwin, D. C.,** Variability within and among species of *Phytophthora,* in *Phytophthora: Its Biology, Taxonomy, Ecology and Pathology,* Erwin, D. C., Bartnicki-Garcia, S., and Tsao, P. H., Eds., American Phytopathological Society, St. Paul, MN, 1983, 149.
9. **Brasier, C. M.,** Problems and prospects in *Phytophthora* research, in *Phytophthora: Its Biology, Taxonomy, Ecology and Pathology,* Erwin, D. C., Bartnicki-Garcia, S., and Tsao, P. H., Eds., American Phytopathological Society, St. Paul, MN, 1983, 351.
10. **Hohl, H. R. and Stössel, P.,** Host-parasite interfaces in a resistant and a susceptible cultivar of *Solanum tuberosum* inoculated with *Phytophthora infestans:* tuber tissue, *Can. J. Bot.,* 54, 900, 1976.
11. **Allen, F. H. E. and Friend, J.,** Resistance of potato tubers to infection by *Phytophthora infestans*: a structural study of haustorial encasement, *Physiol. Plant Pathol.,* 22, 285, 1983.
12. **Storti, E., Pelucchini, D., Tegli, S., and Scala, A.,** A potential defense mechanism of tomato against the late blight disease is suppressed by germinating sporangia-derived substances from *P. infestans, J. Phytopathol.,* 121, 275, 1988.
13. **Trique, B., Ravisé, A., and Bompeix, G.,** Modulation of infections induced by *Phytophthora* spp. in tomato plant, *Agronomie,* 1, 823, 1981.

14. **Beagle-Ristaino, J. E. and Rissler, J. F.,** Histopathology of susceptible and resistant soybean roots inoculated with zoospores of *Phytophthora megasperma* f. sp. *glycinea, Phytopathology,* 73, 590, 1983.

15. **Ward, E. W. B. and Lazarovits, G.,** Temperature-induced changes in specificity in the interaction of soybeans with *Phytophthora megasperma* f. sp. *glycinea, Phytopathology,* 72, 826, 1982.

16. **Stössel, P., Lazarovits, G., and Ward, E. W. B.,** Electron microscope study of race-specific and age-related resistant and susceptible reactions of soybeans to *Phytophthora megasperma* var. *sojae, Phytopathology,* 71, 617, 1981.

17. **Purss, G. S.,** Caloona — stem rot resistant cowpea, *Queensl. Agric. J.,* 89, 756, 1963.

18. **Bateman, K. S., Hinch, J. M., Ralton, J. E., Clarke, A. E., McKenzie, J. A., Imrie, B. C., and Howlett, B. J.,** Inheritance of resistance of cowpea to *Phytophthora vignae* in whole plants, cuttings and stem callus cultures, *Aust. J. Bot.,* 37, 511. 1989.

19. **Zentmyer, G. A.,** *Phytophthora cinnamomi* and the Diseases It Causes, Phytopathol. Monogr. No. 10, American Phytopathological Society, St. Paul, MN, 1980.

20. **Mircetich, S. M. and Matheron, M. E.,** *Phytophthora* root and crown rot of walnut trees, *Phytopathology,* 73, 1481, 1983.

21. **Matheron, M. E. and Mircetich, S. M.,** Relative resistance of different rootstocks of English walnut to six *Phytophthora* spp. that cause root and crown rot in orchard trees, *Plant Dis.,* 69, 1039, 1985.

22. **Julis, A. J., Clayton, C. N., and Sutton, T. B.,** Detection and distribution of *Phytophthora cactorum* and *P. cambivora* on apple rootstocks, *Plant Dis. Rep.,* 62, 516, 1978.

23. **Long, G.,** Apple tree resistance to collar rot disease, *N.Z. Agric. Sci.,* 16, 54, 1982.

24. **Long, P. and Miller, S.,** Rootrot. A cure seems likely, *South. Hortic.,* 25, 26, 1986.

25. **Cameron, H. R.,** Susceptibility of pear roots to *Phytophthora, Phytopathology,* 52, 1295, 1962.

26. **Wilcox, W. F. and Ellis, M. A.,** Phytophthora root and crown rots of peach trees in the eastern Great Lakes region, *Plant Dis.,* 73, 794, 1989.

27. **Hickman, C. J.,** *Phytophthora* — plant destroyer, *Trans. Br. Mycol. Soc.,* 41, 1, 1958.

28. **Afek, U., Sztejnberg, A., and Solel, Z.,** A rapid method for evaluating citrus seedlings for resistance to foot rot caused by *Phytophthora citrophthora, Plant Dis.,* 74, 66, 1990.

29. **Laviola, C., Somma, V., and Evola, C.,** Present status of *Phytophthora* species in the Mediterranean area, especially in relation to citrus, *EPPO Bull.,* 20, 1, 1990.

30. **Cinar, A., Tuzcu, O., and Göksedef, M. O.,** Resistance study of the citrus rootstocks to *Phytophthora citrophthora* (Smith and Smith) Leonian, *J. Turk. Phytopathol.,* 5, 49, 1976.

31. **Tuzcu, Ö.,** Resistance of citrus rootstocks to *Phytophthora citrophthora* during winter dormancy, *Plant Dis.,* 68, 502, 1984.

32. **Kouyeas, H.,** Stone fruit tree apoplexy caused by *Phytophthora* collar rot, *EPPO Bull.,* 7, 117, 1977.

33. **Doster, M. A. and Bostock, R. M.,** Effects of low temperature on resistance of almond trees to *Phytophthora* pruning wound cankers in relation to lignin and suberin formation in wounded bark tissue, *Phytopathology,* 78, 470, 1988.

34. **Chee, K. H.,** Hosts of *Phytophthora palmivora, Rev. Appl. Mycol.,* 48, 337, 1969.

35. **Turner, P. D. and Asomaning, E. J. A.,** Root infection of *Theobroma cacao* by *Phytophthora palmivora, Trop. Agric.,* 39, 339, 1962.

36. **Wilcox, W. F. and Mircetich, S. M.,** Effects of flooding duration on the development of *Phytophthora* root and crown rots of cherry, *Phytopathology,* 75, 1451, 1985.

37. **Wilcox, W. F. and Mircetich, S. M.,** Influence of soil water matric potential on the development of *Phytophthora* root and crown rots of mahaleb cherry, *Phytopathology,* 75, 648, 1985.

38. **Jeffers, S. N. and Aldwinckle, H. S.,** Seasonal variation in extent of colonization of two apple rootstocks by five species of *Phytophthora, Plant Dis.,* 70, 941, 1986.

39. **Graham, J. H.,** Evaluation of tolerance of citrus rootstocks to *Phytophthora* root rot in chlamydospore-infested soil, *Plant Dis.,* 74, 743, 1990.

40. **Weste, G.,** Population dynamics and survival of *Phytophthora,* in *Phytophthora: Its Biology, Taxonomy, Ecology and Pathology,* Erwin, D. C., Bartnicki-Garcia, S., and Tsao, P. H., Eds., American Phytopathological Society, St. Paul, MN, 1983, 237.

41. **Tsao, P. H.,** Why many *Phytophthora* root rots and crown rots of tree and horticultural crops remain undetected, *EPPO Bull.,* 20, 11, 1990.

42. **Phillips, D.,** Responses of Susceptible and Resistant Avocado Cultivars to Infection by *Phytophthora cinnamomi,* Ph.D. thesis, University of Melbourne, Melbourne, Australia, 1989.

43. **Zentmyer, G. A.,** The effect of temperature on growth and pathogenesis of *Phytophthora cinnamomi* and on growth of its avocado host, *Phytopathology,* 71, 925, 1981.

44. **Dakwa, J. T. and Sewell, G. W. F.,** Influence of rootstock type and time of inoculation on the resistance of five apple scion cultivars to collar rot caused by *Phytophthora cactorum, J. Hortic. Sci.,* 56, 357, 1981.

45. **Ho, H. H. and Zentmyer, G. A.,** Infection of avocado and other species of *Persea* by *Phytophthora cinnamomi, Phytopathology,* 67, 1085, 1977.

46. **Hine, R. B., Alaban, C., and Klemmer, H.,** Influence of soil temperature on root and heart rot of pineapple caused by *Phytophthora cinnamomi* and *Phytophthora parasitica, Phytopathology,* 54, 1287, 1964.

47. **Taylor, P. A.,** *Phytophthora* spp. in irrigation water in the Goulburn Valley, Victoria, *Aust. Plant Pathol. Soc. Newsl.,* 6, 41, 1977.

48. **Pegg, K. G., Forsberg, L. I., and Whiley, A. W.,** Avocado root rot, *Queensl. Agric. J.,* 108, 162, 1982.

49. **Tidball, C. J. and Linderman, R. G.,** *Phytophthora* root and stem rot of apple rootstocks from stool beds, *Plant Dis.,* 74, 141, 1990.

50. **Shea, S. R., Shearer, B., Tippett, J., and Deegan, P. M.,** Distribution, reproduction and movement of *Phytophthora cinnamomi* on sites highly conducive to jarrah dieback in South Western Australia, *Plant. Dis.,* 67, 970, 1983.

51. **Marks, G. C., Smith, I. W., and Kassaby, F. Y.,** Trunk infection of *Eucalyptus* species by *Phytophthora cinnamomi* Rands. A preliminary report, *Aust. For. Res.,* 11, 257, 1981.

52. **Weste, G.,** Damage and loss caused by *Phytophthora* species in forest crops, in *Plant Diseases: Infection, Damage and Loss,* Wood, R. K. S. and Jellis, G. J., Eds., Blackwell Scientific, Oxford, 1984, 273.

53. **Hinch, J. and Weste, G.,** Behaviour of *Phytophthora cinnamomi* zoospores on roots of Australian forest species, *Aust. J. Bot.,* 27, 679, 1979.

54. **Hinch, J. M., Wetherbee, R., Mallett, J. E., and Clarke, A. E.,** Response of *Zea mays* roots to infection with *Phytophthora cinnamomi.* I. The epidermal layer, *Protoplasma,* 126, 178, 1985.

55. **Tippett, J. T., Holland, A. A., Marks, G. C., and O'Brien, T. P.,** Penetration of *Phytophthora cinnamomi* into disease tolerant and susceptible eucalypts, *Arch. Microbiol.,* 108, 231, 1976.

56. **Duncan, J. M.,** *Phytophthora* species attacking strawberry and raspberry, *EPPO Bull.,* 20, 107, 1990.

57. **Ploetz, R. C. and Mitchell, D. J.,** Root rot of bamboo palm caused by *Phytophthora arecae, Plant Dis.,* 73, 266, 1989.

58. **Carlile, M. J.,** Motility, taxis, and tropism in *Phytophthora,* in *Phytophthora: Its Biology, Taxonomy, Ecology and Pathology,* Erwin, D. C., Bartnicki-Garcia, S., and Tsao, P. H., Eds., American Phytopathological Society, St. Paul, MN, 1983, 95.

59. **Podger, F. D.,** Aetiology of Jarrah Dieback...A Disease of Dry Sclerophyll *Eucalyptus* Forests in Western Australia, M.Sc. thesis, University of Melbourne, Melbourne, Australia, 1968.

60. **Weste, G. and Taylor, P.,** The invasion of native forest by *Phytophthora cinnamomi.* I. Brisbane Ranges, Victoria, *Aust. J. Bot.,* 19, 281, 1971.

61. **Podger, F. D.,** *Phytophthora cinnamomi,* a cause of lethal disease in indigenous plant communities in Western Australia, *Phytopathology,* 62, 972, 1972.

62. **Van der Merwe, J. J. H. and Matthee, F. N.,** *Phytophthora* crown and root rot of apple trees in South Africa, *Phytophylactica,* 5, 55, 1973.

63. **Keen, N. T. and Yoshikawa, M.,** Physiology of disease and the nature of resistance to *Phytophthora,* in *Phytophthora: Its Biology, Taxonomy, Ecology and Pathology,* Erwin, D. C., Bartnicki-Garcia, S., and Tsao, P. H., Eds., American Phytopathological Society, St. Paul, MN, 1983, 279.

64. **Weste, G. and Cahill, D.,** Changes in root tissue associated with infection by *Phytophthora cinnamomi, Phytopathol. Z.,* 103, 97, 1982.

65. **Cahill, D. M., Weste, G., and Grant, B. R.,** Changes in cytokinin concentrations in xylem extrudate following infection of *Eucalyptus marginata* Donn. ex. Sm. with *Phytophthora cinnamomi* Rands, *Plant Physiol.,* 81, 1103, 1986.

66. **Cahill, D. M. and Ward, E. W. B.,** Rapid localized changes in abscisic acid concentrations in soybean in interactions with *Phytophthora megasperma* f. sp. *glycinea* or after treatment with elicitors, *Physiol. Mol. Plant Pathol.,* 35, 483, 1989.

67. **Deverall, B. J.,** *Defence Mechanisms of Plants,* Cambridge Monogr. in Exp. Biol. No 19, Cambridge University Press, Cambridge, U. K., 1977.

68. **Heath, M. C.,** The absence of active defense mechanisms in compatible host-pathogen interactions, in *Active Defense Mechanisms in Plants,* NATO Advanced Study Institute Series, Wood, R. K. S., Ed., Plenum Press, New York, 1982, 143.

69. **Friend, J.,** Lignification in infected tissue, in *Biochemical Aspects of Plant-Parasite Relationships,* Friend, J. and Threlfall, D. R., Eds., Academic Press, London, 1976, 291.

70. **Misaghi, I. J.,** *Physiology and Biochemistry of Plant-Pathogen Interactions,* Plenum Press, New York, 1982, 287.

71. **Bailey, J. A. and Deverall, B. J., Eds.,** *The Dynamics of Host Defence,* Academic Press, new York, 1983.

72. **Goodman, R. N., Kiraly, Z., and Wood, K. R.,** *The Biochemistry and Physiology of Plant Disease,* University of Missouri Press, Columbia, 1986.

73. **Cahill, D., Legge, N., Grant, B., and Weste, G.,** Cellular and histological changes induced by *Phytophthora cinnamomi* in a group of plant species ranging from fully susceptible to fully resistant, *Phytopathology,* 79, 417, 1989.

74. **Hanchey, P. and Wheeler, H.,** Pathological changes in ultrastructure of tobacco roots infected with *Phytophthora parasitica* var. *nicotianae, Phytopathology,* 61, 33, 1971.

75. **Coffey, M. D. and Wilson, U. E.,** Histology and cytology of infection and disease caused by *Phytophthora,* in *Phytophthora: Its Biology, Taxonomy, Ecology and Pathology,* Erwin, D. C., Bartnicki-Garcia, S. and Tsao, P. H., Eds., American Phytopathological Society, St. Paul, MN, 1983, 289.

76. **Marks, G. C. and Mitchell, J. E.,** Penetration and infection of alfalfa roots by *Phytophthora megasperma* and the pathological anatomy of infected roots, *Can. J. Bot.,* 49, 63, 1971.

77. **Tippett, J. T., O'Brien, T. P., and Holland, A. A.,** Ultrastructural changes in eucalypt roots caused by *Phytophthora cinnamomi, Physiol. Plant Pathol.,* 11, 279, 1977.

78. **Allen, R. N. and Newhook, F. J.,** Chemotaxis of zoospores of *Phytophthora cinnamomi* to ethanol in capillaries of soil pore dimensions, *Trans. Br. Mycol. Soc.,* 61, 287, 1973.

79. **Khew, K. L. and Zentmyer, G. A.,** Electrotaxic response of zoospores of seven species of *Phytophthora, Phytopathology,* 63, 1511, 1974.

80. **Hinch, J. M. and Clarke, A. E.,** Adhesion of fungal spores to root surfaces is mediated by carbohydrate components of the root slime, *Physiol. Plant Pathol.,* 21, 113, 1980.

81. **Grant, B. R., Irving, H. R., and Radda, M.,** The effect of pectin and related compounds on encystment and germination of *Phytophthora palmivora* zoospores, *J. Gen. Microbiol.,* 131, 669, 1985.

82. **Gubler, F., Hardham, A. R., and Duniec, J.,** Characterising adhesiveness of *Phytophthora cinnamomi* zoospores during encystment, *Protoplasma,* 149, 24, 1989.

83. **Hardham, A. R. and Gubler, F.,** Polarity of attachment of zoospores of a root pathogen and pre-alignment of the emerging germ tube, *Cell Biol. Int. Rep.,* 14, 947, 1990.

84. **Wetherbee, R., Hinch, J. M., Bonig, I., and Clarke, A. E.,** Response of *Zea mays* roots to infection with *Phytophthora cinnamomi.* II. The cortex and stele, *Protoplasma,* 126, 188, 1985.

85. **Cahill, D. and Weste, G.,** Formation of callose deposits as a response to infection with *Phytophthora cinnamomi, Trans. Br. Mycol. Soc.,* 80, 23, 1983.

86. **Tippett, J. T.,** The response of eucalypt roots to infection by *Phytophthora cinnamomi,* in *Microbial Ecology,* Loutit, M. and Miles, J. A. R., Eds., Springer-Verlag, New York, 1978, 364.

87. **Byrt, P. N. and Holland, A. A.,** Infection of axenic *Eucalyptus* seedlings with *Phytophthora cinnamomi* zoospores, *Aust. J. Bot.,* 26, 169, 1978.

88. **Phillips, D. and Weste, G.,** Field resistance in three native monocotyledon species that colonize indigenous sclerophyll forest after invasion by *Phytophthora cinnamomi, Aust. J. Bot.,* 32, 339, 1984.

89. **Phillips, D., Grant, B. R., and Weste, G.,** Histological changes in the roots of an avocado cultivar, Duke 7, infected with *Phytophthora cinnamomi, Phytopathology,* 77, 691, 1987.

90. **Malajczuk, N.,** Interactions between *Phytophthora cinnamomi* and Roots of *Eucalyptus calophylla* R. Br. and *Eucalyptus marginata* Donn ex Sm., Ph.D. thesis, University of Western Australia, Perth, 1976.

91. **Stössel, P., Lazarovits, G., and Ward, E. W. B.,** Penetration and growth of compatible and incompatible races of *Phytophthora megasperma* var. *sojae* in soybean hypocotyl tissues differing in age, *Can. J. Bot.,* 58, 2594, 1980.

92. **Goode, P. M.,** Infection of strawberry roots by zoospores of *Phytophthora fragariae, Trans. Br. Mycol. Soc.,* 39, 367, 1956.

93. **Millholland, R. D.,** Pathogenicity and histopathology of *Phytophthora cinnamomi* on highbush and rabbiteye blueberry, *Phytopathology,* 65, 789, 1975.

94. **Shea, S. R., Shearer, B., and Tippett, J.,** Recovery of *Phytophthora* cinnamomi Rands from vertical roots of jarrah (*Eucalyptus marginata* Sm.), *Aust. Plant Pathol.,* 11, 25, 1982.

95. **Tippett, J. T. and Hill, T. C.,** Role of periderm in resistance of *Eucalyptus marginata* roots against *Phytophthora cinnamomi, Eur. J. For. Pathol.,* 14, 431, 1984.

96. **Weste, G.,** Comparative pathogenicity of six root parasites towards cereals, *Phytopathol. Z.,* 93, 41, 1978.

97. **Casares, A., Melo, E. M. P. F., Ferraz, J. F. P., and Ricardo, C. P. P.,** Differences in ability of *Phytophthora cambivora* and *P. cinnamomi* to dephenolize lignin, *Trans. Br. Mycol. Soc.,* 87, 229, 1986.

98. **Hwang, B. K., Kim, W. B., and Kim, W. K.,** Ultrastructure at the host-parasite interface of *Phytophthora capsici* in roots and stems of *Capsicum annuum, J. Phytopathol.,* 127, 305, 1989.

99. **Malajczuk, N., McComb, A. J., and Parker, C. A.,** Infection by *Phytophthora cinnamomi* Rands of roots of *Eucalyptus calophylla* R.Br. and *Eucalyptus marginata* Donn. ex Sm., *Aust. J. Bot.,* 25, 483, 1977.

100. **Paxton, J. D.,** Phytoalexins — a working definition, *Phytopathol. Z.,* 101, 106, 1981.

101. **Bailey, J. A. and Mansfield, J. W.,** *Phytoalexins,* Blackie & Son, Glasgow, 1982.

102. **Darvill, A. G. and Albersheim, P.,** Phytoalexins and their elicitors — a defense against microbial infection in plants, *Annu. Rev. Plant Physiol.,* 35, 243, 1984.

103. **Keen, N. T.,** Evaluation of the role of phytoalexins, in *Plant Disease Control,* Staples, R. C., Ed., John Wiley & Sons, New York, 1981, 155.

104. **Ingham, J. L.,** Phytoalexins from the Leguminosae, in *Phytoalexins,* Bailey, J. A. and Mansfield, J. W., Eds., Blackie & Son, Glasgow, 1982, 21.

105. **Ku'c, J.,** Phytoalexins from the Solanaceae, in *Phytoalexins,* Bailey, J. A. and Mansfield, J. W., Eds., Blackie & Son, Glasgow, 1982, 81.

106. **Yoshikawa, M., Yamauchi, K., and Masago, H.,** Glyceollin: its role in restricting fungal growth in resistant soybean hypocotyls infected with *Phytophthora megasperma* var. *sojae, Physiol. Plant Pathol.,* 12, 73, 1978.

107. **Ward, E. W. B., Cahill, D. M., and Bhattacharyya, M. K.,** Abscisic acid suppression of phenylalanine ammonia-lyase activity and mRNA, and resistance of soybeans to *Phytophthora megasperma* f. sp. *glycinea, Plant Physiol.,* 91, 23, 1989.

108. **Umaerus, V.,** Studies on field resistance to *Phytophthora infestans.* V. Mechanisms of resistance and applications to potato breeding, *Z. Pflanzenzuecht.,* 63, 1, 1970.

109. **Umaerus, V., Umaerus, M., Erjefält, L., and Nilsson, B. A.,** Control of *Phytophthora* by host resistance: problems and progress, in *Phytophthora: Its Biology, Taxonomy, Ecology and Pathology,* Erwin, D. C., Bartnicki-Garcia, S., and Tsao, P. H., Eds., American Phytopathological Society, St. Paul, MN, 1983, 315.

110. **Guest, D. I.,** Evidence from light microscopy of living tissues that Fosetyl-Al modifies the defence response in tobacco seedlings following inoculation by *Phytophthora nicotianae* var. *nicotianae, Physiol. Mol. Plant Pathol.,* 29, 251, 1986.

111. **Beckman, C. H.,** Defenses triggered by the invader: physical defenses, in *Plant Disease. An Advanced Treatise,* Horsfall, J. and Cowling, E., Eds., Academic Press, New York, 1980, 225.

112. **Aist, J. R.,** Structural responses as resistance mechanisms, in *The Dynamics of Host Defence,* Bailey, J. A. and Deverall, B. J., Eds., Academic Press, New York, 1983, 33.

113. **Tippett, J. and Malajczuk, N.,** Interaction of *Phytophthora cinnamomi* and a resistant host, *Acacia pulchella, Phytopathology,* 69, 764, 1979.

114. **Tippett, J. T., Hill, T. C., and Shearer, B. L.,** Resistance of *Eucalyptus* spp. to invasion by *Phytophthora cinnamomi, Aust. J. Bot.,* 33, 409, 1985.

115. **Alt, D.,** Histological studies on the bark of apple trees after infection with *Phytophthora cactorum* (Leb. et Cohn) Schroet, *Angew. Bot.,* 54, 393, 1980.

116. **Struckmeyer, B. E. and Riker, A. J.,** Wound-periderm formation in white-pine trees resistant to blister rust, *Phytopathology,* 41, 276, 1951.

117. **Jewell, F. F. and Spiers, D. C.,** Histopathology of one- and two-year-old resisted infections by *Cromartium fusiforme* in slash pine, *Phytopathology,* 66, 741, 1976.

118. **Jewell, F. F., Jewell, D. C., and Walkinshaw, C. H.,** Histopathology of the initiation of resistance-zones in juvenile slash pine on *Cronartium quercuum* f.sp. *fusiforme, Phytopathol. Mediterr.,* 19, 8, 1980.

119. **Hebard, F. V., Griffin, G. J., and Elkins, J. R.,** Developmental and histopathology of cankers incited by hypovirulent and virulent isolates of *Endothia parasitica* on susceptible and resistant chestnut trees, *Phytopathology,* 74, 140, 1984.

120. **Biggs, A. R.,** Occurrence and location of suberin in wound reaction zones in xylem of 17 tree species, *Phytopathology,* 77, 718, 1987.

121. **Mullick, D. B. and Jensen, G. D.,** New concepts and terminology of coniferous periderms: necrophylactic and exophylactic periderms, *Can. J. Bot.,* 51, 1459, 1973.

122. **Mullick, D. B.,** A new tissue essential to necrophylactic periderm formation in the bark of four conifers, *Can. J. Bot.,* 53, 2443, 1975.

123. **Biggs, A. R.,** Boundary-zone formation in peach bark in response to wounds and *Cytospora leucostoma* infection, *Can. J. Bot.,* 62, 2814, 1984.

124. **Biggs, A. R. and Northover, J.,** Formation of the primary protective layer and phellogen after leaf abscission in peach, *Can. J. Bot.,* 63, 1547, 1985.

125. **Tippett, J. T., Shea, S. R., Hill, T. C., and Shearer, B. L.,** Development of lesions caused by *Phytophthora cinnamomi* in the secondary phloem of *Eucalyptus marginata, Aust. J. Bot.,* 31, 197, 1983.

126. **Bishop, C. D. and Cooper, R. M.,** An ultrastructural study of vascular colonization in three vascular wilt diseases. I. Colonization of susceptible cultivars, *Physiol. Plant Pathol.,* 23, 323, 1983.

127. **Mace, M. E.,** Contributions of tyloses and terpenoid aldehyde phytoalexins to *Verticillium* wilt resistance in cotton, *Physiol. Plant Pathol.,* 12, 1, 1978.

128. **Mace, M. E., Bell, A. A., and Beckman, C. H.,** *Fungal Wilt Diseases of Plants,* Academic Press, New York, 1981.

129. **Charést, P. M., Ouellette, G. B., and Pauze, F. J.,** Cytological observations of early infection process by *Fusarium oxysporum* f.sp. *radicis-lycopersici* in tomato plants, *Can. J. Bot.,* 62, 1232, 1984.

130. **Esau, K.,** *Anatomy of Seed Plants,* 2nd ed., John Wiley & Sons, New York, 1977, 550.

131. **Marks, G. C. and Tippett, J. T.,** Symptom development and disease escape in *Eucalyptus obliqua* growing in soils infected with *Phytophthora cinnamomi, Aust. For. Res.,* 8, 47, 1978.

132. **Kellam, M. K. and Coffey, M. D.,** Quantitative comparison of the resistance to *Phytophthora* root rot in three avocado rootstocks, *Phytopathology,* 75, 230, 1985.

133. **Dolan, T. E. and Coffey, M. D.,** Laboratory screening technique for assessing resistance of four avocado rootstocks to *Phytophthora cinnamomi, Plant Dis.,* 70, 115, 1986.

134. **Gabor, B. K., Guillemet, F. B., and Coffey, M. D.,** Comparison of field resistance to *Phytophthora cinnamomi* in twelve avocado rootstocks, *HortScience,* 25, 1655, 1990.

135. **Gabor, B. K. and Coffey, M. D.,** Comparison of rapid methods for evaluating resistance to *Phytophthora cinnamomi* in avocado rootstocks, *Plant Dis.,* 75, 118, 1991.

136. **Tippett, J. T., Crombie, D. S., and Hill, T. C.,** Effect of phloem water relations on the growth of *Phytophthora cinnamomi* in *Eucalyptus marginata, Phytopathology,* 77, 246, 1987.

137. **Dawson, P. and Weste, G.,** Changes in water relations associated with infection by *Phytophthora cinnamomi, Aust. J. Bot.,* 30, 393, 1982.

138. **Dugger, L. and Zentmyer, G. A.,** Ultrastructural studies of *Phytophthora* root rot of resistant and susceptible species of *Persea, Phytopathology,* 71, 871, 1981.

139. **Mircetich, S. M. and Matheron, M. E.,** *Phytophthora* root and crown rot of cherry trees, *Phytopathology,* 66, 549, 1976.

140. **Mircetich, S. M. and Keil, H. L.,** *Phytophthora cinnamomi* root rot and stem canker of peach trees, *Phytopathology,* 60, 1376, 1970.

141. **Yarwood, C. E.,** Predisposition, in *Plant Pathology. An Advanced Treatise,* Vol. 1, *The Diseased Plant,* Horsfall, J. G. and Dimond, A. E., Eds., Academic Press, New York, 1959, 521.

142. **Schoeneweiss, D. F.,** Water stress as a predisposing factor in plant disease, in *Water Deficits and Plant Growth,* Vol. 5, Kozlowski, T. T., Ed., Academic Press, New York, 1978, 61.

143. **Vanderplank, J. E.,** *Host-Pathogen Interactions in Plant Disease,* Academic Press, New York, 1982.

144. **Cook, R. J. and Papendick, R. I.,** Influence of water potential of soils and plants on root disease, *Annu. Rev. Phytopathol.,* 10, 349, 1972.

145. **Swiecki, T. J. and MacDonald, J. D.,** Histology of chrysanthemum roots exposed to salinity stress and *Phytophthora cryptogea, Can. J. Bot.,* 66, 280, 1988.

146. **Bouchibi, N., van Bruggen, A. H. C., and MacDonald, J. D.,** Effect of ion concentration and sodium: calcium ratio of a nutrient solution on *Phytophthora* root rot of tomato and zoospore motility and viability of *Phytophthora parasitica, Phytopathology,* 80, 1323, 1990.

147. **Sulistyowati, L. and Keane, P. J.,** Effect of soil salinity on accumulation of phytoalexin in roots and stems of citrus seedlings infected by *Phytophthora citrophthora,* in Australasian Plant Pathol. Soc. 8th Conf. Proc., Sydney, October 7 to 11, 1991 (Abstr.).

148. **Blaker, N. S. and MacDonald, J. D.,** The role of salinity in the development of *Phytophthora* root rot of citrus, *Phytopathology,* 76, 970, 1986.

149. **Ristaino, J. B. and Duniway, J. M.,** Effect of preinoculation and postinoculation water stress on the severity of *Phytophthora* root rot in processing tomatoes, *Plant Dis.,* 73, 349, 1989.

150. **Stolzy, L. H., Letey, J., Klotz, L. J., and Labanauskas, C. K.,** Water and aeration as factors in root decay of *Citrus sinensis, Phytopathology,* 55, 270, 1965.

151. **Kuan, T. L. and Erwin, D. C.,** Predisposition effect of water saturation of soil on *Phytophthora* root rot of alfalfa, *Phytopathology,* 70, 981, 1980.

152. **Blaker, N. S. and MacDonald, J. D.,** Predisposing effects of soil moisture extremes on the susceptibility of rhododendron to *Phytophthora* root and crown rot, *Phytopathology,* 71, 831, 1981.

153. **Fraedrich, S. W. and Tainter, F. H.,** Effect of dissolved oxygen concentration on the relative susceptibility of shortleaf and loblolly pine root tips to *Phytophthora cinnamomi, Phytopathology,* 79, 1114, 1989.

154. **Ploetz, R. C. and Schaffer, B.,** Effects of flooding and *Phytophthora* root rot on net gas exchange and growth of avocado, *Phytopathology,* 79, 204, 1989.

155. **Browne, G. T. and Mircetich, S. M.,** Effects of flood duration on the development of *Phytophthora* root and crown rots of apple, *Phytopathology,* 78, 846, 1988.

156. **Drew, M. C. and Lynch, J. M.,** Soil anaerobiosis, Microorganisms, and root function, *Annu. Rev. Phytopathol.,* 18, 37, 1980.

157. **Ouimette, D. G. and Coffey, M. D.,** Phosphonate levels in avocado (*Persea americana*) seedlings and soil following treatment with Fosetyl-Al or potassium phosphonate, *Plant Dis.,* 73, 212, 1989.

158. **De Boer, R. F., Greenhalgh, F. C., Pegg, K. G., Mayers, P. E., Lim, T. M., and Flett, S.,** Phosphorous acid treatments control *Phytophthora* diseases in Australia, *EPPO Bull.,* 20, 193, 1990.

159. **Taylor, P. A. and Washington, W. S.,** Curative treatment for *Phytophthora cactorum* in peach trees using metalaxyl and phosethyl Al, *Aust. Plant Pathol.,* 13, 31, 1984.

160. **Sandler, H. A., Timmer, L. W., and Graham, J. H.,** Effect of fungicide applications on populations of *Phytophthora parasitica* and on feeder root densities and fruit yields of citrus trees, *Plant Dis.,* 73, 902, 1989.

161. **Pegg, K. G., Whiley, A. W., Langdon, P. W., and Saranah, J. B.,** Comparison of phosetyl-Al, phosphorous acid and metalaxyl for the long-term control of *Phytophthora* root rot of avocado, *Aust. J. Exp. Agric.,* 27, 471, 1987.

162. **Darvas, J. M., Toerien, J. C., and Milne, D. L.,** Control of *avocado* root rot by trunk injection with phosetyl-Al, *Plant Dis.,* 68(7), 691, 1984.

163. **Bruin, G. C. A. and Edington, L. V.,** The chemical control of diseases caused by zoosporic fungi, in *Zoosporic Plant Pathogens. A Modern Perspective,* Buczaki, S. T., Ed., Academic Press, London, 1983, 193.

164. **Rohrbach, K. G. and Schenck, S.,** Control of pineapple heart rot, caused by *Phytophthora parasitica* and *P. cinnamomi,* with metalaxyl, fosetyl-Al, and phosphorous acid, *Plant Dis.,* 69, 320, 1985.

165. **Davis, R. M.,** Effectiveness of Fosetyl-Al against *Phytophthora parasitica* on tomato, *Plant Dis.,* 73, 215, 1989.

166. **Utkhede, R. S.,** Chemical and biological control of crown and root rot of apple caused by *Phytophthora cactorum, Can. J. Plant Pathol.,* 9, 295, 1987.

167. **Matheron, M. E. and Mircetich, S. M.,** Control of *Phytophthora* root and crown rot and trunk canker in walnut with metalaxyl and fosetyl Al, *Plant Dis.,* 69, 1042, 1985.

168. **Anderson, R. D., Middleton, R. M., and Guest, D. I.,** Development of a bioassay to test the effect of phosphorous acid on black pod of cocoa, *Mycol. Res.,* 93, 110, 1989.

169. **Fenn, M. E. and Coffey, M. D.,** studies on the *in-vitro* and *in-vivo* antifungal activity of fosetyl-Al and phosphorous acid, *Phytopathology,* 74, 606, 1984.

170. **Bompeix, G.,** Fungicides and host-parasite interactions: the case of phosphonates, *C. R. Acad. Agric. Fr.,* 75, 183, 1989.

171. **Coffey, M. D. and Ouimette, D. G.,** Phosphonates: antifungal compounds against oomycetes, in *Nitrogen, Phosphorus and Sulphur Utilisation by Fungi,* Boddy, L., Marchant, R., and Read, D. J., Eds., Cambridge University Press, Cambridge, U.K., 1989, 107.

172. **Guest, D. I., Upton, J. C. R., and Rowan, K. S.,** Fosetyl-Al alters the respiratory response in *Phytophthora nicotianae* var. *nicotianae*-infected tobacco, *Physiol. Mol. Plant Pathol.,* 34, 257, 1989.

173. **Nemestothy, G. S. and Guest, D. I.,** Phytoalexin accumulation, phenylalanine ammonia lyase activity and ethylene biosynthesis in fosetyl-Al treated resistant and susceptible tobacco cultivars infected with *Phytophthora nicotianae* var. *nicotianae, Physiol. Mol. Plant Pathol.,* 37, 207, 1990.

174. **Smillie, R. H., Dunstan, R. H., Grant, B. R., Griffith, J. H., Iser, J., and Niere, J. O.,** The mode of action of the antifungal agent phosphite, *EPPO Bull.,* 20, 185, 1990.

175. **Saindrenan, P., Barchietto, T., Avelino, J., and Bompiex, G.,** Effects of phosphite on phytoalexin accumulation in leaves of cowpea infected with *Phytophthora cryptogea, Physiol. Mol. Plant Pathol.,* 32, 425, 1988.

176. **Jiang, Y.,** Cytological Alterations in *Phytophthora infestans* in Infected Tomato Leaves Treated with Fosetyl-Al, Ph.D. thesis, University of Hohenheim, Stuttgart, Germany, 1990.

177. **Durand, M. C. and Sallé, G.,** Effect of aluminium tris-o-ethyl phosphonate (TEPA) on *Lycopersicon esculentum* Mill. leaves infected with *Phytophthora capsici* Leon. Cytological and cytochemical study, *Agronomie,* 1, 723, 1981.

178. **Dercks, W. and Buchenauer, H.,** Comparative studies on the mode of action of aluminium ethyl phosphite in four *Phytophthora* species, *Crop Protect.,* 6, 82, 1987.

179. **Cahill, D. M., Bennett, I. J., and McComb, J. A.,** Mechanisms of resistance to *Phytophthora cinnamomi* in roots of *Eucalyptus calophylla* and clonal *E. marginata,* in Australasian Plant Pathol. Soc. 8th Conf. Proc., Sydney, October 7 to 11, 1991 (Abstr.).

180. **Cahill, C. M., Bennett, I. J., and McComb, J. A.,** Resistance of micropropagated *Eucalyptus marginata* to *Phytophthora cinnamoni, Plant Dis.,* 76, 630, 1992.

181. **Taylor, P.,** ICI Australia personal communication.

182. **Wilcox, W.,** New York State Agricultural Experiment Station, Cornell University, Ithaca, personal communication.

183. **Sulistyowati, L., Keane, P. J., and Anderson, J. W.,** Accumulation of the phytoalexin, 6,7-dimethoxycoumarin, in roots and stems of citrus seedlings following inoculation with *Phytophthora citrophthora, Physiol. Mol. Plant Pathol.,* 37, 451, 1990.

Chapter 11

CYTOLOGY, HISTOPATHOLOGY, AND HISTOCHEMISTRY OF CITRUS BLIGHT

Ronald H. Brlansky

TABLE OF CONTENTS

0-8493-2939-6/93/$0.00 + $.50
© 1993 by CRC Press, Inc.

I. INTRODUCTION

A. HISTORY OF THE DISEASE

Citrus blight, a disease of unknown etiology, is one of the most important citrus diseases in Florida, causing estimated yearly losses of $60 million,[38] and speculated to be one of the oldest. The disease was probably present prior to the American Civil War, but it was not considered to be different than the other "dieback" diseases. The first written report of blight was in 1874 by Fowler,[1] and again described in 1883 by Manville.[2] Manville quoted Dr. Z. H. Mason's description of the disease as a "limb blight", from which subsequent authors no doubt appropriated the term "blight" for the disease. Blight occurred in all Florida citrus growing areas by 1894.[3] In 1896, Swingle and Webber[4] provided an excellent description of blight and estimated the annual loss due to the disease to be about $150,000. Blight did not appear in the literature from 1910 to 1924, but Rhoads[5] in 1925 reported that the disease was active and destructive and was spreading into new grove areas. From reports in the literature, the disease has occurred in cycles during the periods from 1926 to 1953 and from the mid-1960s to the present. During the 1960s, an outbreak of blight affecting large acreages of sweet oranges on rough lemon rootstock was reported in the flatwoods area of Florida. Presently, no rootstock is immune to citrus blight. An association appears to exist between the occurrence of the disease and the rootstock. Rough lemon rootstock is considered to be more susceptible than most other rootstocks; however, rootstocks such as rangpur lime, trifoliate orange, and carrizo citrange also are very susceptible.

B. WORLDWIDE DISTRIBUTION

Diseases similar or identical to blight occur in Brazil, Argentina, Australia, South Africa, Cuba, and Uruguay. The disease has been called *declinio* in Brazil, *declinamiento* in Argentina, and such things as young tree decline, sandhill decline, and rough lemon decline in Florida. The disease is not present in California or in countries in the Mediterranean region or in Japan. Little information is available on the disease in Asia. The disease occurs only in fruit-bearing trees older than 5 years.

C. SYMPTOMATOLOGY

Symptoms of the disease include a permanent wilt and decline of the tree canopy from which the trees never recover. Prior to canopy decline, leaves may show zinc deficiency symptoms. These zinc deficiency symptoms may vanish as the tree declines or may occur on any new flushes after initiation of decline. Leaves on affected trees are usually small in size as compared to those of healthy trees and have an off-green to gray color. The wilt symptoms may develop on one or two limbs and the tree may appear sectored. In time, the decline moves throughout the canopy. In some trees, the wilt and decline symptoms may occur throughout the canopy at the same time. The bloom on blight trees may be delayed and few fruit set. The fruit that are set are usually normal in appearance but may have a high acidity and soluble solids. In the early stage of blight decline, the root system is similar to that of healthy trees. Feeder roots are abundant and starch is plentiful throughout the root system. The wilting of blight-affected trees appears to be from a failure of the xylem system to conduct water, rather than failure of the roots to acquire water. Blight is diagnosed by the inability of the tree to uptake water (syringe water injection test), the accumulation of zinc first in the bark and later in the wood, the presence of specific proteins in xylem extracts, and the presence of occlusions in the xylem vessels.

D. TRANSMISSION

Early attempts failed to transmit the causal agent via bud grafts and via reconstituting trees with roots and scion material from blight-affected trees. The unknown causal agent was first transmitted tree-to-tree by root grafting[6] and later by using root pieces from blighted trees grafted onto healthy trees.[7,8] Apparently, the pathogen is located in higher concentrations in the root system of affected trees than in other parts of the tree.

II. HISTOPATHOLOGY: LIGHT MICROSCOPY

A. XYLEM SYSTEM AND WATER TRANSPORT DYSFUNCTION

As mentioned above, blight is primarily a wilt disease caused by a dysfunction of the xylem system. The wilting of blight-affected trees can occur in one sector of the tree, with the remainder of the tree declining later, or the entire tree can wilt rather rapidly. The wilt of a blight-affected tree can occur even with good available soil moisture and does not appear as a typical wilt where the leaves become flaccid and limp.[9] The leaves on declining trees remain turgid, but can curl, and the tips often point upward. Cohen,[10] in 1974, using a gravity flow uptake system, was the first to demonstrate that blight-affected trees reduced water conduction in the trunk. Young and Garnsey[11] found that the water movement in the trees with blight was greatly restricted in the trunks, less so in the major limbs and roots, and seldom in the small limbs and roots. Young[12] also showed with the use of dye infusion that the major reductions in water movement were in the trunk and major scaffold limbs of blight-affected trees, and that the reduction in water movement in the large roots was due to the general reduction in water movement in the trunk. The dye patterns in the trunk of blight trees indicated that only about 6 to 8% of the outer xylem functioned actively in water movement, in contrast to 57 to 73% of the outer xylem of healthy trees.

B. OCCLUSIONS IN BLIGHT-AFFECTED TREES

Two occlusions, filamentous and amorphous plugs, have been described in the xylem of citrus trees with blight. Prior to much of the work on the reduction of water uptake in blight-affected trees, Childs[13] published the first (1954) light micrographs of plugging in the roots of blighted trees. He clearly showed fibers or strands of material in the lumens of xylem vessels in rough lemon roots similar to the micrographs in Figure 1. These filamentous plugs were thought to be the cause for the restriction of water flow through the vessels and were said to resemble masses of actinomycetes. Later, Childs and colleagues[14] presented evidence that these filamentous plugs were a species of the fungus *Physoderma*. Little work followed on this interesting filamentous structure until VanderMolen,[15] using transmission electron microscopy (TEM), reported in 1973 that the interior of the filaments showed no biological structure of fungal mycelia. Childs and Carlysle[16] showed the following year the first scanning electron micrographs (SEM) of unfixed filamentous plugs associated with the blight syndrome. The micrographs showed a material that was described as being similar to fungus hyphae or a fungus-like organism that was coated with an inert substance of what appeared to be a stiff exoskeleton. The nature of the material was unknown. VanderMolen,[17] using both TEM and SEM, provided further evidence that the filamentous plugs were not fungal hyphae, actinomycetes, mycoplasma, or bacteria. He found they were strands of threads or filaments varying in diameter from 250 to 500 nm that appeared to branch and consisted of electron-dense and electron-transparent layers (Figures 2 and 3). The material appeared to originate at pits in the vessel walls and in perforation rims at the junctions of vessel elements. At these areas, the filamentous material was associated with a degeneration and swelling of the primary wall and/or middle lamella. These filamentous plugs also were found in the trunks of both diseased and healthy trees,[18] but were more prevalent in the diseased trees. The filamentous plugs appeared to be present in branches, trunk, and main, primary, pioneer, and fibrous roots, but not in the leaf midribs and petioles, and were generally more numerous in the oldest xylem.[19] Not only were the filamentous occlusions present in both healthy and blighted trees, but they were found present in trees with other diseases such as tristeza, spreading decline, exocortis, and xyloporosis.[19]

FIGURE 1. Filamentous plugging in sections of xylem vessels from blight-affected trees as seen with the light microscope. (Magnification × 225.)

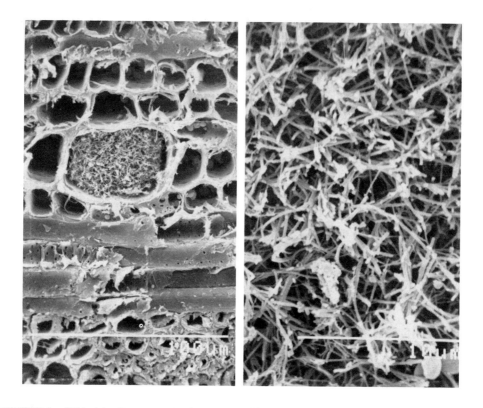

FIGURE 2. SEM of the filamentous plugging in blight-affected trees.

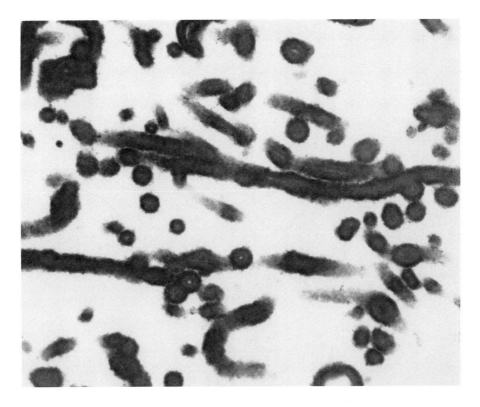

FIGURE 3.   TEM of filamentous plugging from a blighted citrus tree. (Magnification $\times$ 38,000.)

In 1975, the first report of the presence of amorphous plugs appeared in papers by Nemec et al.[20] and VanderMolen et al.[21] Nemec referred to them as gum plugs and described them as being the primary resistance to water movement along the cambium of the trunk in combination with narrow vessels. VanderMolen appears to be the first to use the term ''amorphous plug'' to describe these occlusions and found them along with the filamentous plugs to be significantly greater in the outer functional xylem in roots of blighted trees. Amorphous and filamentous plugs were studied in great detail by Cohen et al.[22] in 1983. The amorphous plug was found to be the most numerous occlusion in the trunk xylem of blighted citrus trees, and appeared by the nature of its structure (Figures 4 to 6) to be the primary blockage to water. The first published photographs of the amorphous plugs appeared in this paper.[22] Filamentous plugs were found usually near the pith in the trunks. This work disagreed with that of Nemec et al.,[20] in that the main resistance to water flow along the cambium region was gum plugs. It was, however, found that high numbers of amorphous plugs were associated with blighted trees. The number of filamentous plugs was similar in both diseased and healthy trees. The amorphous plugs also were found in the major limbs, large roots, but not in the smaller branches and limbs. These observations confirmed the data published by Young[12] that the major areas of reduction in water movement in blighted trees were in the trunks and major scaffold limbs. The relationship between the two types of xylem plugging to reduced water uptake and symptom development was demonstrated by Brlansky et al.[23] Trees with blight in Florida, *declinamiento* in Argentina, and *declinio* in Brazil, which have similar symptoms and diagnostics and are probably the same disease, were shown to contain the same amorphous structures as reported by Cohen et al.[22] Filamentous plugs were also present, but were not consistently associated with the declining

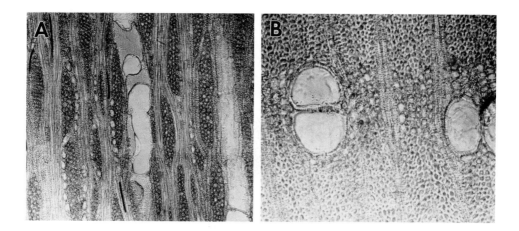

FIGURE 4. Light micrographs of amorphous plugging in the xylem of citrus trees with blight. (Magnification A, × 56; B, × 113).

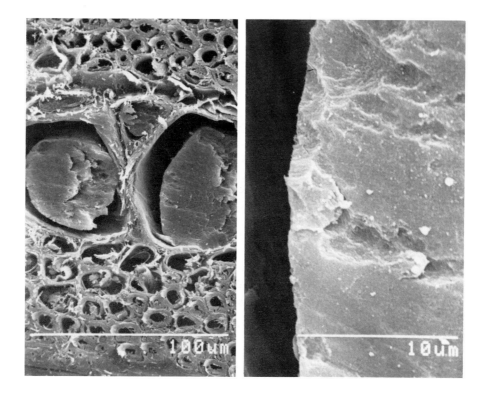

FIGURE 5. SEM of amorphous plugging in the trunk xylem vessels of blight-affected citrus.

trees, and their number was not highly correlated with the reduction in water uptake ($r = -0.43$). The amorphous plugs (Figure 6) were correlated significantly with water uptake ($R = 0.61$) and with symptoms (canopy rating). When the number of amorphous plugs reached 5 to 10/200 vessels, the water uptake was reduced to a level at which the symptoms developed.

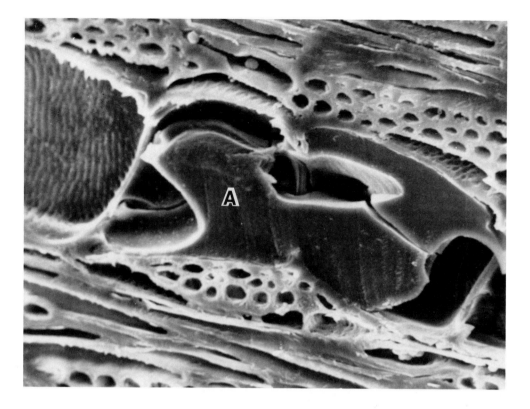

FIGURE 6. SEM of amorphous (A) plugging completely occluding the xylem vessel. (Magnification × 480.)

Timmer et al.[24] showed that the water flow in the first 4 cm of xylem internal to the trunk vascular cambium was high for healthy trees and amorphous plugs were rare in all segments; in blighted trees, little or no water flow occurred deeper than 1 cm from the cambium. The number of amorphous plugs was high in this area of low water flow. In another report by Timmer et al.[25] on the effect of amorphous plugs, filamentous plugs, and depth into the trunk on the water flow of blight-affected citrus trees, it was found that all factors contributed to the reduction of water flow; however, the most important factor in reducing flow was the amorphous plugs. Amorphous plugs, filamentous plugs, and depth into the trunk combined to account for 40% of the reduction in water flow. It has been speculated that vessel size and density are factors that have an effect on the reduced water flow in blight-affected trees. Vasconcellos and Castle[39] recently studied the vessel size and density in blight and healthy trees and reported a significant reduction in vessel size in blighted trees as well as a change in the density of vessels. Blight-affected trees on rough lemon rootstock were found to have a significant reduction in vessel size and an increase in vessel numbers in both the scion and rootstock portions of the trunk. All rootstocks tested had more vessels in the blighted trees.

Psorosis and concave gum diseases of citrus both produce substances in the xylem of diseased trees that have been described as wound gum. Brlansky et al.,[26] in a comparison of the amorphous plugs in blight trees with the wound gum in psorosis and concave gum-affected trees, showed that the amorphous plugs were similar when viewed with light microscopy (Figure 7). By using TEM and SEM, however, the differences between amorphous-like plugs in psorosis and concave gum and blight amorphous plugs were revealed (Figure 8). In the same study, citrus trees with tristeza, stubborn, citrus slump, and

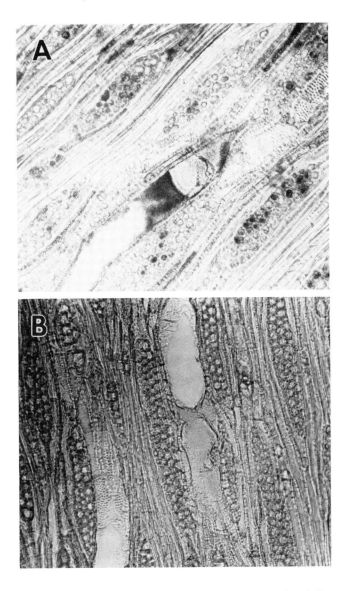

FIGURE 7. Light microscopy of xylem plugging in citrus trees with (A) psorosis and (B) concave gum. (Magnification × 150.) (From Brylansky, R. H. et al., *Phytopathology,* 75, 145, 1985. With permission.)

Phytophthora foot rot and healthy citrus trees all had filamentous plugs similar in structure, as previously reported.[25]

C. PHLOEM AND OTHER TISSUES

Because most of the problem in blight-affected trees is xylem dysfunction, little attention has been given to the phloem and other tissues. Light microscopy of the phloem of the trunk, roots, and small branches has shown no gross abnormalities in blighted trees when compared to healthy citrus trees.[40]

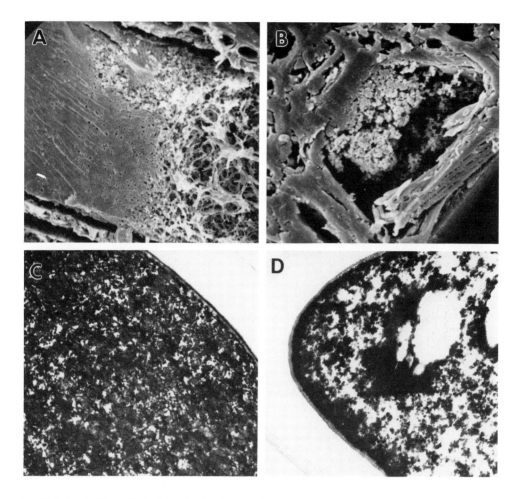

FIGURE 8. SEM and TEM of the plugging found in citrus with psorosis A (A and C) and concave gum (B and D) diseases. (From Brylansky, R. H. et al., *Phytopathology,* 75, 145, 1985. With permission.)

III. CYTOLOGY AND ULTRASTRUCTURE

A. ULTRASTRUCTURE OF BLIGHT-AFFECTED XYLEM AND RELATED OCCLUSIONS

The filamentous plugs are located primarily at the perforation rims that form the junctures between vessel members. The filamentous plugs may or may not completely occlude the vessel lumina. The plugs are a mass of strands or filaments that may vary in size from 250 to 500 nm in diameter and appear to branch.[25] The outside portion of the filaments is usually smaller in diameter than the inner portions and probably led to early reports that the filaments were fungal in nature. The filaments consist of layers of electron-dense and electron-transparent layers (Figure 3) that do not resemble cytoplasm. In cross-section, the filaments appear to have an electron-transparent core and are sometimes coated with other substances. In many vessels, the filaments are visible in the pits along the vessel walls. No filaments are present in the parenchyma and fiber cells adjacent to the vessels, but they can be found in the pits leading from some of these cells, which are often contiguous with the primary wall and the middle lamella. Further observations indicated that the pit aperture was open, but the chamber contained sloughing filamentous material. This may indicate some involvement of the primary wall and middle lamella in the formation of filamentous plugs.

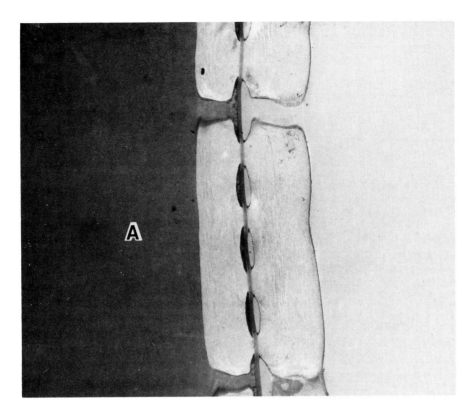

FIGURE 9.   TEM of amorphous plugging (A) in the xylem vessel of a citrus tree with blight. (Magnification ×
4880.) (From Brylansky, R. H. et al., *Phytopathology,* 75, 145, 1985. With permission.)

Via SEM, the amorphous plugging material has been shown to be a dense and nonporous
material that fills all or part of the lumen of the vessels in which they occur (Figure 5). In
some samples examined, vessels can be found that have what appears to be a coating of
amorphous material over the vessel wall. In TEM sections of vessels containing fully formed
amorphous plugs, the plugs appear as dark-staining solid, nonporous, nonparticulate structures that often fill pits, but do not seem to originate from the pits (Figure 9). In early stages
of amorphous plug formation,[41] the first vestiges of amorphous plugs appear as small, round
globules of material formed in the vessel lumen (Figure 10).

Because citrus blight is a disease of xylem dysfunction, the hypothesis has been
studied[27-31] that the causal agent may be a xylem-limited bacterium such as *Xylella fastidiosa*
(the causal agent of Pierce's disease of grape) or a related bacterium. However, microscopic
examinations using light, SEM, and TEM of the xylem of blighted trees in various stages
of decline have yielded negative information on the presence of such a bacterium.[40] *In situ*
immunofluorescence and membrane entrapment immunofluorescence also have been used
to detect the presence of *X. fastidiosa* in blight-affected trees, but have failed to do so. All
of the above methodologies have successfully been used to detect other strains of *X. fastidiosa*
in other plants as well as the xylem-limited bacterium that is associated with citrus variegated
chlorosis disease.[41] From these results, it appears that a xylem-limited bacterium is not
present in citrus trees with blight and the causal agent may be located elsewhere.

B. ULTRASTRUCTURE OF THE PHLOEM

Purcifull et al.[32] examined the phloem for the presence of mycoplasma-like agents and
bacteria, but found no differences in the phloem of the root and leaves of blight-affected

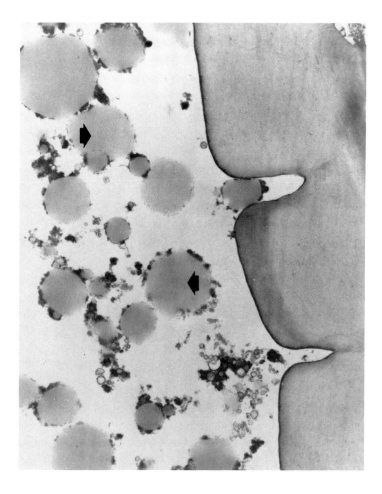

FIGURE 10. Initiation of amorphous plugging (arrows) in the trunk xylem of a citrus tree that eventually was affected by citrus blight. (Magnification × 15,000.)

trees. Mitochondria, nuclei, chloroplasts, endoplasmic reticulum (ER), and dictyosomes were adequately preserved and appeared normal. Brlansky et al.[41] examined the phloem of the trunk of blighted and healthy trees and found some degeneration of the phloem along with normal cellular constituents. As no degeneration was seen with the light microscope, further work is necessary in this area on trees in various stages of decline.

IV. HISTOCHEMISTRY OF XYLEM OCCLUSIONS

The filamentous occlusions were first thought to be the primary blockage structures in the decline of blight-affected trees. Thus, most of the early histochemical studies were performed on these structures. The filamentous or myelin plugs described by Nemec[19] were tested with 31 histochemical stains. The plugs were found to stain positively for unsaturated lipids, phospholipid, myelin, and choline (Table 1). Lipid bodies described in the para-tracheal, ray, and terminal parenchyma cells gave the same reactions, except that they reacted positively with Nile blue sulfate for lipofuscin and gave a weak reaction for protein. The plugs were therefore described as proteolipid with possible free fatty acids (FFA), with phosphatidylcholine as one of the component phosphatides. The strands were interpreted as

TABLE 1
Histochemical Tests on Occlusions in Roots Reported by Nemec,[19] Nemec et al.,[20] and VanderMolen et al.[21]

Compound tested	Histochemical stain	Amorphous	Filamentous
		Blight	
Lipid[19]	Sudan black B, OsO_4	ND[a]	+
Phospholipid[19]	Orange G-aniline blue	ND	
Nyelin[19]	Nile blue sulfate, H_2SO_4	ND	+
Choline[19]	Acid hematein		+
Starch[20]	IKI	−	ND
Lignin[20]	Phloroglucinol-HCl	−	ND
Pectin[20]	Ruthenium red	+	ND
Carbohydrate/polysaccharide[20]	Percodin acid-Schiff's	+	ND
	Heidenhain's iron hematoxylin	+	
Cellulose[21]	Chloriodide-zinc	−	−
Lipid[21]	Sudan black B	−	+
Pectin[21]	Chemical extraction with various procedures	−	−
Lignin[21]	Delignification with sodium chlorite	+	+

[a] ND = not determined.

myelin. VanderMolen et al.[21] found that the filamentous plugs did not contain cellulose or pectin, but were positive for lipid, while the amorphous plugs were positive for pectin (Table 1). Neither plug type was removed with various lipid extraction procedures, but both were completely removed by delignification with sodium chlorite. They concluded that the filamentous plugs were in part lignin and lipid and the amorphous plugs were a pectin-lignin complex. Nemec[19] also studied the histochemistry of the amorphous plugs (Table 1) and found that although they stained red with the nonspecific pectin stain ruthenium red, they did not dissolve in hot water, nor were they positive for pectins after ammonium oxalate extraction. The plugs did not dissolve in 0.1 N trichloroacetic acid (TCA), which dissolves carbohydrates and starch. The amorphous plugs were negative for starch and lignin, but were positive for protein with the stain mercuric bromphenol blue. They did stain blue-black with Heidenhain's iron hemtoxylin. The origin of the amorphous plugs was said to be traced to the parenchyma cells. Nemec[19] concluded that the amorphous plugs were mucopolysaccharide, composed in part of pectin, other sugars, and minor organic components. The unpublished X-ray diffraction data of W. A. Cote was cited and showed that both types of plugs had traces of Si, S, Cl, K, and Ca.

The most recent work on the histochemistry of both filamentous and amorphous plugs was performed by Beretta et al.[33] (Table 2). This work utilizing 15 different histochemical stains, compared blight and *declinio* from Brazil and found that filamentous plugs contain lignin, pectin, gum, protein, neutral lipids, and acidic lipids. Amorphous plugs stained positively for callose, lignin, pectin, gum, protein, neutral lipids, acidic lipids, and total lipids. Plugs from both *declinio* and blight trees stained identically.

V. CONCLUDING REMARKS

Citrus blight currently is a graft-transmissible disease of unknown etiology. Histological and histochemical studies have characterized the disease as being distinct from other known citrus tree declines. The disease is apparently one of xylem dysfunction in which two types of occlusions are formed and xylem vessels are of smaller size but greater in number. The overall consequence is a reduction in the ability of the xylem system to conduct water

TABLE 2
Histochemical Tests on Amorphous and Filamentous Plugs in Blight- and *Declinio*-Affected Trees

Compound detected	Histochemical stain	Blight[a]		Declinio	
		Amorphous plugs	Filamentous plugs	Amorphous plugs	Filamentous plugs
Cellulose	Iodide sulfuric acid	− [b]	−	−	−
	Zinc chloroiodide	−	−	−	−
Callose	Aniline blue (visible light)	+	−	+	−
	Aniline blue (UV light)	+	−	+	−
Starch	IKI reactions	−	−	−	−
Lignin	Phloroglucinol	+	+	+	+
	Maule stain	+	+	+	+
	Potassium permanganate	+	+	+	+
Phenolic compounds and tannins	Ferric chloride	−	−	−	−
Pectic substances	Reuthenium red	+	−	+	−
Gums	Orcinol	+	+	+	+
Proteins	Ninhydrin-Schiff's reagent	+	+	+	+
Neutral lipids	Sudan IV	+	+	+	+
Total lipids	Nile blue	+	− [c]	+	− [c]
Acidic lipids	Sudan black B	+	+	+	+

[a] Tests were based on one horizontal core sampled from each of six blight-affected sweet orange trees (*Citrus sinensis*) on rough lemon rootstock (*C. jambhiri*) and six *declinio*-affected sweet orange trees on rangpur lime (*C. limonia*) rootstock (*C. jambhiri*) and six *declinio*-affected sweet orange trees on rangpur lime (*C. limonia*) rootstock. Cryostat sections 10 to 12 were used from each core sample for staining.

[b] + = Positive color reaction; − = no color reaction.

[c] No red staining for neutral lipids nor blue staining for FFA and phospholipids was observed.

From Beretta, M. J. G. et al., *Plant Dis.*, 72, 1058, 1988. With permission.

resulting in blighted trees being irreversibly wilted and declining. It appears from the data that the amorphous occlusions are unique to blight and are the main obstruction to water movement.

The amorphous and filamentous occlusions are complex in composition, as shown in Tables 1 and 2, and their exact chemical structure is still under study. The filamentous plugs are structures with strands that tend to branch, and contain a central core that is visible in electron micrographs; however, they do not appear to be fungal or bacterial in nature. The amorphous plugs are very solid and electron dense, and often completely fill the vessel lumen. In early stages of development, amorphous material appears as small globules in the vessel lumen.

No gross abnormalities are apparent in the phloem or other tissues, although some small amounts of phloem degeneration may be found. The hypothesis that currently known pathogens such as bacteria, mycoplasmas, or fungi are the cause of blight remains unfounded, since none have consistently been associated in the xylem or other tissues of affected trees. The ability to cause the disease by the use of root grafts (either tree to tree or using root pieces) does suggest the involvement of a causal agent. The involvement of a virus or undescribed pathogen that occurs in low concentration is possible. Since the discovery of a number of pathogenesis-related proteins in blighted trees,[34,35] it may be possible to diagnose trees prior to the presence of high zinc levels, amorphous plug formation, and symptoms of decline.[36,37] A careful study of these predecline trees may aid in the discovery of a causal agent.

Other histological, histochemical, and histopathological events probably occur in trees with citrus blight, and may become better known after the causal agent is discovered.

ACKNOWLEDGMENTS

This chapter is Florida Agricultural Experiment Station Journal Series No. N-00434. The author gratefully acknowledges the assistance of D. Howd for literature surveys, C. Davis for photographic assistance, and P. Hicks for manuscript preparation.

REFERENCES

1. **Fowler, J. H.,** *Orange Culture in Florida,* C. H. Walton, Jacksonville, FL, 1874, 22.
2. **Manville, A. H.,** *Practical Orange Culture; Including the Culture of the Orange, Lemon, Lime, and Other Fruits, as Grown in Florida,* Ashmead Press, Jacksonville, FL, 1883.
3. **Swingle, W. T.,** Blight, *Proc. Fla. State Hortic. Soc.,* 7, 73, 1894.
4. **Swingle, W. T. and Webber, H. J.,** The Principal Diseases of Citrus Fruits in Florida, Pathol. Bull. No. 8, U.S. Department of Agriculture, Washington, D.C., 8, 1896.
5. **Rhoads, A. S.,** Observations on citrus wilt, *Proc. Fla. State Hortic. Soc.,* 32, 26, 1925.
6. **Tucker, D. P. H., Lee, R. F., Timmer, L. W., Albrigo, L. G., and Brlansky, R. H.,** Experimental transmission of citrus blight, *Plant Dis.,* 68, 979, 1984.
7. **Lee, R. F., Brlansky, R. H., Timmer, L. W., Tucker, D. P. H., and Graham, J. H.,** Graft transmission of citrus blight, *Phytopathology,* 78, 1572, 1988.
8. **Timmer, L. W., Brlansky, R. H., Derrick, K. S., and Lee, R. F.,** Transmission of citrus blight by root graft inoculation, in *Proc. 11th Conf. Int. Organ. Citrus Virol.,* 1989, 244.
9. **Smith, P. F.,** History of citrus blight in Florida, *Citrus Ind.,* 55(9), 13, 1974.
10. **Cohen, M.,** Diagnosis of young tree decline, blight and sandhill decline of citrus by measurements of water uptake using gravity injection, *Plant Dis. Rep.,* 58, 801, 1974.
11. **Young, R. H. and Garnsey, S. M.,** Water uptake patterns in blighted citrus trees, *J. Am. Soc. Hortic. Sci.,* 102, 752, 1977.
12. **Young, R. H.,** Water movement in limbs, trunks and roots of healthy and blight affected Valencia orange trees, *Proc. Fla. State Hortic. Soc.,* 92, 64, 1980.
13. **Childs, J. F. L.,** Observations on citrus blight, *Proc. Fla. State Hortic. Soc.,* 66, 33, 1954.
14. **Childs, J. F. L., Knopp, L. E., and Johnson, R. E.,** A species of *Physoderma* present in citrus and related species, *Phytopathology,* 55, 681, 1965.
15. **VanderMolen, G. E.,** Preliminary electron microscope observations of citrus with young tree decline, *Fla. Sci.,* 36, 92, 1973.
16. **Childs, J. F. L. and Carlysle, T. C.,** Some scanning electron microscope aspects of blight disease of citrus, *Plant Dis. Rep.,* 58, 1051, 1974.
17. **VanderMolen, G. E.,** Electron microscopy of vascular obstructions in citrus roots affected with young tree decline, *Physiol. Plant Pathol.,* 13, 271, 1978.
18. **Nemec, S. and Kopp, D.,** Extent of lipid vessel plugs in citrus with and without sandhill and young tree decline symptoms, *Proc. Fla. State Hortic. Soc.,* 87, 107, 1974.
19. **Nemec, S.,** Vessel blockage by myelin forms in citrus with and without rough-lemon decline symptoms, *Can. J. Bot.,* 53, 102, 1975.
20. **Nemec, S., Constant, R., and Patterson, M.,** Distribution of obstructions to water movement in citrus with and without blight, *Proc. Fla. State Hortic. Soc.,* 88, 70, 1975.
21. **VanderMolen, G. E., Gennaro, R. N., Peeples, T. O., and Bistline, F. W.,** Chemical nature and statistical analysis of the distribution of plugging in blight/YTD-affected citrus trees, *Proc. Fla. State Hortic. Soc.,* 88, 76, 1975.
22. **Cohen, M., Pelosi, R. R., and Brlansky, R. H.,** Nature and location of xylem blockage structures in trees with citrus blight, *Phytopathology,* 73, 1125, 1983.
23. **Brlansky, R. H., Timmer, L. W., Lee, R. F., and Graham, J. H.,** Relationship of xylem plugging to reduced water uptake and symptom development with blight and blight-like declines, *Phytopathology,* 74, 1325, 1984.

24. **Timmer, L. W., Brlansky, R. H., Graham, J. H., Sandler, H. A., and Agostini, J. P.,** Comparison of water flow and xylem plugging in declining and apparently healthy citrus trees in Florida and Argentina, *Phytopathology,* 76, 707, 1986.

25. **Timmer, L. W., Brlansky, R. H., Graham, J. H., Agostini, J. P., and Lee, R. F.,** Effect of amorphous plugs, filamentous plugs, and depth into trunk xylem on the water flow in declining and healthy citrus trees, in *Proc. 10th Conf. Int. Organ. Citrus Virol.,* 1984, 370.

26. **Brlansky, R. H., Lee, R. F., and Collins, M. H.,** Structural comparison of xylem occlusions in the trunks of citrus trees with blight and other decline diseases, *Phytopathology,* 75, 145, 1985.

27. **Feldman, A. W., Hanks, R. W., Good, G. E., and Brown, G. E.,** Occurrence of a bacterium in Y.T.D.-affected as well as in some apparently healthy citrus trees, *Plant Dis. Rep.,* 61, 546, 1977.

28. **Hopkins, D. L., Adlerz, W. C., and Bistline, F. W.,** Pierce's disease bacterium occurs in citrus trees affected with blight (young tree decline), *Plant Dis. Rep.,* 62, 442, 1978.

29. **Hopkins, D. L., Thompson, C. M., Bistline, F. W., and Russo, L. W.,** Relationship between xylem-limited bacteria and citrus blight, *Proc. Fla. State Hortic. Soc.,* 102, 21, 1989.

30. **Hopkins, D. L.,** Relation of Pierce's disease bacterium to a wilt-type disease in citrus in the greenhouse, *Phytopathology,* 72, 1090, 1982.

31. **Hopkins, D. L.,** Production of diagnostic symptoms of blight in citrus inoculated with *Xylella fastidiosa, Plant Dis.,* 72, 432, 1988.

32. **Purcifull, D. E., Garnsey, S. M., Story, G. E., and Christie, R. G.,** Electron microscopic examination of citrus trees affected with young tree decline (YTD), *Proc. Fla. State Hortic. Soc.,* 86, 91, 1973.

33. **Beretta, M. J. G., Brlansky, R. H., and Lee, R. F.,** A comparison of histochemical staining reactions of the xylem occlusions in trees affected by citrus blight and declinio, *Plant Dis.,* 72, 1058, 1988.

34. **Bausher, M. G.,** Electrophoretic and immunological evidence of unique proteins in leaves of citrus trees: application to citrus blight detection, *Electrophoresis,* 11, 830, 1990.

35. **Derrick, K. S., Lee, R. F., Brlansky, R. H., Timmer, L. W., Hewitt, B. G., and Barthe, G. A.,** proteins associated with citrus blight, *Plant Dis.,* 74, 168, 1990.

36. **Bausher, M. G. and Sweeney, M. J.,** Field detection of citrus blight using immunological techniques, *Plant Dis.,* 75, 447, 1991.

37. **Derrick, K. S., Lee, R. F., Brlansky, R. H., Timmer, L. W., Hewitt, B. G., and Barthe, G. A.,** Proteins and filamentous structures associated with citrus blight, in *Proc. 11th Conf. Int. Organ. Citrus Virol.,* 1989, 265.

38. **Muraro, R.,** personal communication.

39. **Vasconcellos, L. and Castle, W.,** personal communication.

40. **Brlansky, R. H.,** unpublished data.

41. **Brlansky, R. H. et al.,** unpublished data.

Chapter 12

CYTOLOGY, HISTOLOGY, AND HISTOCHEMISTRY OF MLO INFECTIONS IN TREE FRUITS

Sharon M. Douglas

TABLE OF CONTENTS

0-8493-2939-6/93/$0.00 + $.50

I. INTRODUCTION

Since the first report associating a new type of plant pathogen, the mycoplasma-like organism (MLO), with plant disease in 1967,[1] awareness of the importance of these diseases to worldwide agricultural production has increased. Today, diseases of mycoplasmal etiology involve more than 600 plant species and include food, ornamental, forest, and fiber crops. Comprehensive reviews of MLOs and their associated diseases have been published.[2-9]

The impact of MLO diseases on woody plants and tree fruits is more pronounced than for annual crops since once infected, trees usually remain infected for the remainder of their lives. Tree fruits with diseases associated with nonhelical MLOs include apple, apricot, cherry, coconut, nectarine, peach, pear, and plum. Infection of citrus by the helical mycoplasma, *Spiroplasma citri*, is not addressed in this chapter.

MLO infections often cause a general decline of the tree as evidenced by reduced vigor, dieback, reduced fruit production, and eventual tree death. However, in some cases they can cause a quick wilt and sudden death.[2,6,7] Among the more serious tree fruit diseases associated with MLOs are apple proliferation, pear decline, X-disease of stone fruits, and lethal yellowing of coconut palm. Several of these diseases threaten to eliminate the production of certain crops because of their local impact or widespread distribution. Lethal yellowing in the Caribbean is such an example.[17]

In spite of the prevalence and importance of MLO-associated diseases of tree fruits, information on most aspects of their basic biology, including cytology, histology, and histochemistry, is remarkably limited. These limitations are attributed in large part to the difficulty in working with these fastidious organisms. A significant factor in this difficulty is that to date, numerous and systematic attempts to culture MLOs *in vitro* have been unsuccessful.[10-15] Another factor is the nature of the invaded tissue; MLOs appear to be phloem restricted and selective in pathogenesis. Phloem is a complex tissue by definition, especially when secondary growth is involved, as with tree fruits.[16-19]

In this light, the focus of many studies on diseases of presumed mycoplasmal etiology has been on identification and confirmation of the disease agent in host tissues. This involved use of light, fluorescence, and electron microscopy. Since many of these studies specifically targeted the association of a MLO with the disease, outwardly symptomatic tissues were usually selected for examination. Descriptions of pathological anatomy were, therefore, limited to tissues in advanced stages of disease. As a consequence, discerning any sense of a developmental sequence of anatomical changes was not possible since so many interactions or reactions had already occurred.[5]

Tree fruit diseases associated with MLOs can be placed into two categories, those whose pathological anatomy has been examined and those with unknown or uncharacterized pathological anatomies. The pathological anatomy of the diseases that have been studied has generally been characterized by sieve tube necrosis accompanied by hypertrophy and hyperplasia, lysis, callose deposition, starch accumulation, and excessive cambial activity. The objectives of this chapter are to review the common characteristics of these types of diseases and to infer a sequence for disease-induced anatomical changes in tree fruit hosts with respect to symptom expression.

II. CHARACTERISTICS OF MLO-ASSOCIATED DISEASES

Certain nonanatomical information is necessary to enhance understanding of MLO-associated diseases and is essential to interpretation of the pathological anatomy associated with these pathogens.

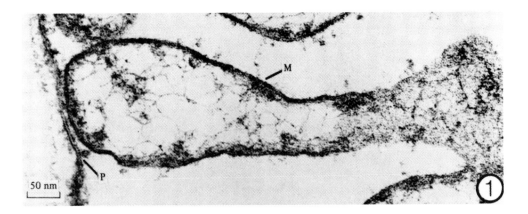

FIGURE 1. Detail of an MLO profile showing the trilamellar membrane (M), DNA material, and ribosomes. The plasmalemma of the host sieve element is also shown (P). (From Waters, H. and Hunt, P., *J. Gen. Microbiol.*, 116, 111, 1980. With permission.)

A. CAUSAL AGENT

The term "mycoplasma-like organism" as used in this chapter refers to the pleomorphic wall-less prokaryotic microorganisms found in phloem of symptomatic plants. This excludes sieve-tube restricted, wall-less Mollicutes with helical morphology (the *Spiroplasmas*, *S. citri* and *S. kunkelii*), sieve-tube-restricted, bacteria-like agents (such as those associated with citrus greening), xylem-limited bacteria (such as *Xylella fastidiosa*, causal agent of Pierce's disease of grape), and nonphytopathogenic Mollicutes isolated from plant surfaces and flowers.

MLOs have pleomorphic cellular morphology. They can appear as small spherical or elongate polymorphic bodies in cross-section (Figure 1) or as filamentous, many-branched, anastomosing bodies when serial sectioned (Figure 2). As seen in electron micrographs, MLOs are typically 60 to 1100 nm in diameter, bounded by a trilaminar unit membrane, electron dense, and slightly granular.[2,5,7,21-23] They often contain ribosome-like particles and fibrillar material presumed to be DNA. Membrane-bound vacuoles or inclusions are typically lacking, although the literature contains reports of such features.[24] MLOs associated with plant disease are generally regarded as morphologically indistinguishable when examined *in situ* by electron microscopy.

Although the aforementioned morphological and ultrastructural features are characteristics of organisms belonging to the class Mollicutes, actual classification of the plant pathogenic MLOs is still tentative.[4,5,7,25] This is because they have not been successfully isolated in pure culture and maintained *in vitro*, and pure cultures have not been reinoculated into plant hosts to provide final proof of pathogenicity in fulfillment of Koch's postulates.

Biological properties have been used to provide additional circumstantial evidence for further characterization of MLOs and for pathogenicity. These properties include symptoms induced in infected plants, antibiotic sensitivity, plant host ranges, and characteristics of transmission by insect vectors, by grafting, and by parasitic plants. For example, tetracycline antibiotics induce symptom remission in plant hosts with MLOs (e.g., peach trees with X-disease, coconut palms with lethal yellowing), whereas no remission is evident following therapy with penicillin, an antibiotic that specifically inhibits cell wall synthesis in bacteria. Although these biological characteristics are important to our understanding of disease, they are relatively nonspecific and do not provide the necessary information for complete pathogen characterization.[2,4,5,7,27]

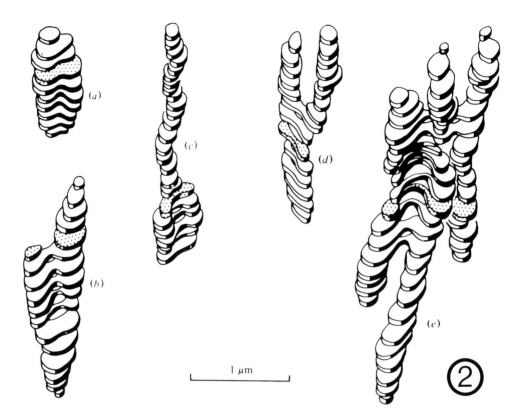

FIGURE 2. Reconstructional drawings of representative erythrocyte-like organisms. The shaded profiles are those which appeared in the middle sections of the appropriate set of serial sections. (From Waters, H. and Hunt, P., *J. Gen. Microbiol.*, 116, 111, 1980. With permission.)

Attempts have been made to classify or group MLOs by the type of symptoms they induce in their host plants (e.g., dieback, decline, witches' brooming, stunting, yellowing, aborted or abnormal fruit, and phyllody, virescence, and sterility of flowers).[5,7] However, problems arise using this approach since some individual MLOs can incite a variety of symptoms in different hosts and several different MLOs can cause the same type of symptom in a common host. Recent and ongoing attempts to develop more specific and sensitive procedures for detection and identification have employed serological and molecular cloning techniques to aid in diagnosis and pathogen characterization. Use of MLO-DNA in nucleic acid hybridization studies has been successful for identification of many MLOs, including those associated with X-disease and apple proliferation.[28,29] Genetic information from this molecular approach also contributes to the development of systems for classifying plant disease MLOs as well as for establishing the relationships among strains of MLOs. For example, these techniques allow one to directly compare MLOs and diseases that induce similar symptoms in the same host, but which may occur in different parts of the world.[30-36] This methodology has been used with X-disease where "strains" have been examined from the eastern U.S., Canada, and California.[29] Development and use of polyclonal and monoclonal antibodies have also aided in detection and have provided insight into the relationships between MLOs in various hosts and from different geographical regions.[37-40]

B. INVADED TISSUES

To date, all evidence indicates that MLOs are tissue-specific phloem inhabitors of plant hosts. They are not distributed in the host as are many other pathogenic agents. Phloem is anatomically and physiologically a complex tissue; this has contributed significantly to the difficulties encountered in dealing with the pathological anatomy of these types of diseases.[5,16,17,20]

The principal components of phloem are sieve tubes, companion cells, phloem parenchyma, phloem fibers, and secondary phloem including ray parenchyma. The presence of secondary phloem in tree fruits and woody species further complicates histopathological studies since woody plants live for many years, during which time secondary growth occurs and secondary phloem accumulates in annual increments. As part of the natural developmental process, phloem loses its primary function and some cells degenerate. Evert[41,42] reported complete degeneration of sieve tubes in the phloem of both healthy apple and pear at the end of the growing season. Therefore, when examining the pathological anatomy of MLO-associated diseases, it is important that these normally degenerated cells not be confused with disease-induced necrotic or distorted cells.[5] In many cases, abnormal sieve elements develop many of the same features as normal, degenerated sieve elements. However, the diseased or abnormal cells develop these features prematurely.

The sieve tube elements in which MLOs are located are living cells that are stacked end to end, forming a pressurized tube.[7,21] The ends of a sieve tube element have sieve plates containing pores. Sieve plate pores of functioning sieve elements are open, have limited callose deposits, and are lined by the plasmalemma and a thin layer of P-protein. Median pore size is 2 μm with a range from 1 to 14 μm. Thus, these pore sizes are large enough to allow for the unimpeded movement and passage of MLOs between functional, healthy sieve tube elements.[7,21,43] However, the cell to cell movement and spread of MLOs throughout plant hosts remains to be determined; it is not known whether MLOs utilize phloem translocation mechanisms or as yet unknown motility or chemotactic factors.

Functional sieve tubes are under positive hydrostatic pressures which range from 8 to 20 bars, and in some cases, 30 bars, and respond physiologically to injury and disruption. Because of this, phloem physiologists and anatomists generally agree that it is almost impossible to preserve the sieve element without physical disruptions.[7,21,41-43] Any cuts made in the process of tissue sampling interrupt this pressurized system; the sudden release of pressure causes a disruption of plastids, plugging of sieve pores with P-protein, mitochondria, starch, and cellular debris, and excessive callose deposition around sieve plates and in their pores. These sampling and fixation artifacts further complicate interpretations of pathogen-induced anatomical changes.[21]

While MLOs have been consistently located and identified in sieve tube elements, some researchers have reported MLOs in companion cells and phloem parenchyma.[44] McCoy[7] presented considerable evidence to discount these observations on the basis of the misidentification of cells (i.e., abnormal or immature sieve elements containing cytoplasmic contents misidentified as parenchyma cells, vesicles in parenchyma cells described incorrectly as MLOs), the physics of MLOs passing from sieve elements to companion cells through plasmadesmata, and the dramatic physiological differences between sieve elements and parenchyma or companion cells. However, this concept of MLO-sieve tube specificity remains open to question. Sears and Klomparens[45] recently reported the occasional presence of MLOs in phloem parenchyma cells in addition to sieve tube elements in leaf tip cultures of *Oenothera hookeri*. They speculated that this observation may be a unique feature of their system, possibly associated with MLO invasion of meristematic tissues, since these cultures were specifically developed and used to maintain MLOs. While the relation of MLOs to the protoplast of phloem parenchyma cells merits further investigation, it serves to mark the importance of correct sieve element identification when studying pathological anatomy. Identification can generally be based on wall thickness and type, presence of sieve areas or sieve plate pores, and type of plasmadesmatal connections to adjacent cells.[7,43]

To date, most studies on the histology/pathology of MLO-associated diseases have examined plant tissues in advanced stages of disease and symptom expression. Consequently, it has been difficult to know which stimulus caused which response. This has thwarted attempts to determine specific sequences of events in pathogenesis. Researchers have not been able to accurately determine sites of initial pathology or of primary establishment of MLOs in the host with respect to transmission, movement, multiplication, and symptom expression. Budding and grafting studies have further complicated the issue by supporting electron microscopic observations that show that MLOs can vary significantly in titer and are typically irregularly distributed in their tree fruit hosts during periods of both active growth and dormancy.[46-49] Furthermore, it is a rare occurrence for all sieve elements within a cross-section to contain MLOs. Invaded cells often occur in apparently random distribution and numbers of MLOs in these invaded cells often vary from few to dense aggregations completely filling lumens of sieve tubes.

Although the histopathology of tree fruit diseases believed to be caused by MLOs is characterized by numerous anatomical changes, it has also been noted that there are often no obvious differences in the overall cytology or in the structure of parenchyma or companion cells associated with sieve elements containing MLOs as compared to their counterparts without MLOs.[5,7] It appears that the actual presence of MLOs within sieve tubes may not be a prerequisite for their eventual collapse or for phloem aberrations, as these symptoms can occur in the absence of MLOs.

Light, fluorescence, and electron microscopic investigations have primarily focused on indirect and direct demonstration of MLOs in host tissues.[5,50,52-58] With few exceptions, descriptions and investigations of pathological anatomy have been secondary. Indirect methods for detection of MLOs have relied primarily on autofluorescence of sieve elements in diseased materials when compared to healthy counterparts[59] and on the presence of abnormally high levels of wound callose in diseased sieve tubes when stained for light microscopy by resorcin blue[56,60] or for fluorescence microscopy by aniline blue, a callose-specific fluorochrome.[61] Direct demonstrations of MLOs in tissues have employed numerous histochemical stains for DNA or RNA detection. Light microscopic studies have used the Feulgen reaction and methyl green, pyronin y, complementary DNAse and RNase treatments, and gallocyanine chromalum to demonstrate MLOs (through DNA or RNA staining) in sieve tubes.[62,63] Toluidine blue O has also been used with marginal success as a stain for MLOs in diseased tissues.[21] For fluorescence microscopy, a variety of stains for DNA have been used, including *N,N'*-diethylpseudoisocyanide chlorides, Feulgen and auramine O as Schiff's reagent, fast-green or Azur I, berberine sulfate, and 4',6-diamidino-2-phenylindole-HCl (DAPI) or Hoechst 33288, a benzimide derivative.[65-73] DAPI and Hoechst 33288 were the most successful of these stains. Seemüller[72] first used DAPI to detect MLOs in pear tissues with symptoms of pear decline. Actual verification and confirmation of the ability of this stain to positively detect DNA/RNA of MLOs in sieve elements using fluorescence microscopy was subsequently reported by Sampson and Welvaart[74] and Behnke et al.[71] by electron microscopy. Douglas[48] used DAPI in conjunction with electron microscopy to follow the seasonal distribution of MLOs with respect to symptom expression in peach and chokecherry affected by X-disease.

More recently, indirect immunofluorescence tests using polyclonal and monoclonal antibodies have been used for *in situ* detection of MLOs, including the MLO associated with X-disease of stone fruits.[34,37,75,76] However, these techniques have been used primarily with herbaceous hosts and have not been extensively tested with tree fruit or woody hosts. Immunosorbent electron microscopy (ISEM) using immunogold labeling of tissue sections has also been used as a sensitive method for direct detection of MLOs, although once again it has had limited testing with MLOs associated with tree fruits.[34,77]

C. SYMPTOMATOLOGY

Despite their long history of being called "yellows-type diseases", the outward symptoms associated with MLO infection are quite varied. In fact, given the range of symptoms, it is not uncommon for MLO diseases to have several different characteristic symptoms which can develop at specific stages of pathogenesis or can vary with strains of the pathogen, host plant, environmental factors, and with the age of the host or the tissue at the time of infection. For example, the first symptoms exhibited by coconut palm infected with the lethal yellowing MLO are premature fruit abscision and necrosis of developing inflorescences. This is followed by a progression of other symptoms that include foliar yellowing, root necrosis, and eventual death of the tree within 1 year of the first outward symptoms.[7]

Another illustration of the inaccuracy in assigning specific symptoms to a particular MLO is X-disease of stone fruits. This disease is one of the more extensively documented examples of a tree fruit MLO, with a diverse and variable symptomatology depending upon the particular host.[78] In peach, the MLO typically causes a general decline, in which the leaves develop necrotic blotches and fall prematurely, and fruit production gradually ceases. In sour cherry, those propagated on 'Mahaleb' rootstocks show a sudden wilt and death in late summer, whereas those on 'Mazzard' rootstocks show a general decline, persist for many years, and produce characteristic small, green, misshapen cherries. Thus, outward symptomatology can generally be regarded as an unreliable diagnostic characteristic.

Attempts to classify symptoms of MLO infections of tree fruits as direct and indirect or as primary and secondary with regard to the presence of the pathogen in affected tissues have met with considerable difficulty. These criteria cannot be readily used to assess symptomatology since, as previously mentioned, MLOs cannot be cultured or even reliably detected *in situ* at low levels. Consequently, cause-effect relationships are difficult to understand since it is rarely possible to determine the site or the timing required for transmission, incubation, movement and spread, and symptom expression.[5]

It is generally agreed that MLOs disrupt normal translocation processes in diseased trees. This disruption can involve physical blockage or plugging of sieve tubes by the MLOs themselves, excessive callose formation and deposition on sieve plate pores, phloem degeneration, and cambial hyperactivity. Since studies have shown that the presence of MLOs (i.e., physical blockage) within sieve tubes is not always a prerequisite for the development of symptoms, additional factors such as phytotoxins and other biologically active compounds have been suggested to contribute to symptom expression.[7] Examples of external symptoms that can be readily attributed to or correlated with tissue abnormalities resulting from disruptions in translocation and possible phytotoxin activity are the dieback, general decline, accumulation of starch and fruit abortion or formation of abnormal fruit of X-disease of stone fruits[52,78] and the sudden wilt and general decline of pear decline.[53,79-81]

Symptoms such as witches'-brooming, phyllody, and virescence provide strong evidence for dramatic physiological changes in tree fruit hosts in response to the presence of MLOs in the phloem. Since information on the physiology of diseases with presumed mycoplasmal etiology is limited, specific suggestions for these alterations are still speculative.[7,20] The physiological changes include imbalances of hormones and growth regulators as well as alteration of other fundamental plant processes. For example, trees with apple proliferation show witches' brooming, lack of apical dominance, and premature breaking of axillary buds in late summer, all symptoms suggesting profound pathogen-associated or -induced physiological changes in the host.[82]

III. HOSTS AND ASSOCIATED DISEASES

Although numerous systems have been proposed to classify or categorize diseases associated with MLOs, one of the more widely recognized systems outlined by Nienhaus and

Sikora[2] divides these diseases into four types: aster yellows (leaf yellowing, elongation of internodes), stolbur (apical dwarfing, stunting, leaf roll, epinasty, wilting, virescence), witches'-broom (proliferation of axillary shoots), and decline (degeneration). Tree fruit diseases of mycoplasmal etiology generally fall into the decline category, although all types are often represented by a specific MLO.

Because information on the pathological anatomy of these diseases is scarce and often lacking, MLO diseases can be divided into those with characterized and those with limited or uncharacterized pathological anatomy.

A. DISEASES WITH CHARACTERIZED PATHOLOGICAL ANATOMY

Of the four diseases that have been most extensively studied with regard to histopathology, the common characteristics include sieve tube necrosis, hyperactivity of the cambium (or procambium), callose accumulation, and formation of excessive replacement phloem. The following section considers these diseases and discusses the information on anatomical/histochemical changes in the tree fruit hosts in response to the presence of MLOs as influenced by rootstock/scion combinations, seasonal development, and developmental stage of disease.

1. Pear Decline

Pear decline was first reported in British Columbia in 1948, and is now found throughout North America and Europe.[79] Common hosts include cultivars and/or rootstocks of *Pyrus communis* (French pear), *P. pyrifolia* (Japanese pear), *P. ussuriensis* (Chinese pear), *P. variolosa*, and *Cydonia oblonga* (quince). A mycoplasmal etiology for this disease was first suggested by Hibino and Schnedier,[81] following electron microscopic studies. This suggestion was further strengthened by studies that demonstrated symptom remission after tetracycline antibiotic therapy[83] and by studies of insect transmission and grafting.[84,85]

Pear decline has been considered a bud-union disorder since symptoms vary with rootstock and scion combinations.[86,87] Symptoms of this disease have been grouped into three categories: quick decline, slow decline, and leaf curl.[79] Trees with quick decline symptoms show a sudden wilt which can occur at any time during summer or fall. These trees usually die the following winter. It is not uncommon for these trees to be asymptomatic prior to the sudden wilt, although some trees do show slow decline or leaf curl symptoms. Quick decline is most common on trees with oriental rootstocks (*P. pyrifolia* or *P. ussuriensis*), but can occur with certain scion grafts onto *P. communis* rootstocks. Slow-decline symptoms include a progressive weakening of the tree as measured by a gradual reduction in crop size and quality, reduced terminal growth, small, off-colored leaves, and premature drop of abnormally red leaves in fall. Trees in this category of decline can survive for many years or they can die within a short time. Slow decline occurs on oriental and French rootstocks. Trees with the leaf curl type of decline are characterized by a downward curling of leaves that are often thickened and that develop a diagnostic red-purple color in late summer and fall. These symptoms can vary in intensity and occurrence from year to year and are often associated with what are considered tolerant rootstocks (*P. communis, P. calleryana,* and *P. betulifolia*).

Anatomical studies on trees with pear decline have been fairly extensive, although many were conducted before the causal agent was identified as a probable MLO. Following examination of woody tissue from the area immediately below the bud union, researchers reported a series of anatomical changes thought to be induced by necrosis of the sieve tubes.[53,86,87] Because no disease agents were observed and scion wood appeared unaffected, it was speculated that a translocatable toxin might be involved. This speculation was not fully substantiated. Abnormal deposits of callose on sieve plate pores followed by loss or breakdown of the cellular contents of he sieve tubes and the callose were observed. Once sieve tube necrosis became evident below the bud union, degeneration of some of the older sieve tubes above the union was also observed. This was attributed to the girdling effect caused by the nonfunctional sieve tubes in the rootstock. Sieve tube necrosis also appeared to stimulate the cambium. Cambium hyperactivity resulted in the formation of replacement phloem. The anatomy of this replacement phloem varied from what appeared to be nearly normal to extremely pathological. These researchers noted that the sieve tubes in this replacement phloem either remained functional for a while or they rapidly degraded. Degradation was usually associated with immediate depositions of callose. Schnedier[53] also noted that the rays in the replacement phloem underwent hyperplasia and hypertrophy, thereby crushing the tissues in that year's annual ring. Sieve tube necrosis in the bud union was thought to interfere with translocation since starch subsequently accumulated in tissues above the union and was deficient in the roots below the union.[87] When extended over several years, the buildup of abnormal replacement phloem often resulted in noticeable swelling at the bud union (Figure 3). The girdling effect and impaired translocation could be correlated with the outward symptoms of general decline, loss of vigor, leaf reddening, and wilt exhibited by the scion.

All of the anatomical symptoms appeared to vary with the type of rootstock and were apparently not affected by the scion or budwood. Symptoms ranged from slight to severe as follows: *P. communis* var. 'Bartlett', *P. communis* (French), *P. calleryana, P. ussuriensis,* and *P. serotina.*[86]

From fluorescence microscopy and DAPI staining of stem tissues, the number of MLOs in the stem was found to vary throughout the year, reaching a maximum in summer and fall and usually disappearing completely during the winter.[46,88-91] However, MLOs were consistently present in root tissue during the winter. Since survival and persistence of MLOs seemed to be linked to the presence of functional sieve tubes, Schaper and Seemüller[91] proposed that the inability to detect MLOs in stems during winter was due to the presence of nonfunctional sieve tubes. They reported that sieve tubes in secondary phloem degenerated in both healthy and pear decline stems as winter approached. This was consistent with what Evert[41,42] had found to occur naturally in healthy pear by the end of the growing season. Evert reported complete degeneration of the phloem under his conditions. In diseased trees, Schaper and Seemüller[91] noted that sieve tubes that had been produced in spring were usually damaged, appearing necrotic and collapsed by the end of the season. A ring of replacement phloem was occasionally present in declined tissues and the degeneration of sieve cells in this tissue was often less pronounced. By midwinter, few functional sieve tubes were present in either healthy or declined stem tissues and few, if any, MLOs were observed.[46] The midwinter lack of functional sieve cells able to sustain MLOs spawned the idea that MLOs must recolonize above-ground tissues each year as new growth began in spring.

Anatomically, no significant differences were observed between diseased or healthy pear roots.[92,93] Both contained functional sieve tubes throughout the winter, although phloem necrosis and nonfunctional sieve tubes were evident. Diseased roots contained relatively high numbers of MLOs during the winter. Functional sieve tubes that contained MLOs were also reported at graft unions and appeared to persist throughout the winter. These observations further substantiated the idea the MLOs overwinter in the roots or graft unions and recolonize the scions each year.[46,92,93]

An extensive histological study of major veins of leaf tissues from healthy and declining 'variolosa' pear was conducted by Schneider and co-workers.[51,94,95] They reported that differences between diseased and healthy tissues were not evident until after the beginning of secondary growth. Starch accumulation was the first abnormality observed. The first structural change was sieve tube and companion cell necrosis. Sieve tubes lost internal structural integrity, which was characterized by general deterioration of the ground, endo-

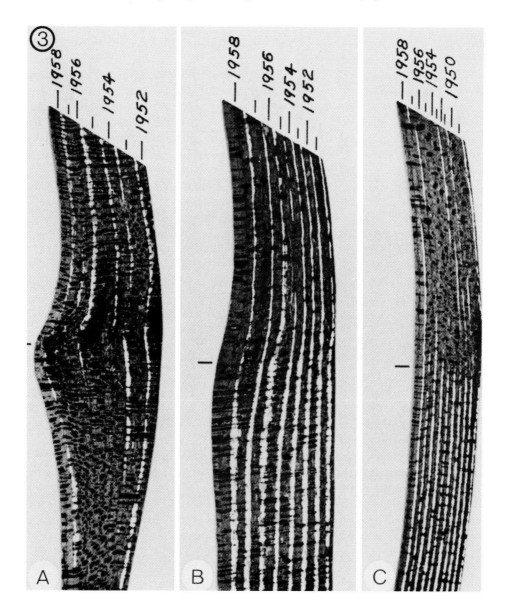

FIGURE 3. Radial sections (×4.2) of phloem. Location of the bud union is indicated by the short horizontal lines at the center of the figures. Sections were photographed with polarized light which caused fiber and crystal idioblast layers between annual rings to photograph white. (A) Bartlett on oriental rootstock in decline stage 4 from Yakima Valley, Washington, where pear decline appeared about 1950. Abnormal replacement phloem occurs in all the annual rings in the section; the oldest shown was formed in 1952. (B) Bartlett on *P. ussuriensis* rootstock in decline stage 2 from Medford, OR. Abnormal replacement phloem occurred in the 1957 and 1958 annual rings. (C) Bartlett on *P. ussuriensis* rootstock not affected by decline from Marysville, CA. No pathological condition is present. (From Batjer, L. P. and Schneider, H., *Proc. Am. Soc. Hortic. Sci.,* 76, 85, 1960. With permission.)

plasmic reticulum (ER), and organelle cytoplasm, membranes become fuzzy, and many sieve tube walls collapsed. With the onset of symptoms, the overall structure of the phloem also changed as the cambium became hyperactive (Figure 4). This resulted in the formation of uneven amounts of excessive phloem. Some areas of the cambium were apparently more active than others and produced more phloem. Many of the sieve tubes in this replacement

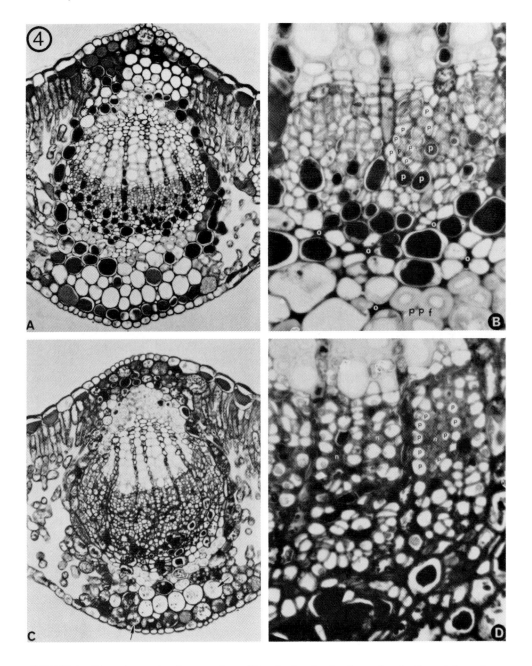

FIGURE 4. Photomicrographs of cross-sections of lateral veins of mature leaves from healthy and diseased pear. (A) Healthy vein. (B) Higher magnification of part of the phloem in A. Sieve tubes and companion cells are interspersed among parenchyma cells (p). The sieve tubes have thick, crenulated, nacreous walls that nearly fill the lumens. Obliterated sieve tubes (o) occur in the nonfunctional primary phloem. (C and D) Veins from a diseased leaf. D is a higher magnification of a portion in C. Sieve tubes and companion cells are necrotic (n), collapsed, and interspersed among parenchyma cells (p). Starch is accumulating in the bundle sheath extensions (arrows in C). (Magnification, A and C, × 140; B and D, × 557.) (From Soma, K. and Schnedier, H., *Hilgardia*, 40, 471, 1971. With permission.)

phloem failed to maintain structural integrity and necrosed and collapsed. Hypertrophy and hyperplasia of parenchyma cells were observed. Schneider[94] also reported the presence of brown oxidized polyphenols in parenchyma cells of the bundle sheath, palisade and spongy mesophylls, and phloem in diseased leaves. However, because these studies were performed before MLOs were known to be associated with the disease, no mention was made of a relationship between the causal agent and the pathological anatomy.

Schneider and co-workers[81,94] also noted that in symptomatic leaves, MLOs were limited to mature sieve tubes and were not found in young sieve tubes or companion or parenchyma cells. They observed starch accumulation, sieve tube necrosis and collapse, as previously reported for the 'variolosa' variety, but they also reported that some sieve tubes remained unaffected and were apparently functional since MLOs were abundant in these cells. They noted that the distribution of MLOs in leaves could be remarkably variable, especially when fine minor veins were compared to major veins. MLOs were frequently associated with tissues exhibiting no significant pathological anatomy and were often not found in tissues exhibiting considerable pathosis. These observations were consistent with other MLO-associated tree fruit diseases.

2. Apple Proliferation

Apple proliferation, a relatively new disease (first described in Italy in 1950), is now found throughout mainland Europe.[82] Economically important woody hosts include apple *(Malus domestica)* and ornamental *Malus* species. Giannotti et al.[96] first proposed a mycoplasmal etiology for this disease in 1968. Successful transmissions by grafting and dodder and symptom remission resulting from treatment with tetracycline subsequently provided further support for this etiology.[82,91]

Symptoms of apple proliferation appear on shoots, leaves, fruits, or roots and often first appear as witches'-brooms, small fruit, and late growth of terminal buds in the fall.[82] Witches'-brooms, distinguishing symptoms of this disease, can appear at the tips of main branches, in the crown of the tree, or on root sprouts. Fruit on infected trees is smaller than normal and often have longer peduncles. Leaves can have enlarged stipules and show premature fall coloration. Terminal buds on shoots of infected trees often break late in the season and result in tufts of new growth at shoot tips. Roots are characteristically undersized, necrotic, and can form subterranean witches'-brooms. Unusual but important features of apple proliferation are spontaneous partial recovery and occasional complete natural remission of symptoms on infected trees.[82]

The pathological anatomy of stems infected with the apple proliferation MLO is very similar to that discussed for pear decline.[46,55,88-90] As with pear, sieve tubes in phloem of both healthy and infected apple stems degenerate as winter approaches. Sieve tubes produced in spring were usually the first damaged and were often necrotic or collapsed (Figure 5). The major observable anatomical difference between pear decline and apple proliferation was the significant amount of replacement phloem produced in apple as a response to infection when compared to the sporadic and limited amount produced in pear. The average width of phloem was 535 and 200 μm for diseases and healthy apple tree, respectively. As a consequence of the large quantity of replacement phloem, trees with apple proliferation had greater numbers of apparently functional sieve tubes during the winter. These were usually located within several layers of the youngest parts of the replacement phloem. Researchers speculated that they probably remained functional longer than other sieve tubes because they were produced later in the season and were therefore younger cells. MLOs were present in some of these cells.

As with pear decline, no significant anatomical differences have been observed between diseased or healthy apple roots.[46,55,89,90,92] Functional sieve tubes were observed throughout

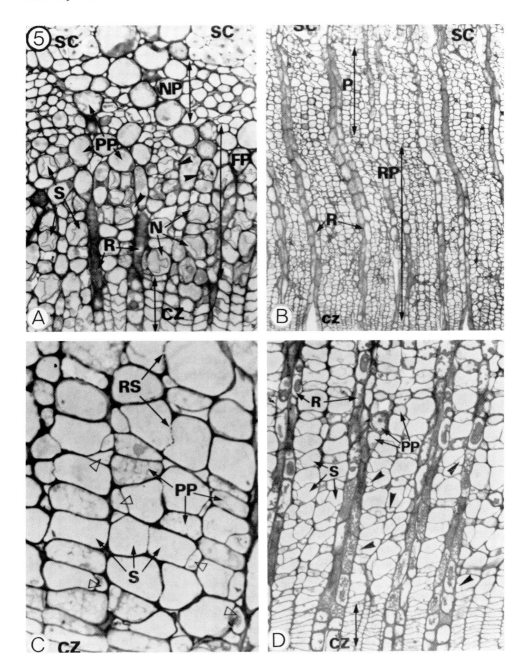

FIGURE 5. Transverse sections of the secondary phloem of the stem of apple and pear in the fall. (A) Cambial zone and phloem of a 1-year-old shoot of a healthy apple tree at the beginning of October. The largest part of the phloem shows functional sieve tubes (FP), most of which have nacreous walls (N). In the oldest part of the phloem, the sieve tubes are collapsed (NP). (Magnification × 322.) (B) Current season's phloem (collected in November) of the trunk of a diseased apple tree consisting of a ring of spring-formed phloem with necrotic sieve tubes (P) and a broad ring of replacement phloem (RP). (Magnification × 84.) (C) Phloem with degenerating sieve tubes of the trunk of a healthy apple tree toward the end of November. Indications of degeneration are the plasmolysis and the large vacuoles of the companion cells (arrow). RS, radial sieve areas. (Magnification × 518.) (D) Phloem with degenerating sieve tubes of the trunk of a diseased pear tree, collected in December. The condition of the phloem is similar to that given in Figure 3 for healthy apple. (Magnification × 203.) Arrows, companion cell; CZ, undifferentiated cambial zone; PP, phloem parenchyma; R, ray; S, sieve tube; SC, fiber sclereids. (From Schaper, U. and Seemüller, E., *Phytopathology, 72,* 736, 1982. With permission.)

the winter, although as previously noted, phloem necrosis and nonfunctional sieve tubes were evident in both healthy and diseased roots. Diseased roots consistently contained relatively high numbers of MLOs.

Functional sieve tubes containing MLOs were also observed at graft unions throughout the winter. However, since these trees were grafted in the fall, the increased cambial activity may have been associated with the time of grafting and not the disease.[92] These sieve tubes were produced late in the season so they were relatively young and appeared to remain functional longer, a phenomenon similar to that found in late-produced replacement phloem; here, functional sieve tubes were still present until late in the season. Again, these results contributed to the theory that the majority of stem tissues of trees infected with apple proliferation are recolonized by MLOs each spring from their overwintering sites at graft unions or in roots. However, with apple proliferation, some MLOs apparently do overwinter in stem tissue. This is documented by observations of apparently normal MLOs in sieve tubes of replacement phloem during winter and occasional positive transmissions from scion wood grafts.[55]

Recent work has shown that *Malus* taxa and hybrids react differently to the apple proliferation MLO. When various taxa were used as rootstocks with scions grafted with two cultivars of *M. pumila*, Kartte and Seemüller[97,98] were able to divide these taxa into five groups based on recovery rate, witches'-broom formation, mortality, and MLO populations. Histopathological studies demonstrated that the extent of phloem necrosis, starch content, and MLO titer also varied significantly with rootstock taxa and developmental stage of disease. They reported the extent of necrosis of the current season's secondary phloem could be quite variable, but that it initially involved sieve tube collapse in defined areas. In extreme cases the necrosis appeared to spread until all elements, including the newly differentiating cells near the cambium, were involved. Necrosis of companion and phloem parenchyma cells was also observed. Starch accumulation in the scion (especially in the leaves) and depletion in the rootstock followed the same trend for the various taxa studied and was indicative of impaired translocation.

Samples from *Malus* taxa in Kartte and Seemüller's group IV[98] (highly susceptible selections) in late stages of decline were examined in developmental sequence as the growing season progressed. At the beginning of the season, phloem of scion and roots was usually only slightly damaged and was not different than healthy tissue. However, as the season progressed, the diseased trees showed significant increases in phloem necrosis. This was first observed in root and then in scion tissues. Replacement phloem was not observed in trees with severe symptoms of decline. Healthy trees exhibited low levels of necrosis in both root sand scions.

Starch content appeared to vary with the condition of the phloem (Figure 6). In roots, as disease increased, the amount of starch decreased, indicative of impaired translocation and nonfunctional phloem. In roots of healthy tree, starch levels appeared to be highest at the end of the season. In scions, starch levels in both diseased and healthy trees fluctuated, but at the end of the season, no significant differences were detectable. MLO populations were consistently low in these highly susceptible taxa.

When decline-tolerant taxa were examined, Kartte and Seemüller[98] reported that the condition of the phloem and starch content did not appear to be affected by the presence of the MLO since no real differences were observed between healthy and diseased samples, although some replacement phloem was evident. MLO populations were always high in both stock and scion and they speculated that the sieve tubes were tolerant to infection.

In the resistant taxon studied, phloem damage was higher and starch content was reduced when compared to the decline-tolerant taxa. MLO titers were lower in the roots when compared to the scion and they decreased over the years, often to the point of complete elimination.

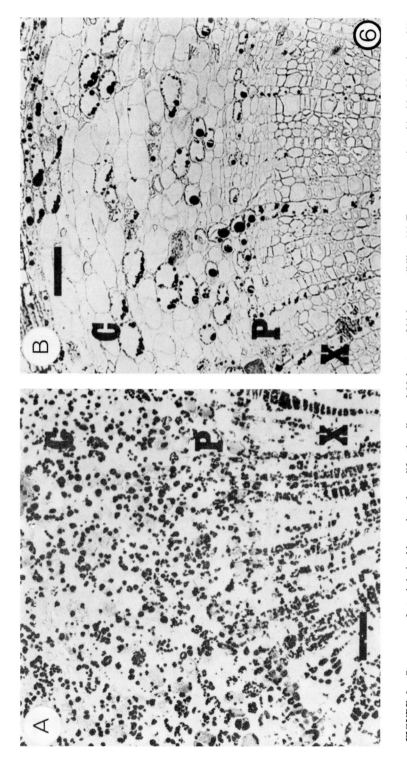

FIGURE 6. Occurrence of starch in healthy and apple proliferation-diseased *Malus* taxa of high susceptibility. (A) Transverse section of healthy *M. tschonoskii* root. Large amounts of starch are present in the parenchyma of xylem (X), phloem (P), and cortex (C). IKI staining. Bar = 100 μm. (B) As A, but diseased root with no detectable amount of starch. The dark, round structures are phenolics stained with osmium tetroxide. (From Kartte, S. and Seemüller, E., *J. Phytopathol.*, 131, 149, 1991. With permission.)

3. X-Disease of Stone Fruits

X-disease of stone fruits is an economically important disease of tree fruits.[78] This name was coined for an unknown disorder of peach that was first reported by Stoddard in Connecticut in 1933[99] and was probably the same disease reported in California on cherry, called cherry buckskin in 1931.[100] Today, X-disease is found throughout most of the U.S. and the stone fruit producing areas of Canada, but it has never been reported from the peach-growing areas of the southern U.S.[78] Even in the early reports, this disease was described as highly variable since it involved a wide host range, a confusing number of common names (e.g., Western X, Eastern X, Cherry Buckskin, Peach Yellow Leafcurl), and was transmitted by a number of species of leafhopper.[78]

The association of a MLO with X-disease was first reported in 1971.[101] Additional lines of evidence have subsequently supported these initial electron microscopic observations, including symptom remission after tetracycline therapy and budding, grafting, and leafhopper transmission studies.[75,102-107] Because of the highly variable nature of the host range and symptoms associated with this MLO, a number of "strains" were proposed. Early publications maintained a distinction between eastern and western forms of X-disease, but additional strains have since been reported from California alone (e.g., Green Valley X, Peach Yellow Leafroll). Recent work using molecular probes and serological techniques have provided additional insight into the relationships of these strains from the various geographical regions.[29,32,33,36-39]

The key economic hosts of X-disease are *Prunus persica* (peach), *P. persica* var. *nectarina* (nectarine), *P. salicina* (Japanese plum), *P. avium* (sweet cherry), and *P. cerasus* (sour cherry). *P. virginiana* (chokecherry) is an important wild reservoir host. The host range of this MLO also extends to over ten other *Prunus* species and a considerable number of herbaceous hosts.[78]

Symptoms of X-disease are highly variable and depend upon strain, host plant, and rootstock.[78,108] In peach, leaves roll longitudinally along the midvein and develop red, irregular blotches which become necrotic and fall out, giving leaves a tattered appearance. Affected leaves fall prematurely and fruit, if produced, are misshapen and bitter in flavor. A diagnostic feature late in the season is a persistent rosetted tuft of leaves at the terminals. Other strains of this MLO produce similar symptoms on peach, but these can vary from excessive leaf rolling and yellowing without significant defoliation to distinct, premature fall coloration and excessive leaf drop.

The variability of symptoms associated with X-disease is further supported by those produced on sweet cherry grafted onto 'Mazzard' (*P. avium*) rootstocks. Such trees survive for many years with chronic symptoms: leaves develop slightly reddened midveins and diagnostic small, often distinctly pointed pale red or green fruit. In contrast, when grafted onto 'Mahaleb' (*P. mahaleb*) rootstocks, trees are short-lived and develop a quick decline or sudden collapse, dying abruptly in late summer.[108]

The basic anatomical features in stems of peach and cherry infected with X-disease are consistent with those reported for other mycoplasmal diseases.[5,52,54] However, since the early studies were conducted before the causal agent was determined, there is limited information on the presence of the MLO with regard to these anatomical changes. In symptomatic peach, the first pathological changes were observed in primary and early secondary phloem. These changes consisted of necrosis and collapse of sieve tubes and companion cells, first in the older phloem, then in the youngest sieve tubes. Typically, the sieve tubes lost their nacreous walls and callose appeared on sieve plate pores; these sieve tubes eventually collapsed from apparent pressure form surrounding cells. Histochemical staining of these tissues with phloroglucinol-HCl indicated the presence of what was interpreted as wound gum in the secondary phloem. This was usually confined to walls and contents of collapsed sieve tubes. Wound gum was not observed in primary phloem.[52,54]

These anatomical changes were often followed by the production of new replacement phloem and wide zones of secondary phloem often resulted. However, in some cases, sieve tubes in this newly formed tissue also became necrotic and collapsed. The cambium, which normally exhibited an orderly arrangement, was often disorganized and some parenchyma cells were hypertrophied.

Petioles of leaves from infected stems also showed evidence of sieve tube necrosis in the primary phloem and a wide band of secondary phloem (Figure 7). In contrast, healthy petioles had very little secondary phloem.[52,54] Many of the sieve tubes in infected petioles were crushed and their end walls contained callus, as determined by lacmoid staining.

When symptomatic leaves were examined using electron microscopy, MLOs were observed in mature phloem elements.[108] MLOs were not present in all sieve tubes of a cross-section and appeared to be randomly and discontinuously distributed in the tissues. These observations were consistent with reports of other tree fruits infected with MLOs. Jones et al.[109] noted that chances of finding MLOs were best in leaves with early symptoms of disease, since severely symptomatic leaves contained many collapsed and necrotic phloem cells. Douglas[48] reported that in peach and chokecherry leaves, MLOs were present in sieve tubes of all stages of maturity before the onset of symptoms and before any anatomical changes due to pathogenesis were evident. Using DAPI and electron microscopy, MLOs were not detected in newly formed leaves at the start of the growing season, but they increased in titer and distribution as the season progressed and with the onset of symptom expression. Hyperplasia and hypertrophy of parenchyma cells were frequently observed. The cambium was disorganized and hyperactive, as evidenced by significant and often uneven amounts of replacement phloem. Many of the sieve tubes in the replacement phloem did not remain functional for very long. Douglas also observed significant depositions of starch in the parenchyma cells of these affected tissues.[110] Sinha and Chiykowski[47] reported a similar situation in infected chokecherry leaves. They observed MLOs in young sieve tubes containing mitochondria, ribosomes, and rough ER. Although X-diseased tissues were not sampled or examined during the winter in these studies, the phenomenon of seasonality and fluctuating MLO titer was consistent with that reported for apple proliferation and pear decline. For these diseases, survival of MLOs in stem tissues appeared to be limited and was attributed to the lack of functional sieve tubes in the stem during the winter. Thus, recolonization of the stem was observed to occur each year.

The histopathology of leaves from cherry grafted onto 'Mazzard' rootstocks was similar to that reported for peach.[52] Leaves typically developed swollen veins which were caused by the production of excessive amounts of new secondary xylem and phloem. Sieve tubes had thick nacreous walls and the metaphloem sieve tubes were usually crushed.

Studies on sweet cherry grafted onto Mahaleb rootstocks showed that extensive sieve tube necrosis and phloem degeneration occurred in rootstock but not in scion wood.[52] Distinct reactions for wound gum stained with phloroglucinol were observed, although they were not as extensive as those reported for peach. Recently, Uyemoto[111] observed significant anatomical changes in the xylem, specifically at the scion/rootstock interface. This was the first report of MLO-induced changes in the histology of xylem tissues. The changes appeared as union aberrations consisting of pits and grooves.

It is interesting to note that Schneider[52] reported that many of the symptoms and anatomical changes in trees infected with X-disease could be produced by girdling a tree. Tissue above the girdle showed callose deposition and necrosis of sieve tubes and companion cells starting with the older cells. Necrosis was followed by cambial hyperactivity; however, dissimilar to X-diseased tissues, both xylem and phloem were produced in reaction to girdling.

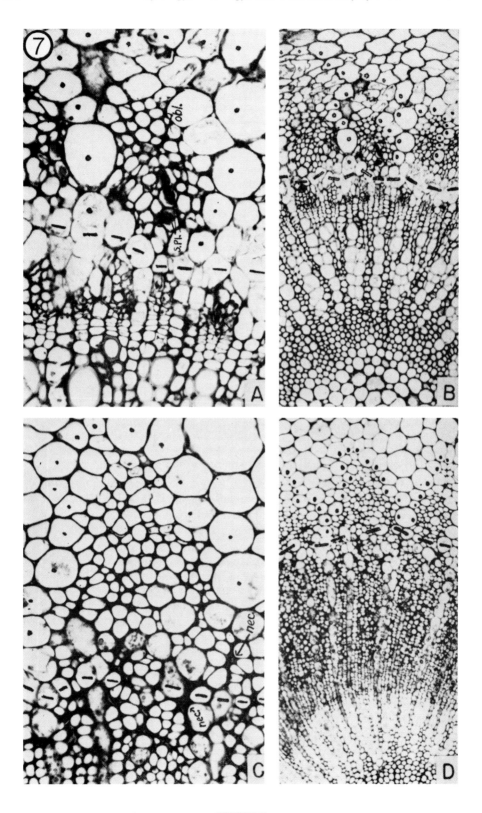

FIGURE 7.

4. Lethal Yellowing of Coconut Palm

Lethal yellowing of coconut palm, *Cocos nucifera*, is a disease that threatens a leading agricultural crop in the humid tropics.[7,49,112-116] This name was first used in 1955 for a specific "unknown disease" of coconut palms in Jamaica.[112] Disease symptoms span a wide range and include premature fruit drop, necrosis of developing inflorescences, foliar yellowing and desiccation, root degeneration, and necrosis. Death usually occurs within 1 year of the first observed symptoms. Besides coconut palm, over 30 other species of palm serve as hosts for this MLO worldwide.[115]

Identification of a MLO as the etiologic agent was first reported in 1972.[113,114] These electron microscopic observations, in conjunction with symptom remission following applications of tetracycline and lack of remission with penicillin, provided strong evidence for the mycoplasmal etiology.[117,118]

Studies on the pathological anatomy of diseased coconut palms have focused on symptomatic tissues. Parthasarathy[49] reported MLOs in mature protophloem sieve elements and in recently matured metaphloem sieve elements in tissues sampled from the rachilla (flowering axis) just below the necrotic portions of unexpanded inflorescences. No MLOs were found in differentiating or mature late metaphloem sieve elements of fully expanded inflorescences, leaves, or stems. Parthasarathy observed that MLOs, when present, were always confined to mature sieve elements and never occurred in parenchyma cells. While still preliminary, he reported that the sieve elements that contained MLOs did not have significantly different structural features than those without them (Figure 8). This relatively "normal" appearance of sieve elements containing MLOs was also reported by Thomas.[115,119]

Although some sieve elements appeared normal, Parthasarathy[49] also observed that others had fibrillar, amorphous substances in the cell lumen, and yet others had sieve plate pores with obvious callose deposits. Thomas,[119] observing a similar phenomenon, defined two matrix types associated with sieve elements containing MLOs. One type had variable electron density and an amorphous to granular structure. The second type had moderate electron density, a fibrillar structure, and a distinct parietal zone of ribosomes that surrounded the matrices. Thomas went on to suggest that the presence of ribosomes in mature sieve elements could be interpreted as a component of abnormal phloem differentiation in diseased trees. He also reported that sieve elements of symptomatic palms degenerated in a fairly regular manner. This is evidenced by a progression of characteristics: cells first lost ultrastructural detail and showed increased electron density; in advanced stages sieve elements were crushed and electron opaque.[115]

Based on the observation that many sieve elements containing MLOs appeared relatively normal, Parthasarathy[49] speculated that the presence of these organisms did not have a significant impact on the structure or apparent function of recently matured sieve elements. Parthasarathy also suggested MLOs move along sieve tubes with assimilates in the "source to sink stream", moving to young, developing inflorescences and leaves. This idea was supported by observing MLOs predominantly in mature protophloem and early metaphloem

FIGURE 7. Cross-section of peach petioles. (A and B) healthy. (B lower magnification of A for orientation.) Dashes set off the primary phloem strands from the secondary phloem, of which only a trace is formed. The inner side of the primary phloem strands contain functioning metaphloem sieve tubes. The outer portion is composed of protophloem cells that enlarged and lengthened as the protophloem sieve tubes were obliterated. In stems, such cells have secondary walls and are the primary phloem fibers. S.Pl., sieve plates; obl., obliterated cells. Dots in parenchyma cells adjoining the primary phloem strands indicate the approximate division between the cortex and primary phloem. (C) Buckskin-diseased petiole. The metaphloem sieve tubes are necrotic (outer margin of the metaphloem is indicated by dots in adjoining parenchyma cells). No large parenchyma cells separate the primary phloem from the secondary phloem. (D) Lower magnification of C showing a wide band of secondary phloem that was produced and in which the sieve tubes became necrotic. Dashes indicate the approximate division between the primary and secondary phloem. nec, Necrosis. (Magnification, A and C, × 280; B and D, × 106). From Schneider, H., *Phytopathology*, 49, 550, 1959. With permission.)

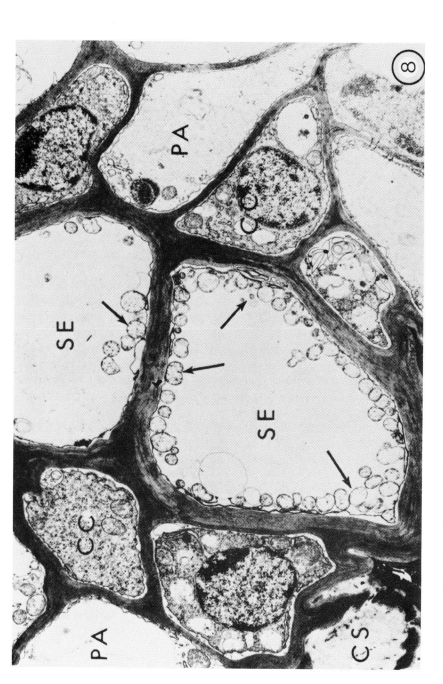

FIGURE 8. Transverse section of a recently matured sieve element of a young inflorescence axis of *Cocos nucifera* (Jamaican material) affected by lethal yellowing. Two relatively undisturbed, recently matured metaphloem elements (SE) containing MLO (arrows). Note the parietal distribution of the MLO in the sieve elements. Crushed protophloem elements (CS), companion cells (CC), and parenchyma cells (PA) are also visible. (Magnification × 6300.) (From Parthasarathy, M. V., *Phytopathology*, 64, 667, 1974. With permission.)

sieve elements as opposed to mature late metaphloem elements of expanded leaves and inflorescences. Subsequent observations by Thomas[115] were in agreement with Parthasarthy's suggestion that MLOs are transported from older to younger tissues in the translocation stream since he reported that sieve elements that contained MLOs occasionally contained dictyosomes and ribosomes, observations indicating that MLOs invaded when the sieve elements were fairly young.

In a study using DAPI staining and fluorescence microscopy, Deutsch and Nienhaus[116] reported that fluorescent particles (equated with MLO-DNA) varied with the age of the tree and the stage of disease. In nonbearing palms, MLOs were highest in samples taken from trees exhibiting advanced stages of disease, whereas in bearing palms, MLOs were present in higher numbers in tissues showing early stages as compared to those in advanced stages of disease. All of the studies agreed with reports of other tree fruit diseases: that it was not uncommon to find vascular bundles completely free of MLOs, while others were packed with MLOs.[113,114] The observations of sieve elements packed with MLOs provided evidence for physical blockage of the translocation stream which could be correlated with the type of outward symptoms associated with this disease.

B. DISEASES WITH LIMITED OR UNDETERMINED PATHOLOGICAL ANATOMY

A number of additional diseases of tree fruits with proposed mycoplasmal etiology have been reported, but available information is extremely sketchy. In most cases, studies have been limited to identification and confirmation of MLOs *in situ* using electron microscopy. These studies reported no significant deviations from previous studies with regard to the apparent phloem specificity of MLOs, nor die they report any obvious or unusual anatomical abnormalities in the phloem or any of the tissues examined.

With Peach Yellows, MLOs were observed in mature sieve tubes of peach leaves and were also reported in phloem parenchyma, companion cells, and immature sieve tubes.[120,121] Tissues from leaves that had exhibited symptoms for several weeks were reported to contain necrotic phloem cells and what appeared to be collapsed MLOs.

Apricot Chlorotic Leafroll, Plum Chlorotic Leafroll, and Declining Plum are thought to be caused by the same or closely related strains of a MLO.[122-125] Electron microscopic studies of symptomatic host tissues revealed that MLOs were restricted to mature sieve cells and never found in companion or parenchyma cells. They were often present in very low concentrations. Researchers observed that host protoplasts of sieve cells could be disorganized and sieve tubes collapsed before normal maturity. Studies on Little Peach,[109] Apple Chat Fruit,[126-128] Albino Disease of Cherry,[23] and Rubbery Wood of Apple[128-130] reported the presence of MLOs in phloem, but presented no information on the histopathology of these tissues. With the exception of rubbery wood, outward symptoms fall into the basic types previously described for MLOs.

IV. CONCLUSION: FUTURE DIRECTIONS

The histopathology of most tree fruit diseases with mycoplasmal etiology is characterized by sieve tube necrosis, hyperactivity of the cambium (or procambium), callose and/or starch accumulation, and the formation of replacement phloem. Hyperplasia and hypertrophy of phloem parenchyma cells have also been reported. In addition to these observable changes in the cytology of the plant host, MLOs also appear to respond and are altered by changes in the delicate environment of the sieve tubes in which they are limited. MLOs respond to these changes by exhibiting degeneration and loss of ultrastructural detail.

The difficulties encountered in assessing the pathological anatomy of these diseases have hampered studies aimed at a direct, straightforward, and systematic approach for determining pathogenesis. New approaches require dynamic rather than static sampling. Tissue would be examined just prior to the onset of symptoms, during initial symptom expression, and at specific intervals in disease development. It is only from studies of this type that degenerative sequences of the histopathology can be developed.

In addition to the aforementioned multiple sampling techniques, it would also be necessary to incorporate and interpret information on host/pathogen interactions and/or altered host physiology into these anatomical studies in order to ascertain primary from secondary and direct from indirect effects of MLOs on plant hosts. For example, overproduction of sieve elements (and replacement phloem) suggests a disturbance of a fundamental process of differentiation in the host. This could be initiated as a direct effect of the MLO in the phloem or as a response of the plant to the presence of the MLO in the phloem. Similarly, starch accumulation could be interpreted as either a primary effect initiated by the physical presence of the MLO in the phloem or by a direct effect of the MLO on starch metabolism, or it could be viewed as a secondary effect simply related to physical blockage or impaired movement of metabolites associated with dysfunctional phloem.

In order to understand the developmental histopathology of these diseases, it will be necessary to utilize an arsenal of existing and newly developed techniques borrowed from other areas of plant pathology, anatomy, and physiology. The complexity of these diseases demands a multifaceted approach. Among the most promising of the techniques are methods for tissue sampling and microscopy that minimize disruption of the phloem, in addition to techniques that employ molecular and serological probes for identification, localization, and quantification of MLOs in tissues. A variety of fixation and sampling techniques have been described that take into account the pressurized environment of functional sieve tubes and help to eliminate artifacts associated with the disruption of this environment which have complicated many previous reports and interpretations of pathological anatomy.

Serological techniques employing both mono- and polyclonal antibodies have been demonstrated for direct and specific detection of MLOs and/or of specific chemicals associated with them. Similarly, molecular probes containing cloned MLO-DNA have been used successfully to detect and differentiate various MLOs in plant hosts. Both types of approaches need to be expanded and modified so they can locate the pathogen in the host or trace specific toxic products of MLOs or MLO-induced plant products associated with the host/parasite interaction. This type of information is essential to understand cause-effect relationships resulting in pathological anatomy.

The challenge of elucidating the histopathology of tree fruit diseases with mycoplasmal etiology is a difficult one compounded by many factors. However, efforts utilizing and modifying currently available technology promise to further our understanding of these important and intriguing plant diseases.

ACKNOWLEDGMENTS

I wish to thank Drs. D. W. Dingman, D. E. Aylor, and R. K. Kiyomoto for their helpful suggestions for improving this manuscript. I am also grateful for the assistance of Mrs. D. Savo in the preparation of the manuscript.

REFERENCES

1. **Doi, Y., Teranaka, M., Yora, K., and Asuyama, H.,** Mycoplasma- or PTL group-like microorganisms found in the phloem of plants infected with mulberry dwarf, potato witches' broom, aster yellows or pawlownia witches' broom, *Ann. Phytopathol. Soc. Jpn.,* 33, 259, 1967.
2. **Nienhaus, F. and Sikora, R. A.,** Mycoplasmas, spiroplasmas, and rickettsia-like organisms as plant pathogens, *Annu. Rev. Phytopathol.,* 17, 37, 1979.
3. **Bove, J. M.,** Wall-less prokaryotes of plants, *Annu. Rev. Phytopathol.,* 22, 361, 1984.

4. **Davis, R. E. and Whitcomb, R. F.**, Mycoplasmas, rickettsiae, and chlamydiae: possible relation to yellows diseases and other disorders of plants and insects, *Annu. Rev. Phytopathol.,* 9, 19, 1971.
5. **Schneider, H.**, Cytological and histological aberrations in woody plants following infection with viruses, mycoplasmas, rickettsias, and flagellates, *Annu. Rev. Phytopathol.,* 11, 119, 1973.
6. **Seliskar, C. E. and Wilson, C. L.**, Yellows diseases of trees, in *Mycoplasma Diseases of Trees and Shrubs,* Maramorosch, K. and Raychaudhuri, S. P., Eds., Academic Press, New York, 1981, 35.
7. **McCoy, R. E.**, Mycoplasmas and yellows disease, in *The Mycoplasmas,* Vol. 3, Whitcomb, R. F. and Tully, J. G., Eds., Academic Press, New York, 1979, 229.
8. **McCoy, R. E.**, Chronic and insidious disease: the fastidious vascular pathogens, in *Phytopathogenic Prokaryotes,* Vol. 1, Mount, M. S. and Lacy, G. H., Eds., Academic Press, New York, 1982, 475.
9. **Saglio, P. H. M. and Whitcomb, R. F.**, Diversity of wall-less prokaryotes in plant vascular tissue, fungi, and invertebrate animals, in *The Mycoplasmas,* Vol. 3, Whitcomb, R. F. and Tully, J. G., Eds., Academic Press, New York, 1979, 1.
10. **Caudwell, A., Kuszala, C., and Larrue, J.**, Sur la culture in vitro des agents infectieux responsables des jaunisses des plantes (MLO), *Ann. Phytopathol.,* 6, 173, 1974.
11. **Hayflick, L. and Arai, S.**, Failure to isolate mycoplasmas from aster yellows-diseased plants and leafhoppers, *Ann. N.Y. Acad. Sci.,* 225, 494, 1973.
12. **Giannotti, J., Vago, C., Marchoux, G., Devauchelle, C., and Czarnecky, D.**, Characterisation par la culture in vitro des souches de mycoplasmes correspondant a huit maladies differentes des plantes, *C. R. Acad. Sci. Paris, Ser. D,* 274, 330, 1972.
13. **Raju, B. C. and Nyland, G.**, Effects of different media on the growth and morphology of three newly isolated plant spiroplasmas, *Phytopathol. News,* 12, 216, 1978.
14. **Smith, A. J., McCoy, R. E., and Tsai, J. H.**, Maintenance in vitro of the aster yellows mycoplasmalike organism, *Phytopathology,* 71, 819, 1981.
15. **Nasu, S., Jensen, D. D., and Richardson, J.**, Primary culturing of the western X mycoplasmalike organism from *Collandonus montanus* leafhopper vectors, *Appl. Entomol. Zool.,* 9, 115, 1974.
16. **Esau, K.**, Development and structure of the phloem tissue, *Bot. Rev.,* 5, 373, 1989.
17. **Evert, R. F.**, Phloem structure and histochemistry, *Annu. Rev. Plant Physiol.,* 28, 199, 1977.
18. **Schneider, H.**, The anatomy of peach and cherry phloem, *Bull. Torrey Bot. Club,* 72, 137, 1945.
19. **Esau, K.**, Phloem degeneration in celery infected with yellow leafroll virus of peach, *Virology,* 6, 348, 1958.
20. **Esau, K.**, An anatomist's view of virus diseases, *Am. J. Bot.,* 43, 739, 1956.
21. **Waters, H.**, Light and electron microscopy, in *Plant and Insect Mycoplasma Techniques,* Daniels, M. J. and Markham, P. G., Eds., John Wiley & Sons, New York, 1982, 101.
22. **Waters, H. and Hunt, P.**, The in vivo three-dimensional form of a plant mycoplasma-like organism by the analysis of serial ultrathin sections, *J. Gen. Microbiol.,* 116, 111, 1980.
23. **Florance, E. R. and Cameron, H. R.**, Three-dimensional structure and morphology of mycoplasmalike bodies associated with albino disease of *Prunus avium, Phytopathology,* 68, 75, 1978.
24. **Hirumi, H. and Maramorosch, K.**, Ultrastructure of the aster yellows agent: mycoplasmalike bodies in sieve tube elements of *Nicotiana rustica, Ann. N.Y. Acad. Sci.,* 225, 201, 1973.
25. **Razin, S.**, The mycoplasmas, *Microbiol. Rev.,* 42, 414, 1978.
26. **Razin, S., Glaser, G., and Amikam, D.**, Molecular and biological features of mollicutes (Mycoplasmas), *Ann. Microbiol. (Inst. Pasteur),* 135A, 9, 1984.
27. **Davis, R. E. and Lee, I.-M.**, Pathogenicity of spiroplasmas, mycoplasmalike organisms, and vascular-limited fastidious walled bacteria, in *Phytopathogenic Prokaryotes,* Vol. 1, Mount, M. S. and Lacy, G. H., Eds., Academic Press, New York, 1982, 491.
28. **Bonnet, F., Saillard, C., Kollar, A., Seemüller, E., and Bove, J. M.**, Detection and differentiation of the mycoplasmalike organism associated with apple proliferation disease using cloned DNA probes, *Mol. Plant-Microbe Interacts.,* 3, 438, 1990.
29. **Kirkpatrick, B. C., Stenger, D. C., Morris, T. J., and Purcell, A. H.**, Cloning and detection of DNA from a nonculturable plant pathogenic mycoplasma-like organism, *Science,* 238, 197, 1984.
30. **Lee, I.-M., Davis, R. E., and Dewitt, N. D.** Nonradioactive screening method for isolation of disease-specific probes to diagnose plant diseases caused by mycoplasmalike organisms, *Appl. Environ. Microbiol.,* 56, 1471, 1990.
31. **Davis, R. E., Lee, I.-M., Douglas, S. M., and Dally, E. L.**, Molecular cloning and detection of chromosomal and extrachromosomal DNA of the mycoplasmalike organism associated with little leaf diseases in periwinkle *(Catharanthus roseus), Phytopathology,* 80, 789, 1990.
32. **Kuske, C. R., Kirkpatrick, B. C., Davis, M. J., and Seemüller, E.**, DNA hybridization between western aster yellows mycoplasmalike organism plasmids and extrachromosomal DNA from other plant pathogenic mycoplasma-like organisms, *Mol. Plant-Microbe Interact.,* 4, 75, 1991.

33. **Davis, R. E., Lee, I.-M., Dally, E. L., Dewitt, N., and Douglas, S. M.,** Cloned nucleic acid hybridization probes in detection and classification of mycoplasmalike organisms (MLOs), *Acta Hort.,* 234, 115, 1988.

34. **Hiruki, C.,** Rapid and specific detection methods for plant mycoplasmas, in *Mycoplasma Diseases of Crops,* Maramorosch, K. and Raychaudhuri, S. P., Eds., Springer-Verlag, New York, 1988, 51.

35. **Kollar, A., Seemüller, E., Bonnet, F., Saillard, C., and Bove, J. M.,** Isolation of the DNA of various plant pathogenic mycoplasmalike organisms from infected plants, *Phytopathology,* 80, 233, 1990.

36. **Kuske, C. R., Kirkpatrick, B. C., and Seemüller, E.,** Differentiation of virescence MLOs using western aster yellows mycoplasmalike organism chromosomal DNA probes and restriction fragment length polymorphism analysis, *J. Gen. Microbiol.,* 137, 153, 1991.

37. **Jiang, Y. P., Che, T. A., Chiykowski, L. N., and Sinha, R. C.,** Production of monoclonal antibodies to peach eastern X-disease agent and their use in disease detection, *Can. J. Plant Pathol.,* 11, 329, 1989.

38. **Sinha, R. C. and Chiykowski, L.,** Purification and serological detection of mycoplasmalike organism from plants affected by peach eastern X-disease, *Can. J. Plant Pathol.,* 6, 200, 1984.

39. **Kirkpatrick, B. C. and Garrot, D. G.,** Detection of X-disease in plant host by enzyme-linked immunosorbent assay, *Phytopathology,* 74, 825, 1984.

40. **Garnier, M., Iskra, M. L., and Bove, J. M.,** Towards monoclonal antibodies against the apple proliferation mycoplasmalike organism, *Isr. J. Med. Sci.,* 23, 691, 1984.

41. **Evert, R. F.,** The cambium and seasonal development of the phloem in *Pyrus malus, Am. J. Bot.,* 50, 149, 1963.

42. **Evert, R. F.,** Ontogeny and structure of the secondary phloem in *Pyrus malus, Am. J. Bot.,* 50, 8, 1963.

43. **Parthasarathy, M. V.,** Sieve-element structure, in *Encyclopedia of Plant Physiology,* Vol. 1, *Transport in Plants I. Phloem Transport,* Zimmermann, M. H. and Milburn, J. A., Eds., Springer-Verlag, New York, 1975.

44. **Kralik, O. and Brcak, J.,** Invaginations in plant cell protoplasts containing mycoplasma-like bodies (MLOs), *Biol. Plant.,* 17, 214, 1975.

45. **Sears, B. B. and Klomparens, K. L.,** Leaf tip cultures of the evening primrose allow stable, aseptic culture of mycoplasma-like organisms, *Can. J. Plant Pathol.,* 11, 343, 1989.

46. **Seemüller, E., Schaper, U., and Zimbelmann, F.,** Seasonal variation in the colonization patterns of mycoplasmalike organism associated with apple proliferation and pear decline, *Z. Pflanzenkr. Pflanzenschutz,* 91, 371, 1984.

47. **Sinha, R. C. and Chiykowski, L. N.,** Transmission and morphological features of mycoplasmalike bodies associated with peach X-disease, *Can. J. Plant Pathol.,* 2, 119, 1980.

48. **Douglas, S. M.,** Detection of mycoplasmalike organisms in peach and chokecherry with X-disease by fluorescence microscopy, *Phytopathology,* 76, 784, 1986.

49. **Parthasarathy, M. V.,** Mycoplasmalike organisms associated with lethal yellowing disease of palms, *Phytopathology,* 64, 667, 1974.

50. **Esau, K., Magyarosy, A. C., and Breazeale, V.,** Studies of the mycoplasma-like organism (MLO) in spinach leaves affected by the aster yellows disease, *Protoplasma,* 90, 189, 1976.

51. **Soma, K. and Schneider, H.,** Developmental anatomy of major lateral veins of healthy and of peardeclined diseased pear trees, *Hilgardia,* 40, 471, 1971.

52. **Schneider, H.,** Anatomy of buckskin-diseased peach and cherry, *Phytopathology,* 35, 610, 1945.

53. **Schneider, H.,** Anatomy of bud-union bark of pear trees affected by decline, *Phytopathology,* 49, 550, 1959.

54. **Rawline, T. E. and Thomas, H. E.,** The buckskin disease of cherry and other stone fruits, *Phytopathology,* 31, 916, 1941.

55. **Schaper, U. and Seemüller, E.,** Condition of phloem and the persistence of mycoplasmalike organisms associated with apple proliferation and pear decline, *Phytopathology,* 72, 736, 1982.

56. **Braun, E. J. and Sinclair, W. A.,** Histopathology of phloem necrosis in *Ulmus americana, Phytopathology,* 66, 598, 1976.

57. **Schneider, H.,** Sectioning and staining pathological phloem, *Stain Technol.,* 35, 123, 1960.

58. **Esau, K.,** Cytological and histologic symptoms of beet yellows, *Virology,* 10, 73, 1960.

59. **Cousin, M. T.,** L'enroulement violace de la pomme de terre, *Ann. Phytopathol.,* 7, 167, 1975.

60. **Cheadle, V. I., Gifford, E. M., and Esau, K.,** A staining combination for phloem and contiguous tissues, *Stain Technol.,* 28, 49, 1953.

61. **Seemüller, E.,** Demonstration of mycoplasmalike organisms in the phloem of trees with pear decline or proliferation symptoms by fluorescence microscopy?, *Phytopathol. Z.,* 85, 368, 1976.

62. **Petzold, H. and Marwitz, R.,** The gallocyanine-chromalum staining for the detection of mycoplasmalike organisms, *Z. Pflanzenkr. Pflanzenschutz,* 87, 46, 1980.

63. **Giannotti, J., Devauchelle, G., and Vago, C.,** Micro-organismes de type mycoplasme chez une cicadelle et une plante infectees par la phyllodie, *C. R. Acad. Sci. Paris, Ser. D,* 266, 2168, 1968.

64. **Cousin, M. T., Darpoux, H., Faivre-Amiot, A., and Staron, T.,** Sur la presence de micro-organismes de type mycoplasme dons le parenchyme cortical de feverdles presentant der syptomes de virescence, *C. R. Acad. Sci. Paris, Ser. D,* 271, 1182, 1970.

65. **Petzold, H. and Marwitz, R.,** Fluorescence microscopical investigations for the detection of mycoplasma-like organisms with the aid of berberine sulphate, *Z. Pflanzenkr. Pflanzenschutz,* 86, 745, 1979.

66. **Petzold, H. and Marwitz, R.,** The fluorescence microscopical detection of mycoplasmalike organisms by Feulgen reaction with auramine O as Schiffs reagent, *Z. Pflanzenkr. Pflanzenschutz,* 86, 577, 1979.

67. **Petzold, H. and Marwitz, R.,** About the suitability of the fluorescence dye *N,N'*-diethylpseudoisocyanine chloride for the detection of mycoplasma-like organisms, *Z. Pflanzenkr. Pflanzenschutz,* 86, 670, 1979.

68. **Petzold, H. and Marwitz, R.,** An improved fluorescence microscopical detection of mycoplasma-like organisms in plant tissue, *Phytopathol. Z.,* 97, 327, 1980.

69. **Goszdziewski, M. and Petzold, H.,** Fluorescent techniques to determine presence of mycoplasma-like organisms in plants, *Phytopathol. Z.,* 82, 63, 1975.

70. **Marwitz, R. and Petzold, H.,** A simple fluorescence microscopical method for the detection of infections by mycoplasma-like organisms in plants, *Phytopathol. Z.,* 97, 302, 1980.

71. **Behnke, H.-D., Schaper, U., and Seemüller, E.,** Association of mycoplasmalike organisms with pear decline symptoms in the Federal Republic of Germany, *Phytopathol. Z.,* 97, 89, 1980.

72. **Seemüller, E.,** Investigations to demonstrate mycoplasmalike organisms in diseased plants by fluorescence microscopy, *Acta Hort.,* 67, 109, 1976.

73. **Hiruki, C.,** Fluorescence microscopy of yellows diseases associated with plant mycoplasma-like organisms, in *Mycoplasma Diseases of Crops,* Maramorosch, K. and Raychaudhuri, S. P., Eds., Springer-Verlag, New York, 1988, 51.

74. **Samyn, G. and Welvaert, W.,** The use of 4'-6-diamidino-2-phenyl-indole-HCl (DAPI) compared with other stains for a quick diagnosis of mycoplasma infections in ornamental plants, *Meded. Fac. Landbouwwet. Rijksuniv. Gent,* 44, 623, 1979.

75. **deRocha, A., Ohki, S. T., and Hiruki, C.,** Detection of mycoplasmalike organisms *in situ* by indirect immunofluorescence microscopy, *Phytopathology,* 76, 864, 1986.

76. **Lin, C. P. and Chen, T. A.,** Comparison of monoclonal antibodies and polyclonal antibodies in detection of the aster yellows mycoplasma-like organism, *Phytopathology,* 76, 45, 1986.

77. **Mouches, C., Candresse, T., McGarrity, G. J., and Bove, J. M.,** Analysis of spiroplasma proteins: contribution to the taxonomy of group IV spiroplasma protein antigen, *Yale J. Biol. Med.,* 56, 451, 1983.

78. **Gilmer, R. M. and Blodgett, E. C.,** X-disease, in *Virus Disease and Noninfectious Disorders of Stone Fruits in North America,* Handb. No. 437, U.S. Department of Agriculture, Washington, D.C., 1976, 145.

79. **Seemüller, E.,** Pear decline, in *Virus and Viruslike Diseases of Pome Fruits and Simulating Noninfectious Disorders,* Cooperative Extension SP003, Fridlund, P. R., Ed., Washington State University, Pullman, 1989, 188.

80. **Tsao, P. W. and Schneider, H.,** Pathological anatomy of pear tissues sensitive to pear decline virus, *Phytopathology,* 57, 103, 1967.

81. **Hibino, H. and Schneider, H.,** Mycoplasmalike bodies in sieve tubes of pear trees affected with pear decline, *Phytopathology,* 60, 499, 1970.

82. **Kunze, L.,** Apple proliferation, in *Virus and Viruslike Diseases of Pome Fruits and Simulating Noninfectious Disorders,* Cooperative Extension SP003, Fridlund, P. R., Ed., Washington State University, Pullman, 1989, 99.

83. **McIntyre, J. L., Dodds, J. A., Walton, G. S., and Lacy, G. H.,** Declining pear trees in Connecticut: symptoms, distribution, symptom remission by oxytetracycline, and associated mycoplasmalike organisms, *plant Dis. Rep.,* 62, 503, 1978.

84. **Kaloostian, G. H., Hibino, H., and Schneider, H.,** Mycoplasmalike bodies in periwinkle: their cytology and transmission by pear psylla from pear trees affected with pear decline, *Phytopathology,* 61, 1177, 1971.

85. **Schnedier, H.,** Graft transmission and host range of the pear decline causal agent, *Phytopathology,* 60, 204, 1970.

86. **Blodgett, E. C., Schneider, H., and Aichele, M. D.,** Behavior of pear decline disease on different stock-scion combinations, *Phytopathology,* 52, 679, 1962.

87. **Batjer, L. P. and Schneider, H.,** Relation of pear decline to rootstocks and sieve-tube necrosis, *Proc. Am. Soc. Hortic. Sci.,* 76, 85, 1960.

88. **Seemüller, E.,** Colonization patterns of mycoplasma-like organisms in trees affected by apple proliferation and pear decline, in *Tree Mycoplasmas and Mycoplasma Diseases,* Canada, Hiruki, C., Ed., University of Alberta Press, Edmonton, 1988, 179.

89. **Schaper, U. and Seemüller, E.,** Recolonization of the stem of apple proliferation and pear decline-diseased trees by the causal organisms in spring, *Z. Pflanzenkr. Pflanzenschutz,* 91, 608, 1984.

90. **Seemüller, E., Kunze, L., and Schaper, U.,** Colonization behavior of MLO and symptom expression of proliferation-diseased apple trees and decline-diseased pear trees over a period of several years, *Z. Pflanzenkr. Pflanzenschutz,* 91, 525, 1984.

91. **Schaper, U. and Seemüller, E.,** Effects of tetracycline treatments on apple proliferation disease, *Z. Pflanzenkr. Pflanzenschutz,* 89, 641, 1982.

92. **Brunner-Keinath, S. and Seemüller, E.,** Overwintering of the apple proliferation and pear decline causing MLOs at the graft union, *Z. Pflanzenkr. Pflanzenschutz,* 94, 457, 1987.

93. **Seemüller, E., Schaper, U., and Kunze, L.,** Effect of pear decline on pear trees on Quince A and *Pyrus communis* seedling rootstocks, *Z. Pflanzenkr. Pflanzenschutz,* 93, 44, 1986.

94. **Schnedier, H.,** Indicator hosts for pear decline: symptomatology, histology, and distribution of mycoplasmalike organisms in leaf veins, *Phytopathology,* 67, 592, 1977.

95. **Tsao, P. W., Schneider, H., and Kaloostian, G. H.,** A brown leaf-vein symptom associated with greenhouse-grown pear plants infected with pear decline virus, *Plant Dis. Rep.,* 50, 270, 1966.

96. **Giannotti, J. G., Morvan, E., and Vago, C.,** Microorganismes de type mycoplasme dans les cellules liberiennes de *Malus sylvestris* L., atteint de la maladie des proliferations, *C. R. Acad. Sci. Paris, Ser. D,* 276, 76, 1968.

97. **Kartte, S. and Seemüller, E.,** Susceptibility of grafted *Malus* taxa and hybrids to apple proliferation disease, *J. Phytopathol.,* 131, 137, 1991.

98. **Kartte, S. and Seemüller, E.,** Histopathology of apple proliferation in *Malus* taxa and hybrids of different susceptibility, *J. Phytopathol.,* 131, 149, 1991.

99. **Stoddard, E. M.,** The X-disease of peach, *Conn. Agric. Exp. Stn. Circ.,* 122, 54, 1938.

100. **Rawlins, T. E. and Horne, W. T.,** "Buckskin," a destructive graft-infectious disease of cherry, *Phytopathology,* 31, 916, 1931.

101. **Granett, A. L. and Gilmer, R. M.,** Mycoplasmas associated with X-disease in various *Prunus* species, *Phytopathology,* 61, 1036, 1971.

102. **MacBeath, J. H., Myland, G., and Spurr, A. R.,** Morphology of mycoplasmalike bodies associated with peach X-disease in *Prunus persica, Phytopathology,* 64, 935, 1972.

103. **Nasu, S., Jensen, D. D., and Richardson, J.,** Electron microscopy of mysoplasma-like bodies associated with insect and plant hosts of peach western X-disease, *Virology,* 41, 583, 1970.

104. **Chiykowski, L. N. and Sinha, R. C.,** Herbaceous host plants of peach eastern X-disease agent, *Can. J. Plant Pathol.,* 4, 8, 1982.

105. **Gilmer, R. M., Moore, J. D., and Keitt, G. W.,** X-disease virus. I. Host range and pathogenesis in chokecherry, *Phytopathology,* 44, 180, 1954.

106. **Sands, D. C. and Walton, G. S.,** Tetracycline injections for control of eastern X-disease and bacterial spot of peach, *Plant Dis. Rep.,* 59, 573, 1975.

107. **Rosenberger, D. A. and Jones, A. L.,** Symptom remission in X-diseased trees as affected by date, method, and rate of application of oxytetracycline-HCl, *Phytopathology,* 67, 277, 1977.

108. **Rawlins, T. E. and Parker, K. G.,** Influence of rootstocks on the susceptibility of sweet cherry to the buckskin disease, *Phytopathology,* 24, 1029, 1934.

109. **Jones, A. L., Hooper, G. R., and Rosenberger, D. A.,** Association of mycoplasmalike bodies with little peach X-disease, *Phytopathology,* 64, 755, 1974.

110. **Douglas, S. M.,** unpublished data.

111. **Uyemoto, J. K.,** Union aberration of sweet cherry on *Prunus mahaleb* rootstock associated with X-disease, *Plant Dis.,* 73, 899, 1989.

112. **Nutman, F. J. and Roberts, P. M.,** Lethal yellowing: the "unknown disease" of coconut palms in Jamaica, *Emp. J. Exp. Agric.,* 23, 257, 1955.

113. **Plavsic-Banjac, B., Hunt, P., and Maramorosch, K.,** Mycoplasmalike bodies associated with lethal yellowing disease of coconut palms, *Phytopathology,* 62, 298, 1972.

114. **Beakbane, A. B., Slater, C. H. W., and Posnette, A. F.,** Mycoplasmas in the phloem of coconut, *Cocos nucifera* L., with lethal yellowing disease, *J. Hortic. Sci.,* 47, 265, 1972.

115. **Thomas, D. L.,** Mycoplasmalike bodies associated with lethal declines of palms in Florida, *Phytopathology,* 69, 928, 1979.

116. **Deutsch, E. and Nienhaus, F.,** Further studies on the distribution of mycoplasmalike organisms in different tissues of lethal diseased coconut palms in Tanzania, *Z. Pflanzenkr. Pflanzenschutz,* 90, 278, 1983.

117. **McCoy, R. E.,** Remission of lethal yellowing in coconut palms treated with tetracycline antibiotics, *Plant Dis. Rep.,* 56, 1019, 1972.

118. **McCoy, R. E.,** Effect of oxytetracycline dose and stage of disease development on remission of lethal yellowing in coconut palm, *Plant Dis. Rep.,* 59, 717, 19175.

119. **Thomas, D. L.,** Phloem ultrastructure of palms with lethal yellowing, in *Proc. 4th Meet. Int. Counc., Lethal Yellowing,* Thomas, D. L., Howard, F. W., and Donselman, H. M., Eds., Agricultural Research Center IFAS, Fort Lauderdale, FL, 1980, 8.

120. **Pine, T. S. and Gilmore, R. M.,** Peach yellows, in *Virus Diseases and Noninfectious Disorders of Stone Fruits in North America,* Handb. No. 437, U.S. Department of Agriculture, Washington, D.C., 1976, 91.

121. **Jones, A. L., Hooper, G. R., Rosenberger, D. A.,and Chevalier, J.,** Mycoplasmalike bodies associated with peach and periwinkle exhibiting symptoms of peach yellows, *Phytopathology,* 64, 1154, 1974.

122. **Rumbos, I. C. and Bosabalidis, A. M.,** Mycoplasmalike organisms associated with declined plum trees in Greece, *Z. Pflanzenkr. Pflanzenschutz,* 92, 47, 1985.

123. **Agrios, G. N.,** Premature foliation, cambial zone discoloration and stem pitting of peach and apricot in Greece, *Plant Dis. Rep.,* 55, 1049, 1971.

124. **Giunchedi, L., Marani, F., and Credi, R.,** Mycoplasma-like bodies associated with plum decline (leptonecrosis), *Phytopathol. Mediterr.,* 17, 205, 1978.

125. **Rumbos, I. C.,** Control of plum chlorotic leaf roll disease by trunk pressure injection of oxytetracycline-HCl, *Z. Pflanzenkr. Pflanzenschutz,* 92, 581, 1985.

126. **Cropley, R.,** Apple chat fruit, in *Virus and Viruslike Diseases of Pome Fruits and Simulating Noninfectious Disorders,* Cooperative Extension SP003, Fridlund, P. R., Ed., Washington State University, Pullman, 1989, 43.

127. **Sharples, R. O.,** The cell characteristics of apple chat fruit, *J. Hortic. Sci.,* 43, 383, 1968.

128. **Beakbane, A. B., Mishra, M. D., Posnette, A. F., and Slater, C. H. W.,** Mycoplasma-like organisms associated with chat fruit and rubbery wood diseases of apple, *Malus domestica* Barkh., compared with those in strawberry with green petal disease, *J. Gen. Microbiol.,* 66, 55, 1971.

129. **Waterworth, H. E. and Fridlund, P. R.,** Apple rubbery wood, in *Virus and Viruslike Diseases of Pome Fruits and Simulating Noninfectious Disorders,* Cooperative Extension SP003, Fridlund, P. R., Ed., Washington State University, Pullman, 1989, 118.

130. **Scurfield, G. and Bland, D. E.,** The anatomy and chemistry of "rubbery" wood in apple var. Lord Lambourne, *J. Hortic. Sci.,* 38, 297, 1963.

Chapter 13

PATHOLOGICAL ANATOMY OF INFECTION AND SYSTEMIC INVASION OF DECIDUOUS FRUIT TREES BY BACTERIAL PATHOGENS

Isabel M. M. Roos, E. Lucienne Mansvelt, and M. J. Hattingh

TABLE OF CONTENTS

I. INTRODUCTION

Successful infection of fruit trees by plant pathogenic bacteria involves movement of the pathogen to the host, attachment to the host surface, entry, multiplication, and establishment in the host. For a general account of these events, the reader is referred to several publications.[1-9] This chapter describes the pathological anatomy of infection and systemic invasion of fruit trees by bacterial pathogens. Much of the information is based on recent scanning electron microscope (SEM) studies.

This chapter is confined to bacterial species or pathovars of *Pseudomonas, Xanthomonas,* and *Erwinia*. The following pathogens, with hosts in parentheses, are considered: *Pseudomonas syringae* pv. *syringae* (stone and pome fruit), *P.s.* pv. *morsprunorum* (stone fruit), *Xanthomonas campestris* pv. *pruni* (stone fruit), and *Erwinia amylovora* (pome fruit). These pathogens primarily cause necroses, manifested as blossom blast, blights, leaf spots, and cankers on stems and trunks. In addition to attacking tender tissue, these bacteria also exploit comparatively mature tissue. They cause extensive damage, but little or not tissue maceration occurs. Where data and research are lacking on deciduous fruit tree bacterial pathogens, other examples have been cited.

It is generally accepted that plant pathogenic bacteria reach favorable entry portals by chance with splashing water. However, on the host surface, flagellar movement[10] and chemotaxis toward exudates and leachates[11-13] may play an important role. Motility and chemotaxis confer survival advantages to certain pathogens and may increase the infection potential by allowing active entry into infection sites.[10] Cell motility and chemotaxis are particularly important in the early stages of infection.[14] However, neither motility nor chemotaxis is a prerequisite for virulence. Electric charges[15] and surface structures of plants and bacteria[16-18] govern recognition and attachment.

A. FLAGELLAR MOTILITY

Motile strains of bacterial pathogens of fruit trees probably occur much more commonly than do nonmotile strains. The role of bacterial motility in infection has been examined extensively with the peritrichously flagellated *E. amylovora*.[19] *In vitro* studies showed that virulent strains of *E. amylovora* had their full complement of normal-sized flagella after 10 h.[20] In contrast, cells of avirulent strains had only a few short flagella. After 24 h, however, these differences were no longer evident. Cells of virulent strains moved more rapidly than did cells of avirulent strains. Motility probably confers invasive advantages to *E. amylovora*, but as mentioned previously, this trait is not required for virulence. For example, when grown at high temperatures (above 33°C) *E. amylovora* loses motility, yet virulence is retained.[21]

It is difficult to determine whether bacterial cells remain motile once infection occurs. Raymundo and Ries[13] stated that *E. amylovora* is not generally motile inside plant tissue. When inoculated directly into apple shoots, virulence of motile and nonmotile strains of *E. amylovora* did not differ.[22] However, when apple blossoms were spray-inoculated, greater rates of infection were obtained with motile strains. Although motility of bacteria seem to be arrested inside host plants, motile cells probably occur immediately beyond the point of entry and where tissue is highly hydrated. Panopoulos and Schroth[10] suggested that motility is hampered in intercellular spaces because of the absence of free water. More recently, Sigee and El-Masry[23] found that flagellation of *P.s.* pv. *tabaci* in tobacco tissue relates directly to the growth phase of the pathogen. They proposed that flagellation is determined by internal factors rather than by external factors within the leaf environment. The pathological advantage of motile over nonmotile strains is accentuated by intercellular water congestion.

B. CHEMOTAXIS

Infection is more likely to occur if plant pathogenic bacteria follow a chemical gradient.[11] *E. amylovora* is attracted chemotactically to nectar extracts of apple blossoms.[21] In this system, chemotaxis is dependent on temperature (20 to 28°C) and pH (6 to 8).[13] *E. amylovora* is attracted primarily to dicarboxylic organic acids such as succinate, malonate, malate, and fumarate in apple nectar as well as to one amino acid, aspartate. A single chemoreceptor site thus appears to be highly specific for three- and four-carbon dicarboxylic acids.[13]

C. RECOGNITION AND CONTACT

Attachment of pathogens to plants, recognition, and specificity of host-pathogen interactions have been considered in several review articles.[12,24-28] It is postulated that early recognition occurs when the host synthesizes compounds which act as molecular signals, thus enabling invading bacteria to recognize susceptible cells. These plant compounds activate bacterial genes that, together with the host genotype, determine the host/pathogen interaction. Genes that are induced by plant compounds have been identified in a number of plant pathogenic bacteria.[29-32]

It is still uncertain exactly how bacteria adsorb to plant cells. Attachment might be based on specific receptor-ligand interactions[33,34] or on nonspecific interactions operating between the surface of the host and the pathogen. Some strains of *P. syringae* possess proteinaceous pili that act as receptor for the lipid-containing bacteriophage φ6.[35] Pili are also involved in the adsorption of bacterial cells to the surface of host plants. Rantala and Romantschuk[35] found that the degree of piliation correlated with the ability of *P. syringae* to adhere to the host plant surface.

Present knowledge of the interaction at the molecular level between bacterial pathogens and fruit trees is based almost exclusively on the *E. amylovora*-apple system. Contact with host cells is required for expression of virulence. The presence of *E. amylovora* in the intercellular spaces of apple tissue leads to electrolyte leakage and death of host cells. Strains of *E. amylovora* that do not produce extracellular polysaccharide (EPS) are avirulent.[36] EPS is therefore required for pathogenicity,[36] but these compounds are believed not to be toxic to plant cells.[37,38] However, Goodman and associates[39-42] indicated that the EPS of *E. amylovora* acts as a host-specific phytotoxin. Beer's group[37,38] questioned this hypothesis and suggested that the putative necrotoxin is an artifact, ascribed to the activity of inorganic salts concentrated during the extraction procedure.

The precise role that bacterial EPS plays in the recognition phase of pathogen establishment within the host is not well known. EPS of some pathogenic pseudomonads causes water soaking in susceptible, but not in resistant, plants.[43-46] The function and mode of action of the outer surface components (EPS and capsule) of *E. amylovora* have received considerable attention in recent years.[36,47-50] The current concept is that capsular EPS protects the bacterial cell against the defense reactions of host cells.

D. PREDISPOSING FACTORS: MOISTURE AND TEMPERATURE

Cells of saprophytic and pathogenic bacteria are dispersed during rainstorms, and arrive on leaves or flowers as a result of splashing. Under these conditions, one may expect that much of the inoculum would be washed off; however, bacteria can effectively attach to host surfaces,[27,51] probably by fibrils resembling the cellulose fibrils produced by *Agrobacterium tumefaciens* on carrot tissue.[52] Fibrils should not be confused with pili. The latter are more likely to be involved in specific adsorption to host cells, rather than with general attachment.[35]

Free water on plant surfaces provides a migratory pathway for bacterial pathogens. Access is gained to substrate and if tissue is water soaked, the endogenous defense mechanism may be compromised. Leben and co-workers[53-55] showed that as a prelude to infection, surface moisture permits the establishment of significant epiphytic populations of bacterial pathogens on plant leaves. Infection requires a high relative humidity in the intercellular spaces of host tissue. Shaw[56] reported that multiplication of *E. amylovora* drops drastically if the relative humidity in intercellular spaces falls slightly below 100%. Under experimental conditions, growth was totally suppressed at 98% relative humidity.

Apart from moisture and nutrients, bacteria require a suitable temperature range for growth and infection. According to Billing,[57,58] temperature is more important than rainfall for primary infection of apple blossoms in spring by *E. amylovora*.

Other predisposing factors involved in pathogenesis are not considered separately in this chapter.

III. ENTRY INTO HOST TISSUE

Pathogenic bacteria are incapable of penetrating intact plant surfaces. Entry, therefore, is restricted to natural openings and wounds.

A. STOMATA AND TRICHOMES

Erwin F. Smith's[59] pioneering work recognized the importance of stomata as points of entry for plant pathogenic bacteria. Stomata occur primarily on leaves and shoots of plants. Each stoma consists of a pair of guard cells with a pore between them.[60] The epidermal cells adjacent to the guard cells frequently differentiate into subsidiary cells. The space formed by the guard cells, the subsidiary cells, and the neighboring cells constitutes the substomatal cavity.

Bacteria most readily enter stomata of young, unfolding, and expanding leaves near shoot tips. *P.s.* pv. *syringae* (on apple and plum),[61,62] *P.s.* pv. *morsprunorum* (on cherry),[63] and *X.c.* pv. *pruni* (on peach)[64] probably enter their respective hosts in the same way. Immediately after inoculation, bacterial cells are randomly dispersed over the leaf surface. Most fail to survive and multiply after inoculation, but the bacteria that do are located on or near stomata. When uninvaded substomatal chambers are reached, bacteria multiply profusely, and masses of these cells are subsequently extruded through stomata. Substomatal chambers are envisaged as important protected sites for establishment of the pathogens during the early phase of the infection process.

Surprisingly, pear stomata seem to be unimportant sites of infection. Instead, *P.s.* pv. *syringae* infects through the open bases of damaged trichomes and possible also through microscopic fissures in depressions of the cuticle.[65]

B. BLOSSOMS

On some plant species, nectar is excreted from flowers through stomata-like nectar-thodes.[60] Nectarthodes on apple, pear, and hawthorn flowers occur in the tissue located between the points of emergence of the styles and stamens.[66,67,75] On pear and hawthorn, this tissue has an open, shallow, fully exposed, saucer-shaped surface. In contrast, the nectarial region on apple flowers supports a ring of stamens and numerous stylar hair-like trichomes and is thus not exposed. Nectarthodes have two guard cells, but these cells do not regulate aperture opening.[60] Nectarthodes and stomata are therefore not directly comparable. A well-developed cuticle shield the surface of nectarial tissue. But nectarthodes *per se* are not covered. Nectar is produced in a zone of tissue 12 to 15 cell layers deep.

In pear, the nectary appears to be the most important site of entry for *E. amylovora*.[66,68,69] Rosen[66] found vast numbers of the pathogen on the nectarial surface 24 to 48 h after inoculating open, nectar-secreting pear blossoms. Bacterial proliferation was generally associated with droplets of nectar. Masses of bacteria were traced from the nectarthodes to underlying tissue. This implies that bacteria had passed through the pores between the guard cells. Invasion of apple nectarial tissue occurs in the same way, but less frequently because their nectaries are well protected.[66] On hawthorn flowers, nectaries are sites for infection only during humid periods. At other times, these nectaries are dry and nectar is absent.[67]

The glandular-like surface of the stigma consists of loose, large, thin-walled cells.[66] These cells may be covered by a cuticle, depending on the fruit cultivar. It is still unclear how infection of apple and pear stigmas by *E. amylovora* leads to stylar invasion and blossom blight, as envisaged by Hildebrand.[68] Rosen[66] and Rundle and Beer[70] found sizeable populations of *E. amylovora* in the lower style of apple after inoculation of stigmas. By contrast, Thomson and associates[71,72] questioned whether infection of stigmas by *E. amylovora* leads to blossom invasion. It was suggested that dew and rain, instead of invasion of stylar tissue, disperse the pathogen to other parts of the flower. Similarly, in hawthorn, *E. amylovora* is believed to be carried externally down the style in stigmatic secretions to infect nectarial tissue.[73]

E. amylovora enters apple anthers and multiplies in the locules.[66] From there, the pathogen makes its way through the intercellular spaces in the filaments, and finally reaches nectary tissue. In hawthorn, the anther locule is invaded via the ruptured dehiscence zone, and possibly also through stomata surrounding the filament insertion.[74] The surface of pollen from infected anthers may contain high numbers of bacteria.[75]

P.s. pv. *syringae* infects blossoms of some apple and pear cultivars more readily than those of other fruit trees. Inherent biochemical factors and morphological properties determine whether infection will occur and how much damage follows successful infection. Although the stigmatic surface of both apple and pear flowers are colonized,[61,75] the tight, circular arrangement of the stamens and the stylar trichomes associated with apple blossoms prevent bacteria from contacting the hypanthium. The open nectaries on pear blossoms are extensively colonized through nectarthodes and nectar-secreting glandular-like trichomes. *P.s.* pv. *syringae* establishes on all parts of the pear flower and also can be associated with pollen.[75,76]

C. WOUNDS

Wounds due to wind, hail, insects, nematodes, and birds[1] allow bacteria to gain direct access to xylem vessels and intercellular spaces. Pruning cuts and damage sustained in orchards during cultivation or harvest might even be more important.[77] Trees are especially vulnerable if damage and water congestion coincide during wet spells. Wind-driven rain disperses inoculum and aids water congestion, whereas strong wind or hail causes wounding. Torn leaves with an exposed vascular system are readily infected.[78] Leakage from wounds enables some bacteria to follow a chemical gradient to the site of injury.

Crosse et al.[78] found that as apple leaves mature, they become more resistant to *E. amylovora* introduced through wounds. They also reported that the percentage of shoots infected increased as inoculum dose (ID) increased. The ID_{50} could in turn be manipulated by altering either the inoculum concentration or the length of time the wound is exposed to inoculum. Infectivity of a wound is thus governed by the number of target cells exposed, number of bacterial cells present, and exposure time of the host cell to the pathogen cell.

1. Frost Injury

Freezing temperatures have a decisive effect on the development of diebacks caused by pseudomonads.[79-84] A decreased sugar content of host tissue[85] or bacterial ice nucleation activity favors the development of dieback symptoms on apricot trees.[86,87] Frost injury due to the presence of ice nucleation activity (INA) *P. syringae* in turn provides avenues for entry of the pathogen to susceptible host tissue.[80,83,88,89]

Süle and Seemüller[90] questioned the hypothesis that infection results from microwounds or plasma membrane damage caused by freezing. Instead, they postulated that INA *P.s.* pv. *syringae* is passively drawn into intercellular spaces via stomata on leaves as a result of a vacuum that develops in the intercellular spaces following air displacement during the process of freezing. More recently, Vigouroux[91] presented evidence that water soaking of tissues caused by freezing and thawing of leaf tissue promotes entry and spread of *Pseudomonas* in fruit trees. Under these conditions, entry of the pathogen appears to be mainly a physical process. This may also explain why winter pruning, especially if done before frost, can lead to a high incidence of infection. Thawing of water-congested, frozen tissue may lead to the resurgence of latent, internal populations of the pathogen[91] or to expansion of existing cankers.[82]

2. Leaf Scars

In England, most cankers caused by *P.s.* pv. *morsprunorum* on cherry trees are believed to originate through leaf scars on fruiting spurs and on extension shoots during autumn.[92] The pathogen is drawn into xylem vessels of the leaf traces by negative pressure and eventually migrates into medullary rays and other living tissue. This type of infection occurs more readily if premature leaf fall exposes a scar without a fully developed abscission layer.[93] The authors[62,94] have been unable to substantiate these findings under South African conditions in which *P.s.* pv. *morsprunorum* or *P.s.* pv. *syringae* probably reaches axillary buds by systemic spread well before leaf fall occurs. Cankers subsequently appear at the base of invaded buds. This also seems to apply for *X.c.* pv. *pruni* on plums.[95,96] It should be emphasized that forcible removal of leaves results in fresh wounds and smooth passage of bacteria to exposed tissue is ensured.[97,98] For example, "unnatural" premature leaf drop during the course of a violent rainstorm could permit bacteria to gain access to xylem cells. By contrast, where suberized protective layers are formed well before normal leaf fall occurs, entry of bacteria is hampered.

IV. ESTABLISHMENT AND COLONIZATION

A. INOCULUM DENSITY

A single bacterial cell can cause infection[78,99] provided conditions for survival and multiplication in the susceptible host are optimal. Neither the method of inoculation nor the inoculum concentration seems to influence the general pattern of tissue colonization by *P. syringae*. Nevertheless, invasion is accelerated if leaves are wounded or if tissue is allowed to become water soaked before being exposed to high inoculum levels of the pathogen.[62,100]

After having entered the tree, extensive multiplication of bacterial pathogens occurs intercellularly, as well as in xylem and phloem cells.[4,42,61-63,94,101-103]

B. INTERCELLULAR SPREAD

SEM is a useful tool to study intercellular spread of bacterial pathogens.[42,61-63,65,75,78,94,95,103,104] We have concentrated part of our investigation on the systemic spread of *P.s.* pv. *syringae* and *P.s.* pv. *morsprunorum* in plum and cherry tissue, respectively. These two host-pathogen systems appear to operate in much the same way. The pathogens probably spread intercellulary through mesophyll (Figure 1) and storage parenchyma cells (Figure 2). Invasion is initially limited to host tissue near the point of inoculation. Once a vein is invaded, infection is no longer localized. Bacteria presumably progress intercellularly through the parenchyma of the bundle sheath into phloem and xylem tissue. We suspect that the physical pressure exerted by the expanding masses of bacteria in the

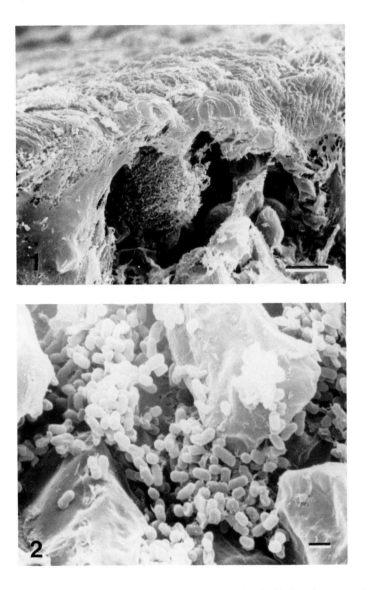

FIGURES 1 and 2. SEMs of cross-sections of cherry leaves inoculated with *Pseudomonas syringae* pv. *morsprunorum* showing intercellular spread through substomatal chamber (Figure 1) and in the spongy parenchyma of the mesophyll (Figure 2). Bars = 10 μm (Figure 1) and μm (Figure 2). (From Roos, I. M. M. and Hattingh, M. J., *Phytopathology*, 77, 1246, 1987. With permission.)

intercellular spaces ruptures parenchyma cells. Intercellular bacterial proliferation is probably fostered by the leakage of both inorganic and organic nutritional substrates.[4]

C. INTRACELLULAR SPREAD

A number of SEM studies indicate that xylem vessels are the most important initial sites of significant intracellular proliferation of *E. amylovora*,[40,42] *X.c.* pv. *pruni*,[95,96,104] *P.s.* pv. *morsprunorum*,[94,103] and *P.s.* pv. *syringae*.[62] In most of these studies, the pathogens gained entry through wounds which had exposed the vascular elements (Figure 3). The bacteria then spread laterally and distally away from the xylem vessels that were colonized first. Long-distance transport of the pathogens occurs in these vessels, but it is uncertain how the xylem lumen is entered. Nelson and Dickey[105] proposed that *P.s.* pv. *caryophylli* passes

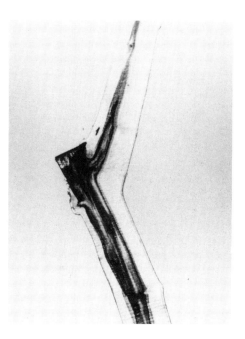

FIGURE 3.   Bacterial canker caused by *Pseudomonas syringae* pv. *syringae* on plum tree. Wood discoloration indicates pathogen entrance via exposed vascular elements of a pruning wound. (From Hattingh, M. J., Roos, I. M. M, and Mansvelt, E. L., *Plant Dis.*, 73, 784, 1989. With permission.)

from an infected vessel member of carnation to adjacent xylem parenchyma through the pit membrane, via the plasmodesmata, to initiate the development of bacterial pockets. Entry of *P.s.* pv. *morsprunorum* into the lumen of xylem vessels might occur readily if pit membranes are ruptured or deteriorated. Alternatively, pressure exerted by bacterial masses on the pit membrane could increase the diameter of plasmodesmata.[94]

V. SYSTEMIC INVASION OF HOST TISSUE

A. LEAVES

After entry through stomata, bacteria can multiply profusely in substomatal chambers, and masses of these cells are extruded through stomata to replenish epiphytic populations on the leaf surface.[61-64,94] The intercellular spaces of the spongy parenchyma are colonized concurrently. The pathogens probably move through the parenchyma of the bundle sheath into the vascular system of a minor vein and from there to the main vein. Apart from colonizing the xylem, bacteria also seem to enter the phloem during the initial stages of pathogenesis. It is unlikely the pathogens are transported passively through the xylem. Although single bacterial cells occur commonly in vessels, large clumps resembling microcolonies have often been seen distant from the site of inoculation. Upon entering a main vein, aggressive strains of the pathogens are virtually assured of passage to axillary buds and to twigs supporting leaves. This promotes long-term survival of the pathogen. In one of our experiments,[94] water-soaked, infected veins (Figure 4) developed on new cherry leaves arising from axillary buds of leaves sprayed the previous spring with a suspension of *P.s.* pv. *morsprunorum*. This agrees with reports on the migration of *P.s.* pv. *syringae*[62] and *X.c.* pv. *pruni* in plum shoots,[96] *P.s.* pv. *morsprunorum* in cherry shoots,[94] and *E. amylovora* in apple shoots.[42,101,102,106]

The authors[62] found that an aggressive strain of *P.s.* pv. *syringae* inoculated into plum

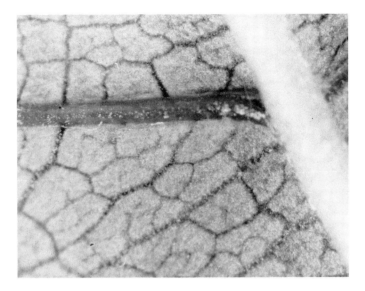

FIGURE 4. A water-soaked vein on a cherry leaf inoculated with *Pseudomonas syringae* pv. *morsprunorum*. (From Hattingh, M. J., Roos, I. M. M., and Mansvelt, E. L., *Plant Dis.*, 73, 784, 1989. With permission.)

petioles in the spring also spread to the xylem and other elements of leaf veins and shoots. Pockets of bacteria occurred in invaded, but symptomless, stems (Figure 5). When present in petioles, leaf blades, and shoots, bacteria were usually found in the spiral or helical vessels of the protoxylem, in scalariform and pitted vessels of the metaxylem, and in the bordering xylem parenchyma (Figures 6 and 7). Bacteria were seen less frequently and in fewer numbers in the phloem and cortex of petioles and shoots. Lateral spread of *P.s.* pv. *syringae* from major xylem vessels into adjacent tissue was more pronounced in leaf veins than in petioles or shoots. Intercellular spread probably occurred in the mesophyll, allowing the pathogen to invade uncolonized minor veins (Figure 8). The pathogen multiplied within the leaf tissue and was exuded from stomata of these invaded leaves. In orchards, internal migration of the pathogen from stems and shoots to leaves may compensate for epiphytic populations lost during unfavorable conditions.

Granular material (Figure 9) found in veins could be composed of gums or gels secreted by the host in response to the pathogen.[62,94] A connecting fibrillar network of bacterial origin might anchor microcolonies to the host cells wall. This agrees somewhat with the situation in xylem vessels of grape petioles invaded by *Xylella fastidiosa*, the causal agent of Pierce's disease.[107] The fibrils seen in plum,[62] especially in xylem vessels (Figure 9), also resemble the cellulose fibrils produced by *A. tumefaciens* during attachment to carrot cells.[52]

The authors[108] investigated systemic spread and pathogenicity of different strains of *P.s.* pv. *syringae* in several plum and apple cultivars. A standardized bacterial suspension was injected into the internode immediately above the petiole of the fifth fully expanded leaf of a vegetative shoot in spring. The extent of shoot invasion in each host/pathogen combination was determined after 2 months. Cells of *P.s.* pv. *syringae* generally spread much further from the site of introduction in plum than in apple shoots. In addition, most strains caused necrotic lesions exceeding 1% of the plum shoot length, whereas lesions in apple shoots were inconspicuous or absent. This strengthens the view that apple is an inhospitable host for *P.s.* pv. *syringae* and that disease is unlikely to develop in apple orchards unless trees are severely stressed or cultural practices favor dissemination and establishment of the pathogen.[77]

Manifestation of host/pathogen interactions on plum cultivars appears to be more varied.

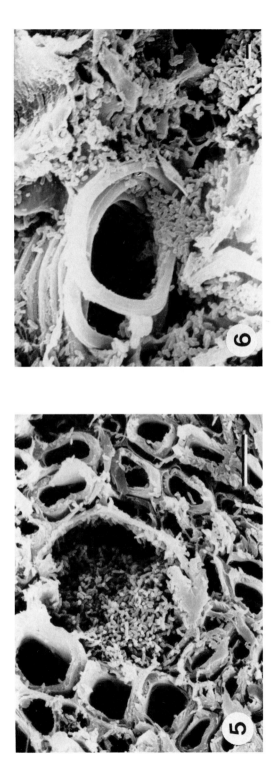

FIGURES 5 to 8. SEMs showing *Pseudomonas syringae* in xylem tissue of plum and cherry trees. **Figure 5.** A systemically invaded symptomless plum stem. (From Hattingh, M. J., Roos, I. M. M., and Mansvelt, E. L., *Plant Dis.*, 73, 784, 1989. With permission.) **Figure 6.** In petioles, leaves, and shoots the bacteria occur in spiral vessels of the protoxylem, and **(Figure 7)** in sclalariform and pitted vessels of the metaxylem and bordering xylem parenchyma. (Figure 6 from Roos, I. M. M. and Hattingh, M. J., *J. Phytopathol. (Berlin)*, 121, 26, 1988. With permission. Figure 7 from Roos, I. M. M. and Hattingh, M. J., *Phytopathology*, 77, 1253, 1987. With permission.) **Figure 8.** Invasion of uncolonized minor veins in leaf tissue via intercellular spread through mesophyll. Bars = 10 μm (Figure 5), 2 μm (Figure 6), and 10 μm (Figure 7).

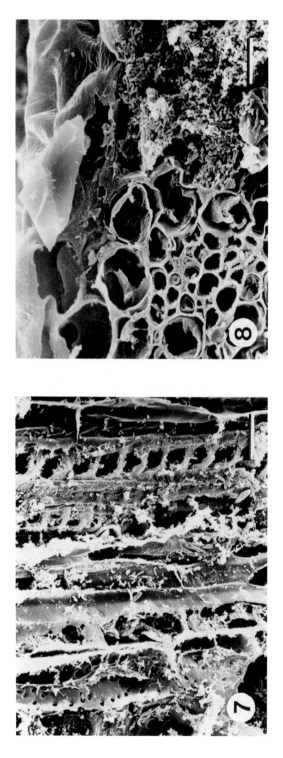

FIGURES 7 and 8.

FIGURE 9. SEM of *Pseudomonas syringae* pv. *morsprunorum* in a cherry vein. Bacteria are embedded in granular material, possibly of plant origin. Note fibrils (arrows) interlacing the bacterial cells. Bars = 2 μm. (From Roos, I. M. M. and Hattingh, M. J., *Phytopathology*, 77, 1246, 1987. With permission.)

Four extremes are mentioned, but many intermediate reactions can occur: (1) no growth of the pathogen and no symptoms; (2) systemic spread but no symptoms; (3) pathogen confined to the lesion (canker); and (4) extensive spread of pathogen beyond the canker. In common with other studies of bacterial diseases on deciduous fruit trees,[109,110] the nature of the response depends on both the bacterial strain and the host cultivar. We concluded that differential interactions occur between strains of *P.s.* pv. *syringae* and plum cultivars and probably also other stone fruit cultivars. Our results are consistent with the general assumption[92] that populations of *P.s.* pv. *syringae* in cankers on deciduous fruit trees decline during summer. However, the rate of decline and the level of the population inside the tree at the end of summer seem to depend on the particular host-pathogen interaction.

B. BLOSSOMS, FRUIT, AND SEED

Cherry blossoms inoculated with an aggressive strain of *P.s.* pv. *morsprunorum* are usually killed.[103] Surviving infected blossoms give rise to infected fruit that have typical dark, sunken necrotic lesions at or near the distal ends. Although these lesions were localized, SEM indicated that bacteria, presumed to have infected the flowers, had invaded the entire pericarp and fruit stalk. High numbers of the pathogen were detected in infected fruit and stalks by dilution plating. The bacteria also spread to adjacent spurs.

The path of entry of the bacteria from inoculated cherry blossoms to developing fruit is not known. Since lesions usually appeared at the blossom end of fruits, the authors hypothesized that the pathogen colonized the stigma and then moved to the developing ovule through the style. The route of invasion followed in the developing fruit was inferred from the cell concentration of the pathogen observed in different tissues, the delayed appearance of spots, and the quantitative results obtained by dilution plating. This would agree with the initial stages of blossom infection by *E. amylovora* of pome fruit trees.[71,111] We have seen *P.s.* pv. *syringae* within styles of apple, but have no direct evidence that *P.s.* pv. *morsprunorum* occurs in stylar tissue of cherry.

After gaining entry through the floral parts on sweet cherry, *P.s.* pv. *morsprunorum* probably multiplies intercellularly for some time in the epidermal and parenchymal tissue of the mesocarp without causing visible damage.[103,108] From there, the pathogen spreads through the rest of the mesocarp and enters the endocarp, fruit stalk, and spur. Large accumulations of bacterial cells were found to be scattered in the intercellular spaces of the pith of the stalk. Much greater numbers were consistently seen in the xylem and phloem vessels. Vascular bundles throughout the stalk were invaded, but bacterial numbers decreased toward the spur. The presence of bacteria in the pits (Figure 10) of the metaxylem suggested lateral movement within the vascular system. Large numbers of *P.s.* pv. *morsprunorum* found in the pericarp of developing fruit as well as in the stony endocarp (Figure 11) led us to believe that the pathogen might be transmitted to seed. The authors[112] subsequently found proof for this, but with *P.s.* pv. *syringae* in peach. Du Plessis[104] also has shown that *X.c.* pv. *pruni* systemically invades plum seeds and fruit through the vascular system of fruit stalks.

In conclusion, strong evidence exists that *P.s.* pv. *syringae, P.s.* pv. *morsprunorum X. c.* pv. *pruni*, and *E. amylovora* spread systemically in shoots, roots, flowers, fruit, and other tissue of deciduous fruit trees without causing symptoms.[101,113] Latent infections probably occur in most of the major fruit-producing regions of the world. Disease development is likely to be triggered by stimuli that alter host or pathogen metabolism. Most researchers regard *P. syringae* as a weak pathogen that causes disease only when the host is stressed. However, the pathogen can aggressively attack weakened trees and is an excellent opportunist by virtue of its ability to colonize the foliar surface and then to spread systemically through the tree. Some of the stress factors that have been recognized in the U.S. and elsewhere include freeze damage, wounds, nematode damage, waterlogging, and dual infections of *P.s.* pv. *syringae* and plant pathogenic fungi such as *Cytospora* and *Nectria*.[108]

C. SYMPTOM DEVELOPMENT

The association between symptom development on fruit trees and systemic invasion by a bacterial pathogen was eloquently demonstrated by Du Plessis[96,104,114] Typical necrotic lesions developed on the main and secondary veins of plum leaves, following systemic migration of *X.c.* pv. *pruni* as far away as 13 cm from the point of inoculation. This agrees with the situation in plum orchards, where necrotic lesions are often situated along main and secondary veins. Lesions on plum fruit extended from the exocarp to the endocarp, and were apparently caused by the pathogen that had moved systemically from leaves and shoots via fruit stalks. Both spring and summer cankers on plums were attributed to systemic movement of *X.c.* pv. *pruni*.

VI. CONCLUSIONS

The most important bacterial pathogens of deciduous fruit trees that cause necroses enter their hosts through natural openings and wounds. They spread through trees systemically and establish a resident phase in/on leaves, shoots, flowers, fruit, and even roots.[62,78,94,96,101,102,106,108,113,114] Trees with latent infections probably occur in most of the fruit-producing regions of the world. The bacterial pathogens have a wide host range, infecting stone and pome fruit trees, as well as ornamentals. The subtle nature of the diseases and the excellent ability of the pathogens to survive on the surface and inside their hosts explain why programs to control bacterial diseases often have been ineffective. The life cycle proposed by Hattingh et al.[108] for bacterial canker of stone fruit illustrates the ramification of resident bacteria inside symptomless tissue. Systemic invasion is of particular concern when one considers that deciduous fruit trees are propagated by grafting vegetative material onto rootstocks.

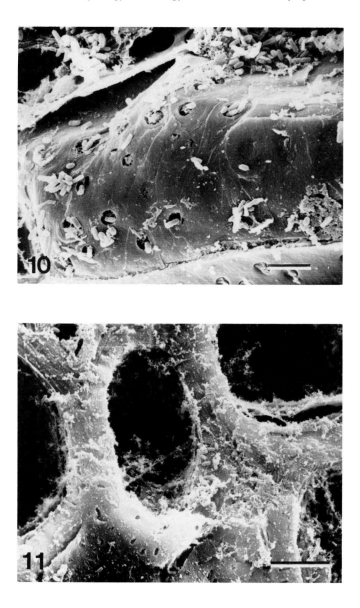

FIGURES 10 and 11.   SEMs of *Pseudomonas syringae* pv. *morsprunorum* in immature cherry fruit. Bacteria in pits of the metaxylem (**Figure 10**), and in the pericarp and stony endorcarp of fruit (**Figure 11**) suggest seed transmission of the pathogen. Bars = 5 μm (Figure 10) and 10 μm (Figure 11). (From Roos, I. M. M. and Hattingh, M. J., *J. Phytopathol. (Berlin)*, 121, 26, 1988. With permission.)

Effective management of bacterial diseases of deciduous fruit trees will in the future rely heavily on selection and breeding for disease resistance. At the same time, factors that tend to aggravate disease need to be neutralized. To achieve this, more information is required on the nature of the host/pathogen interactions. Molecular genetics will be used to determine how genes on chromosomes or plasmids confer pathogenesis and virulence traits.[115] If more becomes known of the genetics of the respective pathogens, development of effective biological control agents or modification of the host to incorporate resistance may become feasible.

REFERENCES

1. **Billing, E.,** Entry and establishment of pathogenic bacteria in plant tissues, in *Bacteria and Plants,* Rhodes-Roberts, M. E. and Skinner, F. A., Eds., Academic Press, London, 1982, 51.
2. **Chatterjee, A. K. and Starr, M. P.,** Genetics of *Erwinia* species, *Annu. Rev. Microbiol.,* 34, 645, 1980.
3. **Goodman, R. N.,** Physiological and cytological aspects of the bacterial infection process, in *Physiological Plant Pathology (Encyclopedia of Plant Physiology),* Vol. 4, Heitefuss, R. and Williams, P. H., Eds., Springer-Verlag, Berlin 1976, 172.
4. **Goodman, R. N.,** The infection process, in *Phytopathogenic Prokaryotes,* Vol. 1, Mount, M. S. and Lacy, G. H., Eds., Academic Press, New York, 1982, 31.
5. **Goodman, R. N., Kiraly, Z., and Wood, K. R.,** *The Biochemistry and Physiology of Plant Disease,* University of Missouri Press, Columbia, 1986, 1.
6. **Hirano, S. S. and Upper, C. D.,** Ecology and epidemiology of foliar bacterial plant pathogens, *Annu. Rev. Phytopathol.,* 21, 243, 1983.
7. **Huang, J. S.,** Ultrastructure of bacterial penetration in plants, *Annu. Rev. Phytopathol.,* 24, 141, 1986.
8. **Kelman, A.,** How bacteria induce disease, in *Plant Disease — An advanced Treatise,* Vol. 4, Horsfall, J. G. and Cowling, E. B., Eds., Academic Press, New York, 1979, 181.
9. **Singh, R. S. and Singh, U. S.,** Pathogenesis and host-parasite specificity in plant diseases, in *Experimental and Conceptual Plant Pathology,* Vol. 2, Hess, W. M., Singh, R. S., Singh, U. S., and Weber, D. J., Eds., Gordon & Breach, Basel, 1988, 140.
10. **Panopoulos, N. J. and Schroth, M. N.,** Role of flagellar motility in the invasion of bean leaves by *Pseudomonas phaseolicola, Phytopathology,* 64, 1389, 1974.
11. **Chet, I., Zilberstein, Y., and Henis, Y.,** Chemotaxis of *Pseudomonas lachrymans* to plant extract and to water droplets collected from the leaf surfaces of resistant and susceptible plants, *Physiol. Plant Pathol.,* 3, 473, 1973.
12. **Gabriel, D. W. and Rolfe, B. G.,** Working models of specific recognition in plant-microbe interactions, *Annu. Rev. Phytopathol.,* 28, 365, 1990.
13. **Raymundo, A. K. and Ries, S. M.,** Chemotaxis of *Erwinia amylovora, Phytopathology,* 70, 1066, 1980.
14. **Klopmeyer, M. J. and Ries, S. M.,** Motility and chemotaxis of *Erwinia herbicola* and its effect on *Erwinia amylovora, Phytopathology,* 77, 909, 1987.
15. **Marshall, K. C.,** Electrophoretic properties of fast- and slow-growing species of *Rhizobium, Aust. J. Biol. Sci.,* 20, 429, 1967.
16. **Graham, T. L. and Sequeira, L.,** Interaction between plant lectins and cell wall components of *Pseudomonas solanacearum*: role in pathogenicity and induced disease-resistance, in *Cell Wall Biochemistry Related to Specificity in Host-Plant Pathogen Interactions,* Solheim, B. and Raa, J., Eds., Universitetsforlaget, Lorentzens Offsettrykkeri, 1977, 417.
17. **Lippincott, B. B. and Lippincott, J. A.,** Bacterial attachment to a specific wound site as an essential stage in tumor initiation by *Agrobacterium tumefaciens, J. Bacteriol.,* 97, 620, 1969.
18. **Sequeira, L.,** Lectins and their role in host-pathogen specificity, *Annu. Rev. Phytopathol.,* 16, 453, 1978.
19. **Huang, P.-Y. and Goodman, R. N.,** Morphology and ultrastructure of normal rod-shaped and filamentous forms of *Erwinia amylovora, J. Bacteriol.,* 102, 862, 1970.
20. **Huang, P.-Y.,** Ultrastructural Modification by and Pathogenicity of *Erwinia amylovora* in Apple Tissues, Ph.D. thesis, University of Missouri, Columbia, 1974.
21. **Raymundo, A. K. and Ries, S. M.,** Motility of *Erwinia amylovora, Phytopathology,* 71, 45, 1981.
22. **Bayot, R. G. and Ries, S. M.,** Role of motility in apple blossom infection by *Erwinia amylovora* and studies of fire blight control with attractant and repellent compounds, *Phytopathology,* 76, 441, 1986.
23. **Sigee, D. C. and El-Masry, M. H.,** Changes in cell size and flagellation in the phytopathogen *Pseudomonas syringae* pv. *tabaci* cultured *in vitro* and *in planta*: a comparative electron microscope study, *J. Phytopathol. (Berlin),* 125, 217, 1989.
24. **Daniels, M. J., Dow, J. M., and Osbourn, A. E.,** Molecular genetics of pathogenicity in phytopathogenic bacteria, *Annu. Rev. Phytopathol.,* 26, 285, 1988.
25. **Dixon, R. A. and Lamb, C. J.,** Molecular communication in interactions between plants and microbial pathogens, *Annu. Rev. Plant Physiol. Plant Mol. Biol.,* 41, 339, 1990.
26. **Lippincott, J. A. and Lippincott, B. B.,** Concepts and experimental approaches in host-microbe recognition, in *Plant-Microbe Interactions. Molecular and Genetic Perspectives,* Vol. 1, Kosuge, T. and Nester, E. W., Eds., Macmillan, New York, 1984, 195.
27. **Pueppke, S. G.,** Adsorption of bacteria to plant surfaces, in *Plant-Microbe Interactions. Molecular and Genetic Perspectives,* Vol. 1, Kosuge, T. and Nester, E. W., Eds., Macmillan, New York, 1984, 215.
28. **Sequeira, L.,** Recognition and specificity between plants and pathogens, in *Challenging Problems in Plant Health,* Kommedahl, T. and Williams, P. H., Eds., American Phytopathological Society, St. Paul, MN, 1983, 301.

29. **Beaulieu, C. and Van Gijsegem, F.,** Identification of plant-inducible genes in *Erwinia chrysanthemi* 3937, *J. Bacteriol.,* 172, 1569, 1990.

30. **Djordjevic, M. A., Redmond, J. W., Batley, M., and Wolfe, B. G.,** Clovers secrete specific phenolic compounds which either stimulate or repress *nod* gene expression in *Rhizobium trifolii, EMBO J.,* 6, 1173, 1987.

31. **Enard, C., Diolez, A., and Expert, D.,** Systemic virulence of *Erwinia chrysanthemi* 3937 requires a functional iron assimilation system, *J. Bacteriol.,* 170, 2419, 1988.

32. **Stadel, S. E., Nester, E., and Zambryski, P.,** A cell plant factor induced *Agrobacterium tumefaciens vir* gene expression, *Proc. Natl. Acad. Sci. U.S.A.,* 83, 379, 1986.

33. **Dazzo, F. B., Truchet, G. L., Sherwood, J. E., Hrabak, E. M., Abe, M., and Pankratz, S. H.,** Specific phases of root hair attachment in the *Rhizobium trifolii-* clover symbiosis, *Appl. Environ. Microbiol.,* 48, 1140, 1984.

34. **De Weger, L. A., Van Loosdrecht, M. C. M., Klaassen, H. E., and Lugtenberg, B.,** Mutational changes in physicochemical cell surface properties of plant-growth-stimulating *Pseudomonas* spp. do not influence the attachment properties of the cells, *J. Bacteriol.,* 171, 2756, 1989.

35. **Rantala, E. and Romantschuk, M.,** Pilus-Expression of *Pseudomonas syringae.* The Role of Piliation in Plant-Microbe Interaction, and Characterization of the Pilus-Specific Genes, Paper P84, 5th Int. Symp. Molecular Genetics of Plant-Microbe Interactions, Interlaken, Switzerland, September 9 to 14, 1990.

36. **Ayers, A. R., Ayers, S. B., and Goodman, R. N.,** Extracellular polysaccharide of *Erwinia amylovora:* a correlation with virulence, *Appl. Environ. Microbiol.,* 38, 659, 1979.

37. **Bauer, D. W. and Beer, S. V.,** Evidence that a putative necrotoxin of *Erwinia amylovora* is an artifact caused by the activity of inorganic salts, *Physiol. Plant Pathol.,* 27, 289, 1985.

38. **Beer, S. V., Sjulin, T. M., and Aldwinckle, H. S.,** Amylovorin-induced shoot wilting: lack of correlation with susceptibility to *Erwinia amylovora, Phytopathology,* 73, 1328, 1983.

39. **Goodman, R. N., Huang, J. S., and Huang, P.-Y.,** Host-specific phytotoxic polysaccharide from apple tissue infected buy *Erwinia amylovora, Science,* 183, 1081, 1974.

40. **Goodman, R. N. and White, J. A.,** Xylem parenchyma plasmolysis and vessel wall disorientation caused by *Erwinia amylovora, Phytopathology,* 71, 844, 1981.

41. **Huang, P.-Y., Huang, J. S., and Goodman, R. N.,** Resistance mechanisms of apple shoots to an avirulent strains of *Erwinia amylovora, Physiol. Plant Pathol.,* 6, 283, 1975.

42. **Suhayda, C. G. and Goodman, R. N.,** Early proliferation and migration and subsequent xylem occlusion by *Erwinia amylovora* and the fact of its extracellular polysaccharide (EPS) in apple shoots, *Phytopathology,* 71, 697, 1981.

43. **El-Banoby, F. E. and Rudolph, K.,** Induction of water-soaking in plant leaves by extracellular polysaccharides from phytopathogenic pseudomonads and xanthomonads, *Physiol. Plant Pathol.,* 15, 341, 1979.

44. **El-Banoby, F. E. and Rudolph, K.,** Purification of extracellular polysaccharides form *Pseudomonas phaseolicola* which induce water-soaking in bean leaves, *Physiol. Plant Pathol.,* 16, 425, 1980.

45. **El-Banoby, F. E., Rudolph, K., and Huttermann, A.,** Biological and physical properties of an extracellular polysaccharide from *Pseudomonas phaseolicola, Physiol. Plant Pathol.,* 17, 291, 1980.

46. **El-Banoby, F. E., Rudolph, K., and Mendgen, K.,** The fate of extracellular polysaccharide from *Pseudomonas phaseolicola* in leaves and leaf extracts from halo-blight susceptible and resistant bean plants (*Phaseolus vulgaris* L.), *Physiol. Plant Pathol.,* 18, 91, 1981.

47. **Hignett, R. C.,** Inhibition of fireblight disease by injection of D-galactose or L-fucose into susceptible apple plants, *Physiol. Mol. Plant Pathol.,* 30, 131, 1987.

48. **Hignett, R. C. and Roberts, A. L.,** A possible regulatory function for bacterial outer surface components in fireblight disease, *Physiol. Plant Pathol.,* 27, 235, 1985.

49. **Romeiro, R. D. S., Karr, A. L., and Goodman, R. N.,** *Erwinia amylovora* cell wall receptor for apple agglutinin, *Physiol. Plant Pathol.,* 19, 383, 1981.

50. **Romeiro, R. D. S., Karr, A. L., and Goodman, R. N.,** Isolation of a factor from apple that agglutinates *Erwinia amylovora, Plant Physiol.,* 68, 772, 1981.

51. **Preece, T. F. and Wong, W. C.,** Quantitative and scanning electron microscope observations on the attachment of *Pseudomonas tolaasii* and other bacteria to the surface of *Agaricus bisporus, Physiol. Plant Pathol.,* 21, 251, 1982.

52. **Matthyse, A. G., Holmes, K. V., and Gurlitz, R. H. G.,** Elaboration of cellulose fibrils by *Agrobacterium tumefaciens* during attachment to carrot cells, *J. Bacteriol.,* 145, 583, 1981.

54. **Leben, C. and Daft, G. C.,** Population variations of epiphytic bacteria, *Can. J. Microbiol.,* 13, 1151, 1967.

55. **Leben, C., Rusch, V., and Schmitthenner, A. F.,** The colonization of soybean buds by *Pseudomonas glycinea* and other bacteria, *Phytopathology* 58, 1677, 1968.

56. **Shaw, L.,** Intercellular humidity in relation to fire-blight susceptibility in apple and pear, *N.Y. (Cornell) Agric. Exp. Stn. Mem.,* 181, 1935.

57. **Billing, E.,** Fireblight in Kent, England in relation to weather (1955–1976), *Ann. Appl. Biol.,* 95, 341, 1980.
58. **Billing, E.,** Fireblight (*Erwinia amylovora*) and weather: a comparison of warning systems, *Ann. Appl. Biol.,* 95, 365, 1980.
59. **Smith, E. F.,** *Bacteria in Relation to Plant Diseases,* Vol. 2, Carnegie Institution, Washington, D.C., 1911, 57.
60. **Esau, K.,** *Anatomy of Seed Plants,* 2nd ed., John Wiley & Sons, New York, 1977, chap. 23.
61. **Mansvelt, E. L. and Hattingh, M. J.,** scanning electron microscopy of invasion of apple leaves and blossoms by *Pseudomonas syringae* pv. *syringae, Appl. Environ. Microbiol.,* 55, 533, 1989.
62. **Roos, I. M. M. and Hattingh, M. J.,** Systemic invasion of plum leaves shoots by *Pseudomonas syringae* pv. *syringae* introduced into petioles, *Phytopathology,* 77, 1253, 1987.
63. **Roos, I. M. M. and Hattingh, M. J.,** Scanning electron microscopy of *Pseudomonas syringe* pv. *morsprunorum* on sweet cherry leaves, *Phytopathol. Z.,* 108, 18, 1983.
64. **Miles, W. G., Daines, R. H., and Rue, J. W.,** Presymptomatic egress of *Xanthomonas pruni* from infected peach leaves, *Phytopathology,* 67, 895, 1977.
65. **Mansvelt, E. L. and Hattingh, M. J.,** Scanning electron microscopy of colonization of pear leaves by *Pseudomonas syringae* pv. *syringae, Can. J. Bot.,* 65, 2517, 1987.
66. **Rosen, H. R.,** Mode of penetration and of progressive invasion of fire blight bacteria into apple and pear blossoms, *Univ. Ark. Agric. Exp. Stn. Bull.,* 331, 1936.
67. **Wilson, M., Sigee, D. C., and Epton, H. A. S.,** *Erwinia amylovora* infection of hawthorn blossom. III. The nectary, *J. Phytopathol. (Berlin),* 128, 62, 1990.
68. **Hildebrand, E. M.,** The blossom-blight phase of fire blight and methods of control, *Cornell Univ. Agric. Exp. Stn. Mem.,* 207, 17, 1937.
69. **Hildebrand, E. M. and MacDaniels, L. H.,** Modes of entry of *Erwinia amylovora* into flowers of the principal pome fruits, *Phytopathology,* 25(Abstr.), 20, 1935.
70. **Rundle, J. R. and Beer, S. V.,** Population dynamics of *Erwinia amylovora* and a biological control agent, *Erwinia herbicola* on apple blossom parts, *Acta Hort.,* 217, 221, 1987.
71. **Thomson, S. V.,** The role of the stigma in fire blight infections, *Phytopathology,* 76, 476, 1986.
72. **Thomson, S. V., Schroth, M. N., Moller, W. J., and Reil, W. O.,** Occurrence of fire blight of pears in relation to weather and epiphytic populations of *Erwinia amylovora, Phytopathology,* 65, 353, 1986.
73. **Wilson, M., Epton, H. A. S., and Sigee, D. C.,** *Erwinia amylovora* infection of hawthorn blossom. II. The stigma, *J. Phytopathol. (Berlin),* 127, 15, 1989.
74. **Wilson, M., Sigee, D. C., and Epton, H. A. S.,** *Erwinia amylovora* infection of hawthorn blossom. I. The anther, *J. Phytopathol. (Berlin),* 127, 1, 1989.
75. **Mansvelt, E. L. and Hattingh, M. J.,** Scanning electron microscopy of pear blossom invasion by *Pseudomonas syringae* pv. *syringae, Can. J. Bot.,* 65, 2523, 1987.
76. **Mansvelt, E. L. and Hattingh, M. J.,** Pear blossom blast in South Africa caused by *Pseudomonas syringae* pv. *syringae, Plant Pathol. (London),* 35, 337, 1986.
77. **Hattingh, M. J., Mansvelt, E. L., and Roos, I. M. M.,** Practical approaches to control apple blister bark, *Decid. Fruit Grower,* 39, 228, 1989.
78. **Crosse, J. E., Goodman, R. N., and Shaffer, W. H., Jr.,** Leaf damage as a predisposing factor in the infection of apple shoots by *Erwinia amylovora, Phytopathology,* 62, 176, 1972.
79. **Klement, Z., Rozsnyay, D. S., and Arsenijevic, M.,** Apoplexy of apricots. III. Relationship of winter frost and the bacterial canker and die-back of apricots, *Acta Phytopathol. Acad. Sci. Hung.,* 9, 35, 1974.
80. **Panagopoulos, C. G. and Crosse, J. E.,** Frost injury as a predisposing factor in blossom blight of pear caused by *Pseudomonas syringae* van Hall, *Nature,* 202, 1352, 1964.
81. **Vigouroux, A.,** Obtention de symptômes de bactériose du pêcher (*Pseudomonas mors-prunorum* f. sp. *persicae*) sur rameaux de pêcher détachés et conservés en survie. Effet du froid, *Ann. Phytopathol.,* 6, 95, 1974.
82. **Vigouroux, A.,** Incidence des basses températures sur la sensibilité du pêcher au dépérissement bactérien (*Pseudomonas mors-prunorum* f. sp. *persicae*), *Ann. Phytopathol.,* 11, 231, 1979.
83. **Weaver, D. J.,** Interaction of *Pseudomonas syringae* and freezing in bacterial canker on excised peach twigs, *Phytopathology,* 68, 1460, 1978.
84. **Montesinos, E. and Vilardell, P.,** Relationships among population levels of *Pseudomonas syringae,* amount of ice nuclei, and incidence of blast of dormant flower buds in commercial pear orchards in Catalunya, Spain, *Phytopathology,* 81, 113, 1991.
85. **Klement, Z., Rozsnyay, D. S., Balo, E., Panczel, M., and Prileszky, Gy.,** The effect of cold on development of bacterial canker in apricot trees infected with *Pseudomonas syringae* pv. *syringae, Physiol. Plant Pathol.,* 24, 237, 1984.
86. **Gross, D. C., Cody, Y. S., Proebsting, E. L., Jr., Radamaker, G. K., and Spotts, R. A.,** Ecotypes and pathogenicity of ice-nucleation-active *Pseudomonas syringae* isolated from deciduous fruit tree orchards, *Phytopathology,* 74, 241, 1984.

87. **Weaver, D. J., Gonzales, C. F., and English, H.,** Ice nucleation by *Pseudomonas syringae* associated with canker production in peach *Phytopathology,* 71(Abstr.), 109, 1981.

88. **Paulin, J.-P. and Luisetti, J.,** Ice nucleation activity among phytopathogenic bacteria, in *Proc. 4th Int. Conf. Plant Pathogenic Bacteria,* Station de Pathologie Végétale et Phytobactériologie, Gibert-Clarey, Ed., INRA, Angers, France, 1978, 725.

89. **Zeller, V. W. and Schmiddle, A.,** The effect of frost on the infection by *Pseudomonas syringae* van Hall on leaves of sour cherries *(Prunus cerasus), Nachrichtenbl. Dtsch. Pflanzenschutzdienst (Berlin),* 31, 97, 1979.

90. **Süle, S. and Seemüller, E.,** The role of ice formation in the infection of sour cherry leaves by *Pseudomonas syringae* pv. *syringae, Phytopathology,* 77, 173, 1987.

91. **Vigouroux, A.,** Ingress and spread of *Pseudomonas* in stems of peach and apricot promoted by frost-related water-soaking of tissues, *Plant Dis.,* 73, 854, 1989.

92. **Crosse, J. E.,** Epidemiological relations of the pseudomonad pathogens of deciduous fruit trees, *Annu. Rev. Phytopathol.,* 4, 291, 1966.

93. **Crosse, J. E.,** Bacterial canker of stone-fruits, III. Inoculum concentration and time of inoculation in relation to leaf-scar infection of cherry, *Ann. Appl. Biol.,* 45, 19, 1957.

94. **Roos, I. M. M. and Hattingh, M. J.,** Systemic invasion of cherry leaves and petioles by *Pseudomonas syringae* pv. *morsprunorum, Phytopathology,* 77, 1246, 1987.

95. **Du Plessis, H. J.,** Scanning electron microscopy of *Xanthomonas campestris* pv. *pruni* in plum petioles and buds, *Phytopathol. Z.,* 109, 277, 1984.

96. **Du Plessis, H. J.,** Systemic migration and establishment of *Xanthomonas campestris* pv. *pruni* in plum leaves and twigs, *J. Phytopathol. (Berlin),* 116, 221, 1986.

97. **Feliciano, A. and Daines, R. H.,** Factors influencing ingress of *Xanthomonas pruni* through peach leaf scars and subsequent development of spring cankers, *Phytopathology,* 60, 1720, 1970.

98. **Gasperini, C., Bazzi, C., and Mazzucchi, U.,** Autumn inoculation of *Xanthomonas campestris* pv. *pruni* through leaf scars in plum trees in the Po valley, *Phytopathol. Mediterr.,* 23, 60, 1984.

99. **Hildebrand, E. M.,** A microsurgical study of crown gall infection in tomato, *J. Agric. Res.,* 65, 45, 1942.

100. **Roos, I. M. M. and Hattingh, M. J.,** Pathogenicity and numerical analysis of phenotypic features of *Pseudomonas syringae* strains isolated from deciduous fruit trees, *Phytopathology,* 77, 900, 1987.

101. **Gowda, S. S. and Goodman, R. N.,** Movement and persistence of *Erwinia amylovora* in shoot, stem and root of apple, *Plant Dis. Rep.,* 54, 576, 1970.

102. **Lewis, S. and Goodman, R. N.,** Mode of penetration and movement of fire blight bacteria in apple leaf and stem tissue, *Phytopathology,* 55, 719, 1965.

103. **Roos, I. M. M. and Hattingh, M. J.,** Systemic invasion of immature sweet cherry fruit by *Pseudomonas syringae* pv. *morsprunorum* through blossoms, *J. Phytopathol. (Berlin),* 121, 26, 1988.

104. **Du Plessis, H. J.,** Systemic invasion of plum seed and fruit by *Xanthomonas campestris* pv. *pruni* through stalks, *J. Phytopathol. (Berlin),* 130, 37, 1990.

105. **Nelson, P. E. and Dickey, R. S.,** Histopathology of plants infected with vascular bacterial pathogens, *Annu. Rev. Phytopathol.,* 8, 259, 1970.

106. **Suhayda, C. G. and Goodman, R. N.,** Infection courts and systemic movement of ^{32}P-labeled *Erwinia amylovora* in apple petioles and stems, *Phytopathology,* 71, 656, 1981.

107. **Tyson, G. E., Stojanovic, B. J., Kuklinski, R. F., DiVittorio, T. J., and Sullivan, M. L.,** Scanning electron microscopy of Pierce's disease bacterium in petiolar xylem of grape leaves, *Phytopathology,* 75, 264, 1985.

108. **Hattingh, M. J., Roos, I. M. M., and Mansvelt, E. L.,** Infection and systemic invasion of deciduous fruit trees by *Pseudomonas syringae* in South Africa, *Plant Dis.,* 73, 784, 1989.

109. **Du Plessis, H. J.,** Differential virulence of *Xanthomonas campestris* pv. *pruni* to peach, plum and apricot cultivars, *Phytopathology,* 78, 1312, 1988.

110. **Norelli, J. L., Aldwinckle, H. S., Beer, S. V., and Lamb, R. C.,** The effects of virulence of *Erwinia amylovora* on the evaluation of fire blight resistance in *Malus, Phytopathology,* 77, 1551, 1987.

111. **Hattingh, M. J., Beer, S. V., and Lawson, E. W.,** Scanning electron microscopy of apple blossoms colonized by *Erwinia amylovora* and *E. herbicola, Phytopathology,* 76, 900, 1986.

112. **Roos, I. M. M., Hattingh, M. J., and Marasas, C. N.,** Transmission of *Pseudomonas syringae* pv. *syringae* by seed of stone fruit trees, in *Proc. 7th Int. Conf. Plant Pathol. Bacteriol.,* Budapest, Hungary, Klement, Z., Ed., Zkademia Kiado, Budapest, 1989, 161.

113. **Cameron, H. R.,** *Pseudomonas* content of cherry trees, *Phytopathology,* 60, 1343, 1970.

114. **Du Plessis, H. J.,** Canker development on plum shoots following systemic movement of *Xanthomonas campestris* pv. *pruni* from inoculated leaves, *Plant Dis.,* 71, 1078, 1987.

115. **Chatterjee, A. K. and Vidaver, A. K.,** Genetics of pathogenicity factors: application to phytopathogenic bacteria, *Adv. Plant Pathol.,* 4, 1, 1986.

Chapter 14

ADAPTATION AND RESPONSE OF FRUIT TREES TO FREEZING TEMPERATURES

Michael Wisniewski and Rajeev Arora

TABLE OF CONTENTS

I. INTRODUCTION

Although injury of fruit tree tissues due to exposure to freezing temperatures is not a pathological disease, the dead tissues that result from freeze injury often serve as a primary infection court for other fruit tree pathogens. Additionally, economic losses can occur as a direct result of severe winter temperatures and/or untimely spring frosts. Thus, it is commonly recognized that low temperatures are a major limiting factor in the production of deciduous fruit crops. For these reasons, cold hardiness is often a selection criterion in breeding programs, and several research programs have placed a primary emphasis on elucidating the mechanisms of freezing injury and temperature acclimation. The response of fruit tree tissues to freezing temperatures has been recently reviewed[1] and a complete treatise on the subject of frost survival of plants has been published by Sakai and Larcher.[2] This chapter attempts to provide a brief overview of the subject and to discuss the various methods that are presently being used to study cold hardiness and cold acclimation.

Strategies that allow plants to survive freezing temperatures have been placed into two major categories:[3] freezing tolerance and freezing avoidance. Mechanisms representing both categories have been well documented in woody perennials.[4] Tissues displaying freeze tolerance respond to a low-temperature stress by the loss of cellular water to extracellular ice. This results in collapse of the cell, including the cell wall (cytorrhysis), which in turn increases the concentration of the cell sap and lowers the freezing point. In contrast, tissues that avoid freezing stress, but are still exposed to freezing temperatures, do so by deep supercooling, a process in which cellular water is isolated from the dehydrative and nucleating effects of extracellular ice.

A distinct seasonality exists for both mechanisms and is marked by a period of cold acclimation in the fall and deacclimation in the spring. This implies a sequence of active processes that results in well-defined changes in cellular composition, both biochemical and structural.[5,6]

Developing an understanding of cold hardiness in fruit trees is a complex issue due to the discovery that different tissues within the same plant respond in quite different ways to exposure to subzero temperatures. This is further complicated by the seasonality of the expression of cold hardiness and by the interplay of other biotic and abiotic stresses (mineral, water, diseases, etc.) on the phenotypic expression of cold hardiness for any individual genotype. This is important to keep in mind when evaluating and interpreting data on cold hardiness.

II. DEEP SUPERCOOLING OF XYLEM TISSUES

Living xylem tissues of several temperate woody species, including species of temperate fruit trees, avoid low temperature stress by deep supercooling.[7-10] A comprehensive list of woody species that exhibit this trait was published by George et al.[11]

During a supercooling event, cellular water remains liquid within xylem parenchyma cells at very low temperatures by remaining isolated from heterogeneous ice nuclei and the nucleating effect of extracellular ice.[12] Supercooled water is in a metastable condition and will form intracellular ice in response to a heterogeneous nucleation event or when the homogeneous nucleation temperature of water ($-38°C$) is reached.[13]

The freezing response of a tissue can be monitored using the technique of differential thermal analysis (DTA), as outlined by Quamme et al.,[7] and since modified to include the use of digital, data acquisition equipment and microcomputers.[14-16] Thermocouples are utilized to detect the heat of fusion produced by the water in the samples as it undergoes a liquid to solid phase change (Figure 1). In DTA, sample temperatures are compared to a

FIGURE 1. Small debarked, twig (arrow) attached to copper-constantan thermocouple. After attachment of thermocouple, the tissue sample is enclosed in aluminum foil and the thermocouple is placed in a test tube. The test tube is then placed in an aluminum heat block inside a low-temperature freezer ($-80°C$) and the tissue sample is subjected to differential thermal analysis. (Magnification $\times$ 0.5).

piece of freeze-dried tissue (reference) undergoing the same rate of cooling. This produces a flat baseline until the freezing of water within the sample tissue results in a difference in temperature between samples and reference. The sample-reference differential is visualized as a peak on the thermogram, hence the term "differential thermal analysis."

In thermograms from woody plants that exhibit deep supercooling (Figure 2), the initial large peak is referred to as a high-temperature exotherm (HTE) and is believed to represent the freezing of bulk water contained within tracheary elements and extracellular spaces, whereas the peak occurring at very low temperatures (deep supercooling) is believed to represent the freezing of intracellular water contained within xylem parenchyma cells. The peak resulting from the freezing of deep supercooled water is referred to as a low-temperature exotherm (LTE). In contrast to deep supercooling, many species exhibit equilibrium freezing, in which no LTE is observed using DTA (Figure 2).

As indicated previously, different tissues of fruit trees within the same plant exhibit contrasting freezing behavior. Evidence for this was first provided in the 1970s.[7-10] In these experiments, the LTE, which occurred near $-40°C$, was correlated with the death of xylem ray parenchyma cells.[7,8,17,18] Since the documentation of this correlation, DTA has been extensively used to evaluate the degree of cold hardiness of stem tissues of fruit trees.

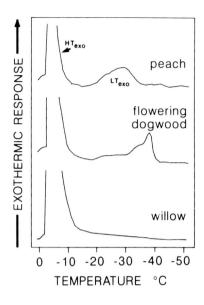

FIGURE 2. Freezing response of debarked, internodal twig section of peach, flowering dogwood, and willow subjected to DTA. The HT_{exo} represents the freezing of the intercellular water and water within lumen of nonliving tracheary elements. The LT_{exo} represents the freezing of intracellular water within living xylem parenchyma cells. Note the absence of any LT_{exo} in willow.

Low-temperature scanning electron microscopy (LTSEM) has also been utilized with great effectiveness to examine plant tissue while in a frozen state. Studies have been conducted to determine the site of ice formation within a tissue and the response of cells to extracellular ice.[19,20] These studies clearly demonstrated that in deep supercooling species, frozen xylem ray parenchyma cells exhibited no evidence of cell collapse or tissue disruption. Regardless of the temperature to which samples were exposed ($-5°$, $-20°$ or $-45°C$), the appearance of cells was similar to unfrozen controls. These observations are consistent with the idea that the cells are deep supercooled and do not exhibit cytorrhysis in response to the presence of extracellular ice.

For deep supercooling to occur, a tissue must exhibit several features:[10] (1) cells must be free of heterogeneous nucleating substances that are "active" at warm freezing temperatures; (2) a barrier must be present that excludes the growth of ice crystals into a cell; and concomitantly (3) a barrier to water movement must exist that prevents a "rapid" loss of cellular water to extracellular ice in the presence of a strong vapor pressure gradient. It is believed that physical properties of the apoplast (i.e., cell wall structure) rather than the symplast (i.e., protoplast) largely account for the ability of xylem parenchyma to deep supercool.[21] In this regard, the porosity and/or permeability of the cell wall plays an essential role and several techniques have been used to examine those properties in species that exhibit deep supercooling.[22-25]

Both lanthanum nitrate[22,23] and colloidal gold manufactured in prescribed sizes[25] have been used as an apoplastic tracer in woody plant tissues. Results of these studies indicated that the pit membrane and associated amorphous layer, not the secondary wall, would play a limiting role in determining the ability of a cell wall to retain water against a strong vapor-pressure gradient and the intrusive growth of external ice crystals (Figures 3 and 4).

Further research[24,26,27] has indicated that modifications in the structure of the pit membrane can result in a loss of deep supercooling as determined by a flattening between the LTEs (Figure 5). Pectinase and oxalic acid (which disrupts Ca bridges of pectin chains)

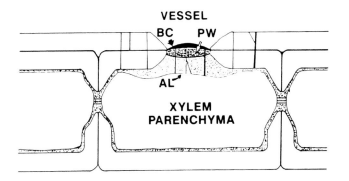

FIGURE 3. Diagrammatic illustration of pit membrane structure in xylem parenchyma cell of peach. The pit membrane consists of three layers: a black cap (BC) of electron-dense material, a primary wall (PW), and an underlying amorphorus layer (AL). Channels indicate the role of pore size and continuity in regulating water flow and ice crystal propagation. (From Wisniewski, M. et al., *Plant Physiol.*, 96, 1354, 1991.)

have proved to have the greatest effect on both pit membrane structure and the character of the LTE. This evidence strongly implicates pectin as playing an integral role in regulating the porosity and/or permeability of the pit membrane. Baron-Epel et al.[28] have also suggested that pectins may play a key role in defining the size of cell wall microcapillaries in soybean suspension cell cultures.

Pectin-mediated regulation of porosity and/or permeability of the pit membrane is an attractive hypothesis because it could account for a loosening of cell wall structure in the spring by disruption of metal ion bridges (primarily Ca^{2+}) or breakage of covalent bonds by intrinsic pectinases. Such modifications readily occur, in fact, during tylose formation when the pit membrane of the cell wall loosens considerably to allow for growth and extension of the xylem parenchyma cell into a neighboring vessel element.[29]

Interestingly, the degree of deep supercooling does not show an annual shift in older wood of peach[29a] and shagbark hickory.[10] It has been reported that as wood ages the pit membrane of xylem parenchyma in many species eventually lignifies.[30,31] The specific timetable of this event varies between species. As the pit membrane lignifies, it exhibits a rigidity and lack of permeability comparable to that of the secondary wall. This can make alterations in wall structure difficult to accomplish. Although this has yet to be documented in relation to deep supercooling, it may account for the characteristics that have been reported for deep supercooling that relate to age structure.

The combined evidence clearly indicates the important role of pit membrane structure in regulating deep supercooling and strongly implicates pectin and the interaction of pectin with other cell wall constituents as playing a role in defining the porosity and/or permeability of the cell wall. There are many questions that must be resolved, however, before a definitive knowledge of the underlying mechanism of deep supercooling of xylem tissues is developed. In apple[32] and peach[24,26] it is clear, for instance, that the xylem tissue does not exhibit a homogeneous freezing response. This is evidenced by multiple LTEs in apple[32] and broad, somewhat bimodal peaks in peach.[24,26] How the complex freezing response of these tissues is regulated is not understood.

Furthermore, recent evidence[33] suggests that contact of the plasmalemma with the cell wall is essential in order to achieve maximum expression of deep supercooling. Why this feature is needed is also not understood. Developing a better understanding of these mechanisms may result in potential new strategies for preventing freezing injury or developing more cold-resistant germplasm.

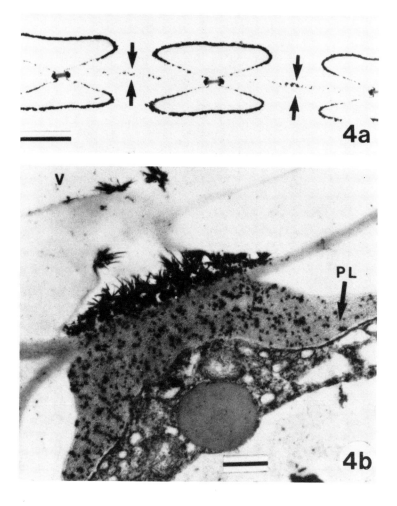

FIGURE 4. Lanthanum deposits in xylem tissue of peach. Note the deposition of lanthanum in intervessel pit membranes (2) and in the pit membrane of xylem ray parenchyma (b) Lanthanum deposits were not observed in secondary cell walls of vessel elements or xylem ray parenchyma. (From Wisniewski, M. et al., *Protoplasma*, 139, 105, 1987. With permission.)

III. EXTRAORGAN FREEZING AND DEEP SUPERCOOLING OF DORMANT BUDS

Research and breeding programs have placed primary emphasis on the cold hardiness of floral buds of fruit trees. The emphasis is justified, not only because of the economic value of the fruit, but also because floral buds tend to be the tissue that is most susceptible to freezing injury. Economic looses occur as a result of severe midwinter temperatures when the floral buds are the most cold hardy and also in early spring when floral bud tissue is the least cold hardy and temperatures drop only a few degrees below freezing.

The response of dormant buds to freezing temperatures is different than in other portions of the tree. Freezing of extracellular water is initiated within the bud scales and subtending stem tissue.[9,34,36] Thus, the shoot apex or floral primordium is isolated from the mechanical damage caused by the presence of large ice crystals. The introduction of ice into the bud tissue results in the establishment of a water potential gradient.

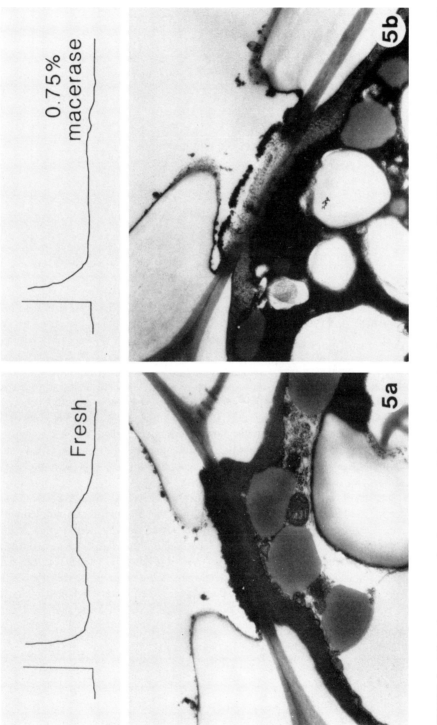

FIGURE 5.   Effect of macerase on freezing response and pit membrane structure of peach. (a) Control tissue. Note LTE and intact pit membrane structure. (b) Macerase-treated tissue. Note flattening of LTE and disruption of pit membrane structure.

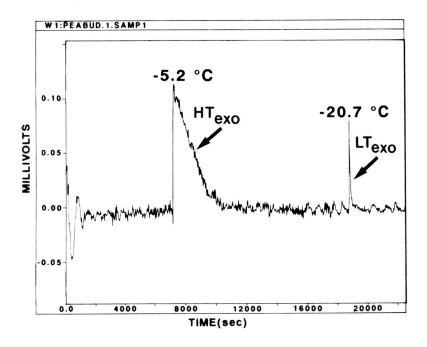

FIGURE 6. Thermogram of peach flower bud obtained using DTA. HTE represents freezing of water in bud scales and subtending axis tissue. LTE represents freezing of water in floral tissues.

In many species of woody plants, water migrates from the shoot or floral apex to the sites of extracellular ice in response to the water-potential gradient. This response to freezing temperatures has been described as extraorgan freezing.[35] When buds are killed, mortality results from the dehydrative stress rather than from low temperature or the presence of ice. Extraorgan freezing is characteristic of the most cold-hardy species, where the ability to withstand severe dehydrative stress has evolved.[2] Among temperate fruit trees, this pattern of freezing is present in floral buds of apple and pear. In addition, vegetative buds of all temperate fruit species respond in this manner.

In other species of woody plants, however, not all the water from the floral tissue migrates to the ice in the bud scales. Instead, a portion of water remains supercooled within the floral tissue. Deep supercooling of flower buds has been documented in grape,[37] blueberry,[38] and several *Prunus* sp.[39] As with deep supercooling of xylem tissues, two distinct exotherms are detected when buds are subject to DTA[34,36] (Figure 6). The HTE is associated with the freezing of water in the bud scales and subtending stem tissue and the LTE is associated with the freezing of intracellular, deep-supercooled water contained within the floral tissue. In species containing multiple flowers within a single flower bud, each floret freezes as an independent unit.[39] This is evidenced by the fact that multiple LTEs are obtained using DTA. This is true for grape, sweet cherry, sour cherry, azalea, and flowering dogwood.[2] The LTE is correlated with the degree of cold hardiness of the tissue[39] and is used extensively as an evaluation tool.

Deep supercooling of peach flower buds has been extensively studied because of the susceptibility of peach to cold injury. As with xylem tissues, in order for deep supercooling to occur in flower buds, a barrier to water movement and ice propagation must exist. In peach floral buds, freezing of the supercooled water occurs at $-20°$ to $-25°C$, which is

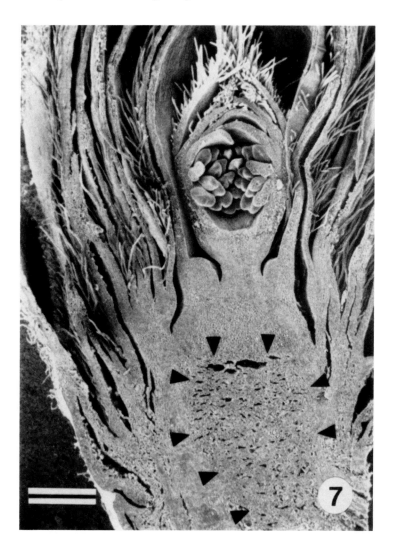

FIGURE 7. SEM of peach flower bud tissue prepared by freeze fixation and freeze substitution. Arrowheads indicate voids in axis tissue subtending the floral buds where ice crystals had formed. Voids are also present in the bud scales. Note the absence of voids in the flower itself. (From Ashworth, E. N. et al., *Plant Cell Environ.*, 12, 521, 1989. With permission.)

well above the homogeneous nucleation point of water; however, the factors responsible for ice nucleation at these temperatures and the nature of the barrier that allows deep supercooling to occur are not completely understood.

A freeze-substitution method developed by Ashworth et al.[40] and used in conjunction with SEM was used to study the distribution of ice in peach flower buds. Their observations confirmed the earlier observations of Quamme[34] and Sakai[35] and demonstrated that large voids are formed in the bud scales and stem tissue subtending the floral bud as a result of the formation of extracellular ice (Figure 7). Although a mechanical disruption of the tissue did occur, no loss in bud viability was observed. After thawing and subsequent refreezing, ice crystals were initiated in these same voids. It was postulated that these voids serve to accommodate the formation of ice throughout the dormant season.

The extent of deep supercooling decreases and cold hardiness is progressively lost as peach flower buds begin to break dormancy.[36,41] Ashworth[42] reported that the loss of deep supercooling was associated with the development of vascular continuity between the flower and the stem axis. Further studies indicated that in the absence of deep supercooling, ice crystals were initiated in the floral tissues rather than just the bud scales and subtending stem tissue.

Recent evidence indicates that both the midwinter and early spring cold hardiness of peach flower buds can be increased with a fall application of ethephon.[43,44] The application of ethephon results in about a 1 to 2-week delay in bloom, and thus prevents the loss of deep supercooling at a critical period.[45,46] Ethephon-treated flower buds deacclimate slower than nontreated flower buds when exposed to warm temperatures (21°C) and prolong dormancy by increasing the chilling requirement.[47] Although the mechanism of increased midwinter cold hardiness is not understood it may result from the smaller floral bud size that is promulgated from the application of ethephon. In general, within the same genus, smaller bud size (i.e., smaller primordia) is associated with increased hardiness.[2] Because of the inherent value associated with flower buds, research will continue to focus on the mechanism regulating extraorgan freezing and deep supercooling of dormant buds, the nature of the barrier in floral tissues that allows deep supercooling to occur, and the development of strategies to increase cold hardiness and/or prevent cold injury. These strategies must be both economically feasible and also conducive to the health of the tree to be of practical value.

IV. EQUILIBRIUM FREEZING: TISSUE RESPONSE TO EXTRACELLULAR FREEZING

In contrast to xylem tissue, which avoids freezing stress by deep supercooling, bark tissue of temperate fruit trees undergoes "equilibrium freezing" and concomitantly tolerates the extracellular ice formation and the dehydrative stress that results from the loss of cellular water to extracellular ice.[18] Equilibrium freezing in plant cells occurs during slow cooling rates (1° to 2°C/h), when ice formation is initiated at high subzero temperatures (0° to −2°C) in the extracellular spaces due to a nucleation event.[3] This occurs because: (1) the extracellular solution has a higher (warmer) freezing point than the intracellular solution and (2) efficient ice nucleators such as dust, bacteria, etc., are prevalent in the extracellular environment. Once the tissue temperature drops below the freezing point of the protoplasm, the internal vapor pressure becomes higher than that of extracellular ice.[3] The formation of this gradient results in the movement of cellular water to extracellular ice crystals which then increase in size. As the temperature decreases further, more water is withdrawn from the cells, cell volume is reduced, and the internal solute concentration is increased.[3,48,49] This also results in collapse of the cell (Figure 8)[3,19,20,50] A gradual or slow cooling allows the diffusion of cellular water to ice at a speed sufficient to increase the solute concentration of the protoplasm as rapidly as the temperature drops. This allows the chemical potential of the protoplasm to be in equilibrium with the chemical potential of the ice, hence, the term "equilibrium freezing". In contrast, if a cell is cooled rapidly, the water cannot diffuse to the extracellular ice at a rate rapid enough for the cell to remain in chemical equilibrium with the ice. If the chemical disequilibrium is maintained and the temperature is sufficiently lowered, the spontaneous formation of intracellular ice may be induced. Unlike extracellular freezing, this method is always lethal.[3,5] "Sunscald", a freezing injury to the trunks, is believed to result from intracellular freezing.[2,3] On a winter day, the sun shining on the south side of the tree may raise tissue temperature high enough (25° to 30°C above the shady side) to instigate thawing. When the sun suddenly disappears behind a cloud, the temperature of the thawed tissue may drop rapidly enough that intracellular freezing and death may occur.

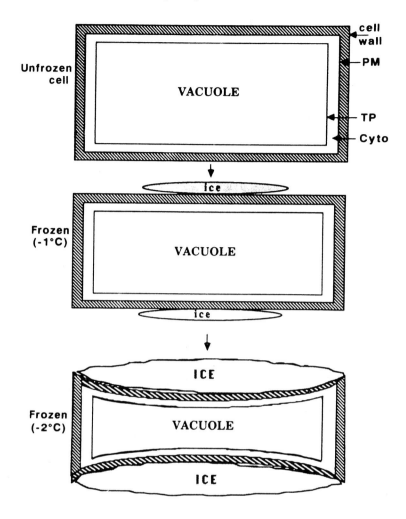

FIGURE 8. Diagrammatic representation of cell collapse during equilibrium freezing. Abbreviations: PM, plasma membrane, TP, tonoplast; Cyto, cytoplasm.

Recent studies with apple, red osier dogwood, and flowering dogwood using low-temperature SEM (LTSEM) and freeze substitution have corroborated the proposal that bark tissue undergoes equilibrium freezing.[19,20] Clear evidence for the collapse of cortical cells and the formation of voids in the tissue from extracellular ice was presented in these studies. Equilibrium freezing may also explain the natural phenomenon of "frost splitting", which is exhibited in some tree trunks during a severe frost.[52]

In contrast to xylem parenchyma cells, which experience primarily low-temperature stress during a frost episode, bark cells experience various stresses: (1) low temperature per se, (2) dehydration, (3) mechanical shear and tear, and (4) concentrated cell sap. Extracellular freezing and subsequent thawing also result in an increased efflux of ions and organic solutes from the cell. This causes the tissue to take on a water-soaked appearance and loose turgor.[53-55] Freezing stress has been shown to result in specific perturbation of membrane-associated calcium.[56,57] Based on the above scenario, several possible mechanisms of injury have been proposed. These include the mechanical stress theory,[58] protein aggregation due to dehydration-induced intermolecular disulfide bonding,[59] protein coagulation due to high salt concentration,[60] alterations in the hydrophilic/hydrophobic interactions within cell membranes,[5] functional alteration of the membrane transport system,[61] perturbation of membrane and/or cellular calcium,[62] and establishment of a minimum critical volume[63] or membrane surface area.[5]

V. PLASMA MEMBRANE ALTERATIONS DURING LOW TEMPERATURE RESPONSE

The water-soaked appearance and loss of turgor of the tissue led early researchers to believe that the cell membrane was a site of injury. This was recognized as early as 1912 by Maximov,[64] however, the nature of injury to the cell membrane at more fundamental level has been investigated only recently. Palta et al.[61] first proposed that freeze-thaw stress results in the alteration of active transport functions (presumably ion transporting ATPases) of the cell membrane. Several subsequent studies have supported this hypothesis.[55,65-67] Although the structural and biochemical alterations in the membrane due to freeze-thaw stress in herbaceous plants have been extensively investigated,[5,68] research on membrane perturbations in woody tissues is very limited.[69-71] This has been, in part, due to the difficulty of isolating purified plasma membranes from woody tissue. Using purified plasma membranes prepared by an aqueous, two-polymer, phase-partition system, Hellergren et al.[70] demonstrated that plasma membrane ATPase was the site of alteration due to freeze-thaw injury in pine needles.

Using purified plasma membranes from mulberry bark cells, Yoshida[72] showed a significant increase in phospholipids and unsaturation of their fatty acids during cold acclimation. The fluidity of the plasma membrane (which is believed to be associated with increased unsaturation level in fatty acids), determined by fluorescent polarization technique, also increased during early autumn to winter.[71] Evidence is also accumulating that plasma membrane ATPase activity increases during cold acclimation in pine needles[73] and apple bark cells.[74]

VI. ADAPTATION TO LOW TEMPERATURE STRESS

Bark and leaves of woody perennials exhibit a distinct seasonality in freezing tolerance.[3] In temperate woody perennials, cold acclimation is generally accompanied by a slowing and cessation of growth culminating in endodormancy. Deacclimation in spring is characterized by the resumption of growth. Several studies have been conducted to understand the structural and/or biochemical changes associated with these seasonal fluctuations and have been recently reviewed.[2,6]

Pomeroy and Siminovitch[75] conducted a seasonal study on the ultrastructural changes in bark tissue of black locust trees and noted a seasonal augmentation of total cytoplasm in late October (i.e., dense cytoplasm containing numerous small vacuoles). This is in contrast to the appearance of cells in spring and summer, when a large central vacuole with a thin band of peripheral cytoplasm is present. These changes were closely associated with seasonal changes in cold hardiness. Niki and Sakai[76] extended these observations to mulberry bark tissues. Wisniewski and Ashworth[77] confirmed these findings with their studies on peach cortical cells. Figure 9 illustrates these ultrastructural changes in peach.

The metabolic shifts associated with cold acclimation and deacclimation have been the subject of much research. Cold acclimation in woody perennials, including deciduous fruit trees, is typically a two-stage process.[2] The first stage after growth has ceased, appears to be induced by short days and proceeds at 10° to 20°C in autumn.[78-80] The prevention or retardation of timely autumn cold acclimation in some of the trees that experience long day conditions near street lights[81] supports this notion. During the first stage, abundant organic substances (primarily starch and neutral lipids) are stored, which serve as essential substrates

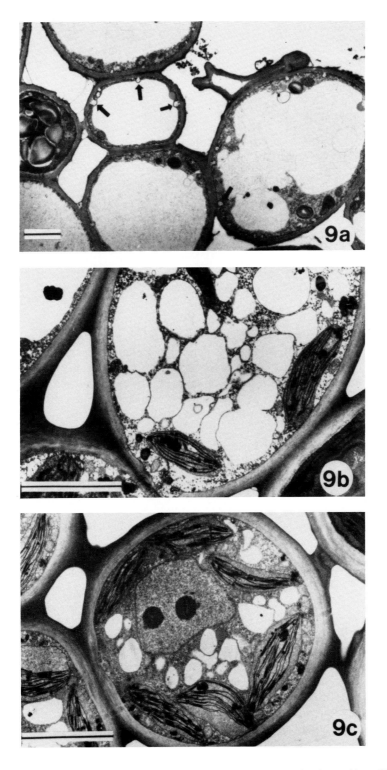

FIGURE 9. (a to c) Ultrastructure of peach bark cells during acclimation. Note breakup and loss of large central vacuole, proliferation of protoplasm, and central location of nucleus at peak of acclimation. (From Wisniewski, M. and Ashworth, E. N., *Bot. Gaz.,* 147, 407, 1986. With permission.)

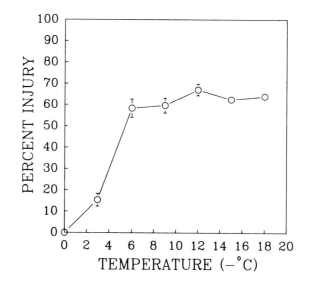

FIGURE 10.   Frost hardiness (LT_{50} of about $-5.5°C$) of peach bark tissue evaluated by the electrolyte leakage method.

and energy source for the metabolic changes occurring during the second stage. The second stage is induced by low ($<5°C$, especially subzero) temperatures. In this stage, proteins and membrane lipids are neosynthesized and/or undergo structural changes, ultimately leading to maximum hardiness.[2] A detailed account of changes in metabolic pathways, carbohydrates, proteins, nucleic acids, phospholipids, and growth regulators during cold acclimation and deacclimation is beyond the scope of this chapter and has been reviewed elsewhere.[2,6,82]

VII. TECHNIQUES TO EVALUATE FREEZE-INDUCED INJURY AND TO STUDY COLD ACCLIMATION

Various techniques have been used to evaluate the freezing injury in herbaceous and woody tissues: regrowth,[83] visual rating of discoloration/browning,[84] vital staining,[85] protoplasmic streaming and ability of cell to plasmolyze,[57,86] triphenyl tetrazolium chloride (TTC) reduction to formazan,[20,87-89] leakage of amino acids,[90] differential thermal analysis,[7] and electrolyte leakage.[18,74,86,91-93] Examples of the data obtained from TTC reduction test on peach cell suspensions and electrolyte leakage test on peach bark tissue are presented in Figures 10 and 11, respectively.

Electrical conductivity in conjunction with visual observations on the browning/water soaking of tissue is by far the most commonly used method to evaluate freezing injury in the bark and leaf tissue. This method, which was first developed by Dexter et al.,[94] is based on the fact that a considerable amount of solutes (ions, sugars, proteins) are lost after cell damage or cell death. It was assumed that the greater the injury to the living tissue, the greater the efflux of ions from the tissue. Electrical conductivity (measure of ion leakage) of the effusate is recorded after freeze-injured tissue is incubated and shaken in deionized water. The conductivity is once again recorded at room temperature after the tissue is heat killed in the same solution, which provides a measurement of total ions present in the tissue. Ion efflux from the injured tissue is calculated as a percentage of the total ions present in the tissue. This method, therefore, has been used to assess the relative injury to the stressed tissue. Freezing tolerance (LT_{50}) of the tissue is generally regarded as the temperature at which 50% ion leakage occurs, however, some drawbacks are inherent in this interpretation:

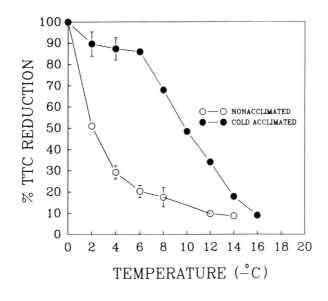

FIGURE 11. Frost hardiness of nonacclimated (LT_{50} of $-2°C$) and cold-acclimated (LT_{50} of $-11°C$) peach cell suspensions estimated by TTC reduction method.

(1) generally, a substantial amount of solutes leak out, even when a nonfrozen tissue is incubated in water and (2) sometimes all the solutes do not leak out (100% relative to the ion leakage from heat-killed tissue), even when the tissue is irreversibly injured due to freeze-thaw stress. Recently, Zhang and Willison[92] modified the conductivity method to overcome these problems.

It is extremely important that experimental tests (in laboratory) to assess relative freezing tolerance of plant tissue are designed to provide accurate simulation of the type and magnitude of stresses associated with frost episode in nature. Cooling rates (in the subfreezing range) during a frost episode in nature are generally 1° to 2°C/h,[3,51] accompanied by slow thawing. Recent studies of Steffen et al.[51] demonstrated that during experimental tests a simple increase in the cooling rates from 1° to 3°C/h meant the difference between cell survival and cell death. In addition, these studies showed that different cellular functions are differentially sensitive to freezing stress, e.g., membrane transport functions are most sensitive, followed by chloroplast functions (photosynthesis), and the mitochondrial functions (respiration) being least sensitive. Since various viability assays routinely used to assess the freeze-induced injury are based on the injury to different cellular functions, caution should be exercised in the interpretation of the experimental results.

One aspect of the physiology of freezing is the ability of injured tissue to recover following moderate stress.[55,95] Contrary to the popular view that freezing injury is a cataclysmic event, it was demonstrated in these studies that depending upon the degree of initial injury, a complete recovery following removal of the stress is possible. Moreover, a complete disappearance of water soaking of the tissue and reduction in the ion leakage was noted during the recovery process.[55,95] True frost hardiness of a plant species is ultimately its ability to survive low temperature stress and maintain or recover its metabolic functions. It is therefore important that experimental tests assessing low temperature tolerance are designed to take recovery into consideration.

In order to study the structural, physiological, or biochemical changes associated with cold acclimation on a year-round basis, various techniques are used to artificially induce cold acclimation in plant tissues. Controlled-environment growth rooms have been used to subject plant tissue to low temperatures (generally 0° to 5°C), and investigators have used short photoperiods for the induction of cold acclimation.[74,96-98] Plants that possess the ability to acclimate increase their freezing tolerance during this treatment, whereas species that do not acclimate exhibit no change in their freezing tolerance. The duration of the time required (days to weeks) to attain maximum freezing tolerance is temperature and species dependent.[99,100] Plants requiring photoperiodic initiation of dormancy and cold acclimation (e.g., deciduous fruit trees) require longer periods to make the transition from cold sensitive to a fully hardy state,[101] while plants strictly dependent on low temperature exposure (primarily herbaceous species) may cold acclimate and reach maximum hardiness more rapidly. The maximum freezing tolerance of species of potato that are able to cold acclimate is reached in about 2 weeks of low temperature (2° to 4°C) exposure, and deacclimation is largely completed in 1 to 2 d when transferred to warmer (noninductive) temperatures.[98] Similar kinetics of cold acclimation and deacclimation have also been reported in *Arabidopsis thaliana*.[102] On the other hand, bark of apple trees reached maximum hardiness when vegetatively mature 1-year-old twigs were subjected to 0°C in darkness for 21 d.[74] In these studies, the onset of vegetative maturity was regarded as the date of defoliation after which no bud regrowth was observed within 14 d.[103]

Increased freezing tolerance upon exposure to inductive temperatures has also been reported in various other model systems such as stem cultures,[99,104] callus tissue,[99,105-107] and cell suspension.[89,108,109]

Low temperature exposure is not the only means to induce freezing tolerance. Exogenous treatment of cell cultures, stem cultures, and seedlings with abscisic acid (ABA) at warmer (nonacclimating) temperatures has been shown to alter freezing tolerance.[110-112] The level of freezing tolerance induced by ABA in these studies was comparable to that developed upon exposure to low temperatures. Interestingly, the induction of maximum freezing tolerance by ABA is extremely rapid compared to that by low temperature treatment. In the cases of wheat and brome grass cell cultures maximum freezing tolerance occurs over a period of 2 to 4 d.[110] The ABA concentration required to evoke freezing tolerance ranges between 10^{-4} and 10^{-5} M. The ability of ABA to rapidly alter freezing tolerance offers two important advantages in the study of cold tolerance mechanisms: (1) experiments need not be confounded by the plant responses to cold that are not related to cold acclimation and (2) the rapidity of tolerance induction allows the inclusion of experimental manipulations that are not suitable over time intervals of weeks and months.

Limited desiccation can also increase freezing tolerance.[113,114] Desiccation for as little as 1 d at nonacclimating temperatures was shown to induce freezing tolerance in winter rye plumules.[115,116] The extent of desiccation-induced freezing tolerance in these experiments correlated with the level of hardiness following 4 weeks of low temperature exposure.

Although reports abound on the mechanism and/or physiology of freezing injury and cold acclimation in herbaceous plants, research on woody plant species, particularly deciduous fruit trees, is more limited. Deciduous fruit trees undergo endodormancy during the autumn at which time they also attain maximum cold hardiness. It is therefore difficult to separate the physiological changes associated with growth cessation and/or defoliation from those that are related to increased cold hardiness. The economic importance of fruit production and the limits placed on it due to problems of cold injury, however, will ensure that research continues in its effort to develop a better understanding of the adaptation and response of fruit trees to cold temperatures.

APPENDICES

FREEZE FIXATION AND FREEZE SUBSTITUTION OF BUD AND STEM TISSUE FROM WOODY PLANTS FOR SEM[40]
Preparation of Plant Material

1. Place twig with six to eight buds (leaf tissue can also be prepared in this manner) in a large test tube (size of test tube depends on size of sample) containing 1 to 2 ml water.
2. Place test tubes with samples in cooling bath and initiate freezing at −1°C with a small chip of ice. Allow 1 to 2 h for tissue to freeze and come to temperature equilibrium with the cooling bath.
3. Cool at 1 to 2°C/h to desired temperature.

Preparation of Fixative

Use the following table to prepare fixative for specific fixation temperatures. Other mixtures can be formulated for other temperatures.

−4°C	16 ml	37% formaldehyde + 84 ml deionized water
−8°C	29 ml	37% formaldehyde + 71 ml deionized water
−12°C	46 ml	37% formaldehyde + 54 ml deionized water

Fixative Performance

Fill a test tube with the volume necessary to completely immerse tissue samples. Allow the solution to equilibrate with the designated fixation temperature. Then add a small chip of ice. The ideal solution is one in which ice crystals are slowly generated from the ice chip, giving a feathery, ice crystal appearance in the tube.

If the ice chip slowly melts, add a small volume of water to the test tube, vortex, and return the tube to the cooling bath to evaluate freezing response of the solution. If ice rapidly fills the entire test tube, let the solution thaw, add a small volume of 37% formaldehyde, vortex, and return test tubes to cooling bath for further evaluation.

Fixation and Dehydration

1. Place appropriate fixative solution in empty test tubes (one for each sample). Place test tubes in cooling bath until temperature is equilibrated.
2. Quickly transfer fixative into test tubes containing samples (enough to completely immerse tissue).
3. Add a small chip of ice to each test tube.
4. Maintain test tubes in cooling bath at the desired temperature for 7 d.
5. After 7 d, remove fixative with a pipette and replace with 50% ethanol (v/v). Ethanol solution should be at the fixation temperature. Tissue should sit in ethanol solution for 30 min.
6. Remove test tubes from cooling bath and continue dehydration in ethanol series 75%, 90%, 95%, and 100% (3×) for 3 min each.
7. Prepare tissue for SEM using standard procedures: critical point dry and sputter coat with gold/palladium.

ELECTROLYTE LEAKAGE METHOD (Based on the Methods of Stefen et al.[51] and Zhang and Wilson[92])

1. Prepare pieces of tissue of uniform size and shape (e.g., remove bark from 1-cm long internodal stem section of peach) which were frozen to various treatment temperatures and subsequently thawed. Nonfrozen tissue stored at about 0°C serves as control.
2. Transfer two pieces into 2.5×20 cm test tubes containing a known volume (e.g., 20 ml) of distilled-deionized water.
3. Vacuum infiltrate the tissue until all the air is excluded. This operation fills extracellular spaces with distilled-deionized water and facilitates the leakage of ions.
4. Shake the tissue in the capped tubes for 1.5 h at about 250 RPM on a gyratory shaker at room temperature.
5. Record the electrical conductivity (C_1) of the leachate using a conductivity salt bridge (YSI Model 32, Yellow Springs Instruments, Yellow Springs, OH).
6. Heat kill the tissue in the same solution and let the solution cool off.
7. Vortex and record the electrical conductivity (C_2) of the solution once again at room temperature.
8. Calculate the percentage of ion leakage by the expression $C_1/C_2 \times 100$.
9. Calculate the percentage of injury using the following expression

$$\% \text{ injury} = \frac{\% \text{ leakage}_{(t)} - \% \text{ leakage}_{(c)}}{100 - \% \text{ leakage}_{(c)}} \times 100$$

where % leakage$_{(t)}$ and % leakage$_{(c)}$ are the measurements of the percentage of ion leakage for freeze treatments and unfrozen controls, respectively.
10. Plot the percentage of injury as a function of treatment temperatures. The temperature at which 50% injury occurs is defined as LT_{50}.

TTC REDUCTION TEST (Based on the Methods of Towill and Mazur[87] and Arora et al.[89])

1. Transfer 100 mg of peach cell suspensions into test tubes (which are covered with aluminum foil) containing 2 ml of 0.16% (w/v) TTC solution made in 0.05 M KH_2PO_4–K_2HPO4 buffer (pH 7.4). The concentration and amount of TTC solution is dependent on species and size of the tissue.
2. Incubate the tissue in TTC solution for 22 to 24 h at room temperature in darkness.
3. Remove the tissue from the solution, rinse it with distilled-deionized water, and extract the reduced TTC in 3 ml of 95% ethanol for 24 h in the dark.
4. Remove the red-colored ethanol extract and read absorbance at 485 nm. Use ethanol extract from nonstressed cells without TTC incubation as a blank. The absorbance values are a measure of TTC reduction to the red-colored formazan compound by mitochondrial respiratory enzymes in viable cells.
5. Using absorbance values for unfrozen control as 100%, calculate the percentage of TTC reduction (relative measure of viability) for treatment temperatures. Plot the percentage of TTC reduction as a function of treatment temperatures. Freezing tolerance is expressed as LT_{50}, the temperature at which a 50% decrease in TTC reduction occurred as compared to unfrozen control tissue.

REFERENCES

1. **Ashworth, E. N. and Wisniewski, M. E.,** Response of fruit tree tissues to freezing temperatures, *HortScience,* 26, 501, 1991.
2. **Sakai, A. and Larcher, W.,** *Frost Survival of plants,* Springer-Verlag, Berlin, 1987.
3. **Levitt, J.,** *Response of plants to Environmental Stresses,* Academic Press, New York, 1980.
4. **Burke, M. J. and Stushnoff, C.,** Frost hardiness: a discussion of possible molecular causes of injury with particular reference to deep supercooling of water, in *Stress Physiology in Crop Plants,* Mussel, H. and Staples, R. C., Eds., Wiley-Interscience, New York, 1979, 197.
5. **Steponkus, P.,** The role of the plasma membrane in freezing injury and cold acclimation, *Annu. Rev. Plant Physiol.,* 35, 543, 1984.
6. **Guy, C.,** Cold acclimation and freezing stress tolerance: role of protein metabolism, *Annu. Rev. Plant Physiol.,* 41, 187, 1990.
7. **Quamme, H. A., Stushnoff, C., and Weiser, C. J.,** The relationship of exotherms to cold injury in apple stem tissue, *J. Am. Soc. Hortic. Sci.,* 97, 608, 1972.
8. **Quamme, H. A., Weiser, C. J., and Stushnoff, C.,** The mechanism of freezing injury in xylem of winter apple twigs, *Plant Physiol.,* 51, 273, 1973.
9. **George, M. F., Burke, M. J., and Weiser, C. J.,** Supercooling in overwintering azalea flower buds, *Plant Physiol.,* 54, 29, 1974.
10. **George, M. F. and Burke, M. J.,** Cold hardiness and deep supercooling in xylem of shagbark hickory, *Plant Physiol.,* 59, 319, 1977.
11. **George, M. F., Becwar, M. R., and Burke, M. J.,** Freezing avoidance by deep supercooling of tissue water in winter-hardy plants, *Cryobiology,* 19, 628, 1982.
12. **Burke, M. J.,** Discussion. Water in plants: the phenomenon of frost survival, in *Comparative Mechanisms of Cold Adaptations,* Underwood, L. S., Tieszen, L. L., Callahan, A. B., and Folk, G. E., Eds., Academic Press, New York, 1979, 259.
13. **Rasmussen, D. H. and Mackenzic, A. P.,** Effect of solute on ice-solution interfacial free energy: calculation from measured homogeneous nucleation temperature, in *Water Structure at the Water Polymer Interface,* Jellnek, H. H. G., Ed., Plenum Press, New York, 1972, 126.
14. **Wisniewski, M., Lightner, G., davis, G., and Schiavone, M.,** System configuration for microcomputer-controlled, low-temperature, differential thermal analysis, *Comp. Elec. Agric.,* 5, 223, 1990.
15. **Wolf, T. K. and Pool, R. M.,** Microcomputer-based differential thermal analysis of grapevine dormant buds, *HortScience,* 21, 1447, 1986.
16. **Rajashekar, C. B.,** Supercooling characteristics of isolated peach flower bud primordia, *Plant Physiol.,* 89, 1031, 1989.
17. **Hong, S. G., Sucoff, E., and Lee-Stadelmann, O. Y.,** Effects of freezing deep supercooled water on the viability of ray cells, *Bot. Gaz.,* 141, 464, 1980.
18. **Ashworth, E. N., Rowse, D. J., and Billmeyer, L. A.,** The freezing of water in woody tissues of apricot and peach and the relationship to freezing injury, *J. Am. Soc. Hortic. Sci.,* 108, 299, 1983.
19. **Ashworth, E. N., Echlin, P., Pearce, R. S., and Hayes, T. L.,** Ice formation and tissue response in apple twigs, *Plant Cell Environ.,* 11, 703, 1988.
20. **Malone, S. R. and Ashworth, E. N.,** Freezing stress response in woody tissues observed using low-temperature scanning electron microscopy and freeze substitution techniques, *Plant Physiol.,* 95, 871, 1991.
21. **George, M. F.,** Freezing avoidance by deep supercooling in woody plant xylem: preliminary data on the importance of cell wall porosity, in *Current Topics in Plant Biochemistry 1983,* Randall, D. D., Blevins, D. G., Larson, R. L., Rapp, B. J., Eds., University of Missouri Press, Columbia, 1983, 84.
22. **Wisniewski, M., Ashworth, E. N., and Schaffer, K.,** The use of lanthanum to characterize cell-wall permeability in relation to deep supercooling and extracellular freezing in plants I. Intergeneric comparisons between *Prunus, Cornus* and *Salix, Protoplasma,* 139, 105, 1987.
23. **Wisniewski, M., Ashworth, E. N., and Schaffer, K.,** The use of lanthanan to characterize cell-wall permeability in relation to deep supercooling and extracellular freezing in woody plants. II. Intrageneric comparisons between *Betula lenta* and *Betula papyrifera, Protoplasma,* 141, 160, 1987.
24. **Wisniewski, M., Davis, G., and Schaffer, K.,** Mediation of deep supercooling of peach and dogwood by enzymatic modifications in cell-wall structure, *Planta,* 184, 254, 1991.
25. **Schaffer, K. and Wisniewski, M.,** Colloidal gold in prescribed sizes as an apoplastic tracer for the determination of pore size in woody plant tissues, *J. Electron Micros. Tech.,* 15, 218, 1990.
26. **Wisniewski, M. and Davis, G.,** Evidence for the involvement of a specific cell-wall layer in regulation of deep supercooling of xylem parenchyma, *Plant Physiol.,* 91, 151, 1989.
27. **Wisniewski, M., Davis, G., and Arora, R.,** The effect of macerase, oxalic acid, and EGTA on deep supercooling and pit membrane structure of xylem parenchyma of peach, *Plant Physiol.,* 96, 1354, 1991.
28. **Baron-Epel, O., Gharyal, P. K., and Schindler, M.,** Pectins as mediators of wall porosity in soybean cells, *Planta,* 175, 389, 1988.

29. **Beckman, C. H.,** The plasticizing of plant cell walls and tylose formation — a model, *Physiol. Plant Pathol.,* 1, 1, 1971.

29a. **Wisniewski, M. and Arora, R.,** unpublished data.

30. **Chafe, S. C.,** Cell wall formation and protective layer development in the xylem parenchyma of trembling aspen, *Protoplasma,* 80, 335, 1974.

31. **Fujii, T., Harada, H., and Sasaki, H.,** Ultrastructure of amorphous layer in xylem parenchyma cell wall of angiosperm species, *Mokuzai Gakkaishi,* 27, 149, 1979.

32. **Ketchie, D. O. and Kammereck, R.,** Seasonal variation of cold resistance in *Malus* woody tissue as determined by differential thermal analysis and viability tests, *Can. J. Bot.,* 65, 2640, 1987.

33. **Wisniewski, M., Arora, R.,and Davis, G.,** Role of the protoplast in deep supercooling of xylem tissue, *HortScience,* 26, 727, 1991.

34. **Quamme, H. A.,** Mechanism of supercooling in overwintering peach flower buds, *J. Am. Soc. Hortic. Sci.,* 103, 57, 1978.

35. **Sakai, A.,** Freezing avoidance mechanism of primordial shoots of conifer buds, *Plant Cell Physiol.,* 20, 1381, 1979.

36. **Ashworth, E. N.,** Properties of peach flower buds which facilitate supercooling, *Plant Physiol.,* 70, 1475, 1982.

37. **Pierquet, P., Stushnoff, C., and Burke, M. J.,** Low temperature exotherms in stem and bud tissues of *Vitis riparia* Michx., *J. Am. Soc. Hortic. Sci.,* 102, 54, 1977.

38. **Biermann, J., Stushnoff, C., and Burke, M. J.,** Differential thermal analysis and freezing injury in cold hardy blueberry flower buds, *J. Am. Soc. Hortic. Sci.,* 104, 449, 1979.

39. **Quamme, H. A.,** An exothermic process involved in freezing injury to flower buds of several *Prunus* species, *J. Am. Soc. Hortic. Sci.,* 99, 315, 1974.

40. **Ashworth, E. N., Davis, G. A., and Wisniewski, M. E.,** The formation and distribution of ice within dormant and deacclimated peach flower buds, *Plant Cell Environ.,* 12, 521, 1989.

41. **Proebsting, E. L.,** The role of air temperature and bud development in determining hardiness of dormant Elberta peach fruit buds, *Proc. Am. Soc. Hortic. Sci.,* 83, 259, 1963.

42. **Ashworth, E. N.,** Xylem development in *Prunus* flower buds and the relationship to deep supercooling, *Plant Physiol.,* 74, 862, 1984.

43. **Gianfagna, T. J., Durner, E. F., and Teiger, G. S.,** Reducing low temperature injury to peach flower buds with ethephon, *Acta Hortic.,* 239, 203, 1989.

44. **Durner, E. F.,** Cryoprotection of deacclimating peach flower buds by ethephon alteration of pistil car-bohydrate content, *Cryobiology,* 26, 290, 1989.

45. **Gianfagna, T. J., Marini, R., and Rachmiel, S.,** Effect of ethephon and GA₃ on time of flowering in peach, *HortScience,* 21, 69, 1986.

46. **Crisosto, C. H., Lombard, P. B. and Fuchigami, L. H.,** Fall ethephon delays bloom in 'Redhaven' peach by delaying flower differentiation during dormancy, *J. Am. Soc. Hortic. Sci.,* 114, 881, 1989.

47. **Durner, E. F. and Gianfagna, T. J.,** Ethephon prolongs dormancy and enhances supercooling in peach flower buds, *J. Am. Soc. Hortic. Sci.,* 116, 500, 1991.

48. **Mazur, P.,** Freezing injury in plants, *Annu. Rev. Plant Physiol.,* 20, 419, 1969.

49. **Burke, M. J., Gusta, L. V., Quamme, H. A., Weiser, C. J., and Li, P. H.,** Freezing and injury in plants, *Annu. Rev. Plant Physiol.,* 27, 507, 1976.

50. **Pearce, R. S.,** Extracellular ice and cell shape in frost-stressed cereal leaves: a low temperature scanning-electron-microscopy study, *Planta,* 175, 313, 1988.

51. **Steffen, K. L., Arora, R., and Palta, J. P.,** Sensitivity of photosynthesis and respiration to a freeze-thaw stress: role of realistic freeze-thaw protocol, *Plant Physiol.,* 89, 1372, 1989.

52. **Ishida, S.,** On the development of frost cracks on "Todumatsu" trunks, *Abies sachalinensis,* especially in relation to their wetwood, *Res. Bull. Coll. Exp. For. Hokkaido Univ.,* 22, 273, 1963.

53. **Palta, J. P., Levitt, J., and Stadelmann, E. J.,** Freezing tolerance of onion bulbs and significance of freeze-induced tissue infiltration, *Cryobiology,* 14, 614, 1977.

54. **Palta, J. P.,** Stress interactions at the cellular and membrane levels, *HortScience,* 25, 1377, 1990.

55. **Arora, R. and Palta, J. P.,** A loss in the plasma membrane ATPase-activity and its recovery coincides with incipient freeze-thaw injury and postthaw recovery in onion bulb scale tissue, *Plant Physiol.,* 95, 846, 1991.

56. **Arora, R. and Palta, J. P.,** Protoplasmic swelling as a symptom of freezing injury in onion bulb cell: its simulation in extracellular KCl and prevention by calcium, *Plant Physiol.,* 82, 625, 1986.

57. **Arora, R. and Palta, J. P.,** *In vivo* perturbation of membrane-associated calcium by freeze-thaw stress in onion bulb cells. Simulation of this perturbation in extracellular KCl and alleviation by calcium, *Plant Physiol.,* 87, 622, 1988.

58. **Iljin, W. S.,** Über den Kaltetod der Pflanzen und seine Ursachen, *Protoplasma,* 20, 105, 1933.

59. **Levitt, J.,** A sulfhydryl-disulphide hypothesis of frost injury and resistance in plants, *J. Theor. Biol.,* 3, 355, 1962.

60. **Lovelock, J.**, The mechanism of the protective action of glycerol against haemolysis by freezing and thawing, *Biochim. Biophys. Acta*, 11, 28, 1953.

61. **Palta, J. P., Jensen, K. G., and Li, P. H.**, Cell membrane alterations following a slow freeze-thaw cycle: ion leakage, injury and recovery, in *Plant Cold Hardiness and Freezing Stress. Mechanisms and Crop Implications*, Li, P. H. and Sakai, A., Eds., Academic Press, New York, 1982, 221.

62. **Arora, R. and Palta, J. P.**, Perturbation of membrane calcium as a molecular mechanism of freezing injury, in *Environmental Stress in Plants: Biochemical and Physiological Mechanisms*, Cherry, J. H., Ed., Springer-Verlag, New York, 1989, 281.

63. **Meryman, H. T.**, Osmotic stress as a mechanism of freezing injury, *Cryobiology*, 8, 489, 1971.

64. **Maximov, N. A.**, Chemische Schutzmittel der Pflanzen gegen Erfrieren, *Ber. Dtsch. Bot. Ges.*, 30, 52, 1912.

65. **Pomeroy, M. K., Pihakaski, J. S., and Andrews, C. J.**, Membrane properties of isolated winter wheat cells in relation to icing stress, *Plant Physiol.*, 72, 535, 1983.

66. **Uemura, M. and Yoshida, S.**, Studies on freezing injury in plant cells. II. Protein and lipid changes in plasma membranes of Jerusalem Artichoke tubers during a lethal freezing *In vivo*, *Plant Physiol.*, 80, 187, 1986.

67. **Iswari, S. and Palta, J. P.**, Plasma membrane ATPase activity following reversible and irreversible freezing injury, *Plant Physiol.*, 90, 1088, 1989.

68. **Palta, J. P.**, Plasma membrane ATPase as a key site of perturbation in response to freeze-thaw stress, in *Current Topics in Plant Biochemistry and Physiology*, Randall, D. D., Ed., University of Missouri Press, Columbia, 1989, 41.

69. **Yoshida, S.**, Freezing injury and phospholipid degradation *in vivo* in woody plant cells. III. Effects of freezing on activity of membrane-bound phospholipase D in microsome-enriched membranes, *Plant Physiol.*, 64, 252, 1979.

70. **Hellergren, J., Widell, S., and Lundborg, T.**, Freezing injury in purified plasma membranes from cold acclimated and non-acclimated needles of *Pinus sylvestris*: is the plasma membrane bound ion stimulated ATPase the primary site of freezing injury?, in *Plant Cold Hardiness*, Li, P. H., Ed., Alan R. Liss, New York, 1987, 211.

71. **Pukacki, P. and Pukacka, S.**, Freezing stress and membrane injury of Norway spruce tissues, *Physiol. Plant.*, 69, 156, 1987.

72. **Yoshida, S.**, Chemical and biophysical changes in the plasma membrane during cold acclimation of mulberry bark cells, *Plant Physiol.*, 76, 257, 1984.

73. **Hellergren, J., Widell, S., Lundborg, T., and Lylin, A.**, Frost hardiness development in *Pinus sylvestris*: the involvement of K^+-stimulated Mg^{2+}-dependent ATPase from purified plasma membranes of pine, *Physiol. Plant.*, 58, 7, 1983.

74. **Mattheis, J. P. and Ketchie, D. O.**, Changes in parameters of the plasmalemma ATPase during cold acclimation of apple tree bark tissues, *Physiol. Plant.*, 78, 616, 1990.

75. **Pomeroy, M. K. and Siminovitch, D.**, Seasonal cytological changes in secondary phloem parenchymia cells in *Robin pseudoacacia* in relation to cold hardiness, *Can. J. Bot.*, 49, 787, 1971.

76. **Niki, T. and Sakai, A.**, Ultrastructural changes related to frost hardiness in the cortical parenchyma cells from mulberry twigs, *Plant Cell Physiol.*, 22, 171, 1981.

77. **Wisniewski, M. and Ashworth, E. N.**, A comparison of seasonal ultrastructural changes in stem tissues of peach (*Prunus persica*) that exhibit contrasting mechanism of cold hardiness, *Bot. Gaz.*, 147, 407, 1986.

78. **Tumanov, I. I., Kuzina, G. V. and Karnikova, L. D.**, Effect of photoperiod on the frost resistance of apricots and black currants, *Fiziol. Rast.*, 12, 665, 1965.

79. **Irving, R. M. and Lamphear, F. O.**, The long-day leaf as a source of cold hardiness inhibitors, *Plant Physiol.*, 42, 1384, 1967.

80. **Weiser, C. J.**, Cold resistance and injury in woody plants, *Science*, 169, 1269, 1970.

81. **Kramer, P. J.**, Photoperiodic stimulation of growth by artificial light as a cause of winter killing, *Plant Physiol.*, 12, 881, 1937.

82. **Graham, D. and Patterson, B. D.**, Responses of plant to low non-freezing temperatures: proteins, metabolism, and acclimation, *Annu. Rev. Plant Physiol.*, 33, 347, 1982.

83. **Palta, J. P., Levitt, J., and Stadelmann, E. J.**, Plant viability assay, *Cryobiology*, 15, 249, 1978.

84. **Kraut, J. L., Walsh, C. S., and Ashworth, E. N.**, Acclimation and winter hardiness patterns in the eastern thornless blackberry, *J. Am. Soc. Hortic. Sci.*, 3, 347, 1986.

85. **Stadelmann, E. J. and Kinzel, H.**, Vital staining of plant cells, in *Methods in Cell Physiology*, Prescott, D. M., Ed., Academic Press, New York, 1972, 325.

86. **Palta, J. P., Levitt, J. and Stadelmann, E. J.**, Freezing injury in onion bulb cells. Evaluation of the conductivity method and analysis of ion and sugar efflux from injured cells, *Plant Physiol.*, 60, 393, 1977.

87. **Towill, L. E. and Mazur, P.**, Studies on the reduction of 2,3,5-triphenyl tetrazolium chloride as a viability assay for plant tissue cultures, *Can. J. Bot.*, 53, 1097, 1975.

88. **Ketchie, D. D .and Kammereck, R.,** Seasonal variation of cold resistance in *Malus* woody tissue as determined by differential thermal analysis and viability tests, *Can. J. Bot.,* 65, 2640, 1987.

89. **Arora, R., Wisniewski, M., and Hershberger, W.,** Low temperature-induced cold acclimation in cell suspension cultures of peach, *Plant Physiol.,* 96, 57, 1991.

90. **Pomeroy, M. K., Simimovitch, D., and Wightman, F.,** Seasonal biochemical changes in the living bark and needles of red pine in relation to adaptation to freezing, *Can. J. Bot.,* 48, 953, 1970.

91. **Wilner, J.,** Relative and absolute electrolytic conductance tests for frost hardiness of apple varieties, *Can. J. Plant Sci.,* 40, 630, 1960.

92. **Zhang, M. I. N. and Willison, J. H. M.,** An improved method for the measurement of frost hardiness, *Can. J. Bot.,* 65, 710, 1987.

93. **Arora, R., Wisniewski, M., and Scorza, R.,** Seasonal patterns of cold hardiness and polypeptides in bark and xylem of genetically related (sibling) deciduous and evergreen peach tress, *HortScience,* 26, 727, 1991.

94. **Dexter, S. T., Tottingham, W. E., and Graber, L. F.,** Investigations of the hardiness of plants by measurement of electrical conductivity, *Plant Physiol.,* 7, 63, 1932.

95. **Palta, J. P., Levitt, J., and Stadelman, E. J.,** Freezing injury in onion bulb cells. II. Post-thawing injury and recovery, *Plant Physiol.,* 60, 398, 1977.

96. **Kushad, M. M. and Yelenosky, G.,** Evaluation of polyamine and proline levels during low temperature acclimation of citrus, *Plant Physiol.,* 84, 692, 1987.

97. **Yelenosky, G. and Guy, C. L.,** Carbohydrate accumulation in leaves and stems of valencia orange at progressively colder temperatures, *Bot. Gaz.,* 138, 13, 1977.

98. **Chen, H. H. and Li, P. H.,** Characteristics of cold acclimation and deacclimation of tuber-bearing *Solanum* species, *Plant Physiol.,* 65, 1146, 1980.

99. **Chen, H. H., Gavinlertvatana, P., and Li, P. H.,** Cold acclimation of stem cultured plants and leaf callus of *Solanum* species, *Bot. Gaz.,* 140, 142, 1979.

100. **Fennell, A. and Li, P. H.,** Rapid cold acclimation and deacclimation in winter spinach, *Acta Hort.,* 168, 179, 1985.

101. **van Huystee, R. B., Weiser, C. J., and Li, P. H.,** Cold acclimation in *Cornus stolonifera* under natural and controlled photoperiod and temperature, *Bot. Gaz.,* 128, 200, 1967.

102. **Thomashow, M. F., Gilmour, S. J., Hajela, R., Horvath, D., Lin, C., and Guo, W.,** Studies on cold acclimation in *Arabidopsis thaliana,* in *Horticultural Biotechnology,* Bennett, A. B. and Sharman, D. O., Eds., Wiley-Liss, New York, 1990, 305.

103. **Fuchigami, L. H., Hotze, M., and Weiser, C. J.,** The relationship of vegetative maturity to rest development and spring bud break, *J. Am. Soc. Hortic. Sci.,* 102, 450, 1977.

104. **Guy, C. L. and Haskell, D.,** Induction of freezing tolerance in spinach is associated with the synthesis of cold acclimation induced proteins, *Plant Physiol.,* 84, 872, 1987.

105. **Tumanov, I. I., Butenko, R. G., and Oglovets, I. V.,** Use of the isolated tissue method for studying hardening of plant cells, *Fiziol. Rast.,* 15, 749, 1968.

106. **Oglovets, I. V.,** Hardening of isolated callus tissue of woody plants with different frost resistances, *Fiziol. Rast.,* 23, 139, 1976.

107. **Sakai, A. and Sugauara, Y.,** Survival of poplar callus at super-low temperatures after cold acclimation, *Plant Cell Physiol.,* 14, 1201, 1973.

108. **Hellergren, J.,** Cold acclimation of suspension cultures of *Pinus sylvestris* in response to light and temperature treatments, *Plant Physiol.,* 72, 992, 1983.

109. **Wallner, S. J., Wu, M.-T., and Anderson-Krengel, S. J.,** Changes in extracellular polysaccharides during cold acclimation of cultured pear cells, *J. Am. Soc. Hortic. Sci.,* 3, 769, 1986.

110. **Chen, T. H. H. and Gusta, L. V.,** Abscisic acid-induced freezing resistance ni cultured plant cells, *Plant Physiol.,* 86, 344, 1983.

111. **Reaney, M. J. T. and Gusta, L. V.,** Factors influencing the induction of freezing tolerance by abscisic acid in cell suspension cultures of *Bromus inermis* and *Medicago sativa, Plant Physiol.,* 83, 423, 1987.

112. **Tanino, K. K., Chen, T. H. H., Fuchigami, L. H., and Weiser, C. J.,** Metabolic alterations associated with abscisic acid-induced frost hardiness in brome grass suspension cultures, *Plant Cell Physiol.,* 31, 505, 1990.

113. **Levitt, J.,** Frost drought and heat resistance, *Annu. Rev. Plant Physiol.,* 2, 245, 1951.

114. **Chen, P., Li, P. H., and Weiser, C. J.,** Induction of frost hardiness in red osier dogwood stems by water stress, *HortScience,* 10, 372, 1975.

115. **Cloutier, Y. and Andrews, C. J.,** Efficiency of cold hardiness induction by desiccation stress in four winter cereals, *Plant Physiol.,* 76, 595, 1984.

116. **Siminovitch, D. and Cloutier, Y.,** Twenty-four hour induction of freezing and drought tolerance in plumules of winter rye seedlings by desiccation stress at room temperature in dark, *Plant Physiol.,* 69, 250, 1982.

INDEX